Next Generation Probiotics

Next Generation Probiotics: Natural Therapeutics, Volume 2 presents a comprehensive and forward-looking exploration of probiotics as scientifically grounded natural therapeutics in modern health and disease management. Building on advances in microbiome science, biotechnology, and computational biology, this volume examines how next-generation probiotics are discovered, designed, evaluated, and applied. The book bridges foundational probiotic research with emerging technologies such as machine learning, genomics, and genetic engineering, highlighting what distinguishes next-generation probiotics from traditional formulations.

This second volume expands its scope to reflect rapid developments in probiotic science and translational applications. It integrates insights from probiotic genomics, CRISPR-Cas9 technology, microbiota-targeted foods, and personalized microbiome-based nutrition. By addressing disease-specific, strain-specific, and host-specific effects, the book provides a holistic view of how probiotics can be optimized for therapeutic efficacy. Contributions from experts across bioinformatics, microbiology, clinical research, and food science ensure a multidisciplinary and evidence-based perspective.

Key Features

- Detailed coverage of machine learning applications in probiotic development and fermented dairy products (new to this edition)
- In-depth discussion of probiotic genomics, cancer associations, and CRISPR-Cas9-based strain improvement
- Exploration of emerging probiotic candidates, including *Akkermansia, Faecalibacterium, Ruminococcus*, and butyrate-producing microbes
- Comprehensive overview of novel probiotic classes such as postbiotics, parabiotics, pharmacobiotics, and immunobiotics (expanded in this edition)
- Practical insights into challenges in probiotic selection, regulation, stress tolerance, and personalized nutrition strategies

Bringing together fundamental science and applied research, *Next Generation Probiotics: Natural Therapeutics, Volume 2* serves as an essential resource for researchers, clinicians, nutritionists, food technologists, and graduate students. It is particularly valuable for those interested in microbiome-based therapeutics, functional foods, and personalized health solutions, offering both conceptual frameworks and real-world relevance for advancing probiotic research and application.

Current Developments in Agricultural Biotechnology and Food Security

Series Editor: Charles Oluwaseun Adetunji, Professor and Director of Research and Innovation, Edo State University Uzairue

This series intends to provide comprehensive coverage of sustainable modern technologies, aimed at improvement of food production via agriculture and food biotechnological techniques. Proposed suite of books focusses on topics in agricultural microbiology, biotechnology, food science, crop production, post-harvest management, aimed at both basic and advance food and agricultural biotechnology. The series seeks to discuss and provide foundational content from bench to bedside in food microbiology, agricultural and food biotechnology. The Series' goal is to enhance knowledge, and present update on hot topics in the field of the Agricultural Biotechnology specifically, the series aims to translate results and recent findings of studies into enhanced food production. This is primarily intended for researchers, students in Food and Agricultural Biotechnology at the graduate level and above, including those working in academic, corporate, or non-profit settings.

Phyto-Oxylipins: Metabolism, Physiological Roles, and Profiling Techniques
Edited by Sheikh Mansoor Shafi, Chukwuebuka Egbuna,
Charles Oluwaseun Adetunji

Agricultural Biotechnology: Food Security Hot Spots
Edited by Charles Oluwaseun Adetunji, Deepak Gopalrao Panpatte,
Yogeshvari Kishorsinh Jhala

Next Generation Probiotics: From Novel Drugs to Future Therapeutics
Edited By Shakira Ghazanfar, Charles Oluwaseun Adetunji, Alessandro Di Cerbo,
Ghulam Muhammad Ali, Wajya Ajmal, Muhammad Naeem Riaz

Chitosan-Based Edible Coatings in the Food Industry
Charles Oluwaseun Adetunji, Raghvendra Raman Mishra, Pragya Mishra,
Dr Juliana Bunmi Adetunji

Biotechnological Relevance of Cytochrome P450 in Boosting Agricultural Production and Food Security
Wadzani Palnam Dauda and Charles Oluwaseun Adetunji

Next Generation Probiotics: Natural Therapeutics, Volume 2
Edited by Shakira Ghazanfar, Charles Oluwaseun Adetunji,
Wajya Ajmal, Adnan Ahmad Ansari

For more information about this series, please visit: https://www.routledge.com/Current-Developments-in-Agricultural-Biotechnology-and-Food-Security/book-series/CRCABFS

Next Generation Probiotics

Natural Therapeutics, Volume 2

Edited by
Shakira Ghazanfar, Charles Oluwaseun Adetunji,
Wajya Ajmal, and Adnan Ahmad Ansari

CRC Press
Taylor & Francis Group
Boca Raton London New York

CRC Press is an imprint of the
Taylor & Francis Group, an **informa** business

First edition published 2027
by CRC Press
2385 NW Executive Center Drive, Suite 320, Boca Raton FL 33431

and by CRC Press
4 Park Square, Milton Park, Abingdon, Oxon, OX14 4RN

CRC Press is an imprint of Taylor & Francis Group, LLC

ISBN: 978-1-032-50822-1 (hbk)
ISBN: 978-1-032-50823-8 (pbk)
ISBN: 978-1-003-39979-7 (ebk)

DOI: 10.1201/9781003399797

Typeset in Times
by KnowledgeWorks Global Ltd.

Contents

Editors' Biographies

Dr. Shakira Ghazanfar has been working as Senior Scientific Officer at PARC, Islamabad, Pakistan, since 2007. She did her PhD in Microbiology from Quaid-i-Azam University, Islamabad, Pakistan. Her area of research is the development and commercialization of the TARGET BASED Probiotic Products based on AI/DL/ML. She is regarded among the pioneers who have initiated Animal Probiotic Research to improve livestock productivity in Pakistan. She is the Main Organizer (100+ training organizer) of the Training/Workshops/Webinars in her department. She published more than 100+ research papers (400+ impact factor), 50+ book chapters, and has been the main editor of CRC Press, USA and Springer Nature publishers' books. She is a Guest Editor in well-known journals. She co/supervised/supervising research of 20+ PhD, 100+ MPhil/BS students. She has published widely in the fields of AI/ML/DL and food microbiology, drug discovery, nutraceuticals, and functional foods with specialization in probiotic-based food/feed and probiotic-related anticancer agents.

Professor Charles Oluwaseun Adetunji is presently a faculty member at the Microbiology Department, Faculty of Science, Edo State University, Uzairue, Edo State, Nigeria, where he utilizes the application of biological techniques and microbial bioprocesses for the actualization of sustainable development goals and agrarian revolution through quality teaching, research, and community development. He was formerly the Acting Director of Intellectual Property and Technology Transfer, the Head of the Department of Microbiology, and the Sub Dean of the Faculty of Science, and is currently the Chairman of the Grant Committee and immediate Dean of the Faculty of Science and now the Director of Research and Innovation at EDSU.

Wajya Ajmal is an accomplished bioinformatics professional who serves as Scientific Officer at the National Institute for Genomics and Advanced Biotechnology (NIGAB), Islamabad, Pakistan, a role she has held since April 2022. With over eight years of experience in the biotechnology industry, including prior work as Assistant Scientific Officer at NIGAB, she possesses expertise in sequence analysis, bioinformatics, Python, systems biology, and Perl. A strong research background supports their work, including a Master of Philosophy (MPhil) in bioinformatics from Quaid-i-Azam University (2012–2014) and a Bachelor of Science (BS) in bioinformatics from COMSATS University Islamabad (2007–2011). Wajya's career reflects a robust commitment to advancing genomics and biotechnology through innovative research and technical proficiency.

Dr. Adnan Ahmad Ansari is an accomplished bioinformatics professional who serves as a faculty member in the Department of Bioinformatics at Quaid-i-Azam University (QAU), Islamabad, Pakistan. With extensive experience in academic research and higher education, he has developed strong expertise in computational biology, sequence analysis, genomics, data analysis, and bioinformatics programming. His work is supported by a solid research and teaching background, contributing to both undergraduate and postgraduate training in bioinformatics. Dr. Ansari holds advanced academic qualifications in bioinformatics and related life sciences, reflecting a sustained commitment to scientific excellence. His career demonstrates a strong dedication to advancing bioinformatics research, capacity building, and innovation through interdisciplinary approaches and technical proficiency.

Contributors

Heba A.M. Abdalla
Department of Botany
Agricultural and Biological Research Institute
National Research Centre
Dokki, Giza, Egypt

Samson Olusola Adeboye
Department of Biological Sciences
College of Science and Technology
Covenant University
Ota, Nigeria

Charles Oluwaseun Adetunji
Department of Microbiology
Edo State University
Uzairue, Edo State, Nigeria

S.A. Ado
Department of Microbiology
Ahmadu Bello University
Zaria, Nigeria

Mohammad Javad Fakkari Afjadi
Department of Biology
Faculty of Science
Islamic Azad University, Ashkezar, Branch
Yazd, Iran

Hadia Afzal
Department of Biotechnology
Quaid-i-Azam University
Islamabad, Pakistan

Muhammad Saad Ahsan
University Institute of Public Health
Lahore, Pakistan

Ali Ajami
Tabriz University of Medical Sciences
Manager of Nobel Medical Laboratory
Esfahan, Iran

Hafiz Muhammad Bilal Akhtar
National Institute for Genomics and Advanced Biotechnology (NIGAB)
National Agricultural Research Center
Islamabad, Pakistan

Hamza Ali
Rawalpindi Medical University
Rawalpindi, Pakistan

Hassan Ali
Institute of Microbiology
Government College University
Faisalabad, Pakistan

Samia S. Alkhalil
Department of Clinical Laboratory Sciences
College of Applied Medical Sciences
Shaqra University
Alquwayiyah, Riyadh, Saudi Arabia

Muhammad Lawal Attanda
Agricultural and Environmental Engineering
Bayero University
Kano State, Nigeria

Furqan Awan
Department of Epidemiology and Public Health
University of Veterinary and Animal Sciences
Lahore, Pakistan

Joy Iziegbe Azeta
Department of Biological Sciences
College of Science and Technology
Covenant University
Ota, Nigeria

Rameesha Azhar
Department of Epidemiology and Public Health
University of Agriculture
Faisalabad, Pakistan

Mohammad Moeen Badar
PAEC Model College
Islamabad, Pakistan

Maryam Bakhtiyari
Department of Medical Laboratory Sciences
Faculty of Allied Medicine
Qazvin University of Medical Sciences
Qazvin, Iran

Abdullah Ahmed Butt
Department of Dairy Technology
University of Veterinary and Animal Sciences
Lahore, Pakistan

Hasan Ejaz
Department of Clinical Laboratory Sciences
College of Applied Medical Sciences
Jouf University
Sakaka, Saudi Arabia

Sherifat Ozavize Enejiyon
Department of Microbiology
Federal University of Technology
Minna, Niger State, Nigeria

Yaseen Galali
Food Technology Department
College of Agricultural Engineering Sciences
Salahaddin University-Erbil
Erbil, Iraq

Lia Gallo-Urrego
Harvard Medical School, Harvard University, USA

Mordecai Gana
Department of Microbiology
Federal University of Technology
Minna, Niger State, Nigeria

Muhammad Hussain Ghazali
School of Food and Biological Engineering
Jiangsu University (JSU)
Zhenjiang, China

Shakira Ghazanfar
National Institute for Genomics and Advanced Biotechnology (NIGAB)
National Agricultural Research Center
Islamabad, Pakistan

Maryam Gohar
Department of Biotechnology
Quaid-i-Azam University
Islamabad, Pakistan

Mehraj Gul
National Institute of Health
Islamabad, Pakistan

Leila Haghshenas
Postdoc Association Member of Harvard Medical School
Boston, MA, USA

Mohammad Al Hemairy
Research Institute of Science and Engineering (RISE), Department of Information Systems, College of Computing and Informatics, Univeristy of Sharjah,
United Arab Emirates

Fiaz Hussain
Department of Agricultural Engineering
Faculty of Agricultural Engineering and Technology
PMAS-Arid Agriculture University
Rawalpindi, Pakistan

Sara Ijaz
Department of Epidemiology and Public Health
University of Veterinary and Animal Sciences
Lahore, Pakistan

Huma Ikram
Neurochemistry and Biochemical Neuropharmacology Research Unit
Department of Biochemistry
University of Karachi
Karachi, Pakistan

Sajid Iqbal
Oujiang Laboratory (Zhejiang Lab for Regenerative Medicine, Vision and Brain Health)
Wenzhou, Zhejiang, P.R. China

Aatka Jamil
National Institute for Genomics and Advanced Biotechnology (NIGAB)
National Agricultural Research Center,
Islamabad, Pakistan

Shayista Jan
Department of Biotechnology
Quaid-i-Azam University
Islamabad, Pakistan

Jumoke Florence Joda
Genetics, Genomic and Bioinformatics Department
National Biotechnology Research and Development Agency (NBRDA)
Abuja, Nigeria

Shamsa Kanwal
Department of Microbiology
Hazara University
Mansehra, Pakistan

Sharjeel Khalid
National Institute for Genomics and Advanced Biotechnology (NIGAB)
National Agricultural Research Center (NARC)
Islamabad, Pakistan

Ayesha Khan
National Institute for Biotechnology and Genetic Engineering
Faisalabad, Pakistan

Mohsin Khurshid
Institute of Microbiology
Government College University Faisalabad
Faisalabad, Pakistan

Khudija Tul Kubra
The University of Haripur
Haripur, Pakistan

Haala Marikar
Department of Biotechnology
Quaid-i-Azam University
Islamabad, Pakistan

Rabia Mehmood
Department of Microbiology
Quaid-i-Azam University
Islamabad, Pakistan

Fatima Mujahid
Institute of Microbiology
Government College University
Faisalabad, Pakistan

Rimsha Munawar
Abasyn University
Islamabad, Pakistan

Abdul Nafeh
Department of Pathology
Sialkot Medical College
Sialkot, Pakistan

Amina Najam
National Institute of Health
Islamabad, Pakistan

Saiqa Noureen
National Institute of Genomics and Advanced Biotechnology (NIGAB)
National Agricultural Research Center
Islamabad, Pakistan

Gbotemi Jerry Oni
Department of Molecular Biology and Bioinformatics
Africa Center of Excellence for Mycotoxins and Food Safety
Federal University of Technology
Minna, Nigeria

Samuel Adeniyi Oyegbade
Department of Biological Sciences
College of Science and Technology
Covenant University
Ota, Nigeria

Oluwafemi Adebayo Oyewole
Department of Microbiology
Federal University of Technology
Minna, Niger State, Nigeria

Mehwish Rafiq
National Institute of Food Science and Technology
University of Agriculture
Faisalabad, Pakistan

Maliha Rashid
University Institute of Biochemistry and Biotechnology
PMAS Arid Agriculture University
Rawalpindi, Pakistan

Noshaba Rashid
Azad Jammu and Kashmir Medical College
Muzaffarabad, Pakistan

Haseeb Raza
Institute of Microbiology
Faculty of Veterinary Science
University of Agriculture
Faisalabad, Pakistan

Abdul Rehman
Animal Quarantine Department
Ministry of National Food Security
and Research
Government of Pakistan

Muhammad Zia ur Rehman
National Institute for Genomics and Advanced
Biotechnology (NIGAB)
National Agricultural Research Center
Islamabad, Pakistan

Nazia Rehman
National Institute for Genomics and Advanced
Biotechnology
National Agricultural Research Centre
Islamabad, Pakistan

Muhammad Naeem Riaz
National Institute of Genomics and Advanced
Biotechnology (NIGAB),
National Agricultural Research Centre,
Islamabad, Pakistan

Evelyn Saba
Department of Veterinary Biomedical
Sciences
Faculty of Veterinary and Animal Sciences
PMAS-Arid Agriculture University
Rawalpindi, Pakistan

Ya Sakina
Pakistan Agricultural Research Council
Islamabad, Pakistan

Faizan Saleem
Faculty of Veterinary Sciences
Baha Ud Din Zakariya University (BZU)
Multan, Pakistan

Samy Selim
Department of Clinical Laboratory Sciences
College of Applied Medical Sciences
Jouf University
Sakaka, Saudi Arabia

Abdul Sammad Ali Khan Shirwany
Department of Epidemiology
and Public Health
University of Veterinary
and Animal Sciences
Lahore, Pakistan

Maria Fareed Siddiqui
University of Lahore
Lahore, Pakistan

Muhammad Tahir
School of Allied Health Sciences
Faculty of Sciences
Multan University of Science
and Technology
Multan, Pakistan

Ahsen Taqveem
School of Allied Health Sciences
Faculty of Sciences
Multan University of Science
and Technology
Multan, Pakistan

Zahid Ullah
Department of Microbiology
Hazara University
Mansehra, Pakistan

Ahmed Umar
University Institute of Public Health
Lahore, Pakistan

Preface

The science of probiotics has advanced far beyond its traditional definition of live microorganisms conferring health benefits to the host. In recent years, the emergence of *next generation probiotics* has transformed the field into a multidisciplinary domain that integrates microbiology, genomics, computational biology, and biotechnology. Probiotics are no longer viewed solely as components of fermented foods but as precision-based natural therapeutics designed to target specific physiological and pathological conditions.

With the expansion of systems biology and big data analytics, machine learning has become an essential tool in microbiological research. Feature engineering, predictive modeling, and artificial intelligence now facilitate the identification of functional microbial traits and the development of precision probiotics. These approaches are particularly valuable in optimizing probiotic applications in fermented dairy foods and tailoring formulations to individual health requirements.

Metagenomics has revolutionized the exploration of microbial diversity, enabling the discovery of novel probiotic strains without reliance on conventional culturing methods. Advances in probiotic genomics further deepen our understanding of molecular mechanisms underlying host–microbe interactions, including the role of probiotics in complex diseases such as cancer. High-throughput sequencing and genome annotation provide insights into functional attributes, safety, and therapeutic potential.

The application of CRISPR-Cas9 technology and other genome-editing tools has opened new avenues for probiotic research. Genetically engineered and bioengineered strains with enhanced functionality, improved stress tolerance, and targeted therapeutic properties are becoming increasingly relevant in probiotic production and clinical application. These innovations bridge natural therapeutics with modern biotechnology.

This volume also addresses the critical importance of genus-, species-, and strain-specific effects, emphasizing disease-condition specificity in probiotic applications. The expanding spectrum of postbiotics, parabiotics, pharmacobiotics, and immunobiotics reflects the evolution of probiotic science beyond live microbial administration. At the same time, researchers and clinicians face practical challenges in evidence-based decision-making when prescribing probiotics for specific health concerns.

Life-stage–specific interventions, from early-life microbial programming to probiotic use in aging populations, highlight the broad applicability of these agents. The modulation of the gut–brain axis, stress tolerance mechanisms in probiotic formulations, and the development of personalized microbiome-based nutritional strategies underscore the shift toward precision and preventive healthcare.

Next Generation Probiotics: Natural Therapeutics, Volume 2 presents a comprehensive overview of contemporary research and emerging innovations in probiotic science. By integrating computational tools, genomic technologies, bioengineering approaches, and clinical perspectives, this book aims to serve as a valuable resource for researchers, clinicians, and industry professionals dedicated to advancing microbiome-based therapeutics.

1 (Part I) Machine Learning and Its Applications in Microbiology

Feature Engineering

Abdul Rehman, Furqan Awan, Ahmed Umar, Abdul Sammad Ali Khan Shirwany, Fiaz Hussain, Muhammad Saad Ahsan, Shakira Ghazanfar, Sara Ijaz, and Saiqa Noureen

LIST OF ABBREVIATIONS USED

Abbreviation	Full Form
A3C	Asynchronous-advantage actor-critic
AUC-ROC	Area under receiver characteristic curve
CV	Cross-validation
DBSCAN	Density-based spatial clustering of applications with noise
DT	Decision tree
LASSO	Least absolute shrinkage and selection operator
LDA	Least discriminate analysis
LR	Logistic regression
LSA	Latent semantic analysis
ML	Machine learning
NN	Neural network
ODA	Oracle digital assistance
ODA	Optimal-discriminate analysis
PCA	Principle component analysis
RMSE	Root mean square error
SARSA	State-action-reward-state-action
SVD	Singular-value-decomposition
SVM	Support vector machine

1.1 INTRODUCTION

Probiotics as microbiome medicine are defined as living microorganisms that can produce health benefits when administered in a sufficient quantity (Moens et al. 2019). The human gut is composed of trillions of microorganisms that have beneficial health effects (Martinez-Guryn, Leone, and Chang 2019). Researchers are interested in investigating the specific strains of probiotics as potential therapeutics with targeted mechanisms, characterized indications, and the capability to improve the gut health (Ghyselinck et al. 2021; Veiga et al. 2020). These therapeutics are called

DOI: 10.1201/9781003399797-1

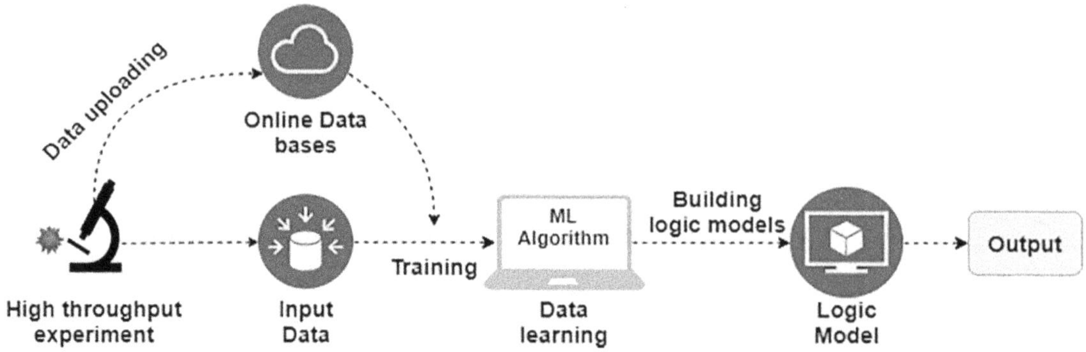

FIGURE 1.1 The graphical illustration of microbiome data collection and mining in constructing ML algorithms.

"next-generation probiotics or live-biotherapeutics" and have the potential to revolutionize disease treatment and prevention in the near future (Yu et al. 2020).

From the advent of the 21st century, substantial research has been focused on the microbiome. Cost-effective and precise DNA sequencing techniques facilitated the start of the Human Microbiome Project: a collaborative effort that resulted in the genomic profiling of microbiota at crucial anatomical locations (2012). Subsequently, the project also evaluated the significance of the microbiome in relation to human health and ailments. Furthermore, various research laboratories across the globe have also made a significant contribution to the extensive body of knowledge surrounding microbiomes. Now, a vast amount of data is globally accessible through publications along with online data sources such as NIH-Human Microbiome Project (Human), MicrobiomeDB (Oliveira et al. 2018), and China-GeneBank (Zhu et al. 2022). These datasets are increasingly being utilized by scientists for the purpose of devising targeted therapies that specifically address the microbiome and forecasting the potential interactions between the microbiome and drugs (Figure 1.1).

1.2 MACHINE LEARNING (ML) CLASSIFICATION

With the recent advancements in computational power, there has been enormous growth in readily available microbiome data (Camacho et al. 2018). Human minds, along with conventional software, cannot analyze and extract meaningful insight from huge databases. Machine learning (ML), a subfield of artificial intelligence, can extract meaningful insights without being specifically programmed (Namkung 2020a; Vujkovic-Cvijin et al. 2020). In ML, programmed algorithms are used to understand, analyze, and process the data patterns to classify or predict new information/outcomes (Azad et al. 2025). Moreover, ML is classified into supervised, semi-supervised, unsupervised, and reinforcement learning (Camacho et al. 2018) (Figure 1.2). Supervised and unsupervised learnings have been extensively applied to analyze the gut microbiome (Manandhar et al. 2021).

1.2.1 Supervised Learning

It mainly involves the training of algorithms on the labeled dataset. Once an algorithm is trained on the feeding or training dataset, it is evaluated on the unseen or test dataset. In supervised learning, the predetermined output variables are used to construct the model (Manandhar et al. 2021). These known variables can predict the outcomes of new samples. Supervised learning is mainly categorized into classification and regression. If the outcome variables are categorical, the model will behave as classification model (Karlsson et al. 2013), whereas for the continuous variables, it will act as regression model (Zhou et al. 2018). The classification supervised models are mostly applied to

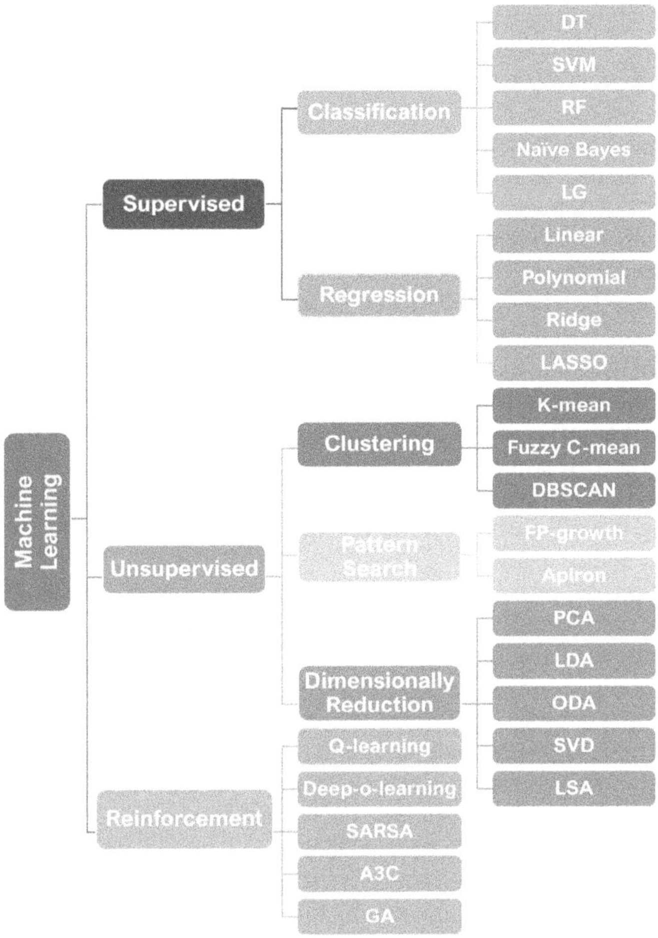

FIGURE 1.2 The hierarchical classification of ML and algorithms.

analyze microbiome data. Classification algorithms include support vector machine (SVM) (Cortes and Vapnik 1995), decision tree (DT), Random Forest (RF) (Breiman 2001), and logistic regression (LR) (Friedman, Geiger, and Goldszmidt 1997). These algorithms are mainly tree-based models that illustrate the combination of various independent variables to predict the outcome variables (Long et al. 2020), whereas the regression algorithms comprise linear regression, multiple regression, non-linear regression, Bayesian linear regression, and polynomial regression. Regression algorithms demonstrate the effect of independent variables to predict the output variable, keeping the input variables constant.

Supervised learning is regarded as data-driven approach to project, optimize, and classify the probiotic strains. Various supervised learning algorithms are applied in probiotics with improved understanding and probiotic formulations (Noor et al. 2024).

Westfall et al. (2021) studied the synergism between exogenous probiotics and microbiota using supervised ML-based multiple-adaptive regression splines (MARS) model. They applied an artificial model of the human gastrointestinal tract known as a bioreactor imitation of microbiota environment (ABIOME). The purpose of this study was to predict and optimize the metabolic activity of probiotic formulations through combining fecal-inoculated ABIOME with probiotics and polyphenol-rich prebiotics. Efficiently, the MARS model was used for exploring the synergetic effects of the inoculated probiotics with the indigenous microbiome. Through the model, the specific probiotic combinations were identified as being involved in the stimulation of bioavailable metabolite production with unique curative capabilities. The study demonstrated the potential of ABIOME in conjunction with the MARS algorithm for designing probiotic formulations intended for therapy.

Active ML, a supervised approach, has been employed to predict the effect of different excipients on the growth of specific probiotics. McCoubrey et al. (2021) utilized an active ML model to predict the growth of *Lactobacillus paracasei*, a key probiotic used as a therapeutic agent to improve health benefits. The model precisely predicted the effects of many excipients, subsequently validated experimentally.

Neural networks (NNs) have recently been involved in differentiating the probiotics and non-probiotic bacteria based on distinct genomics. Bergamini et al. (2022) applied NN to analyze the bacterial genomic data, leading to remarkable performance to differentiate between the probiotics and non-probiotic bacteria. It turns out that features of transfer RNA (tRNA) sequences, such as Chargaff's scores and Shannon's entropy, emerged as important discriminators pointing to the potential evolutionary adaptations relating to symbiosis within the human gut environment. The study resulted not only in identifying the new probable probiotics but also in highlighting the efficacy of NN to correctly differentiate probiotic and non-probiotic bacteria based on genomic features.

Recently, supervised ML models have been applied to develop quantum-inspired differential evolution (QDE) algorithm to analyze whole genomic data of bacterial strains to identify the distinct probiotic marker genes. Kamarudin, Ong, and Tan (2022) demonstrated QDE model proved to be an effective approach for discriminating the probiotic bacterial strains with a relatively high classification accuracy. The study singled out genetic features that could serve as promising marker genes signifying probiotic traits in *Bacillus* spp. strains. The QDE has proven its effectiveness in correctly identifying the genomic markers that point to probiotic traits and are related to certain strains.

SVMs are used to analyze whole-genomic (WG) data of bacterial strains to accurately predict the potential probiotics strains. Sen, Hajra, and Ghosh (2019) used SVM algorithm along with k-mer composition as a feature to differentiate between the probiotic and non-probiotic strains of bacteria. It was noted that the SVMs can deliver reasonably high-accuracy predictions of probiotic-associated genes. However, several limitations were highlighted in this study, including relying on genomic data and the fact that the probiotics' functionality may have multiple genes involved.

A study by Lam et al. (2020) used an RF algorithm to examine the structural properties of beta-glucans and their associated probiotic potential. In the research, RF was applied to determine structural factors that significantly influenced probiotic growth parameters (lag phase, maximum growth rate, and population increase). The key findings focused on characteristics such as low molecular weight, high water solubility, and the lack of carboxymethyl functional groups that contribute to the optimization of the prebiotic beta-glucans. The detailed flowchart of supervised learning is given in Figure 1.3.

Overall, the integration of supervised ML models was found to be significantly helpful in guiding decisions for identifying the potential probiotics and enhancing their current potential.

1.2.2 Unsupervised Learning

In this type of ML, the models are not trained on the labeled dataset. Instead, models learn hidden and novel data patterns by themselves, so-called "data-driven learning" (Ghahramani 2004). Unsupervised learning is further categorized into clustering analysis and dimension reduction. Clustering algorithms consist of K-means (Hartigan and Wong 1979), Fuzzy C-means, and self-organizing map (SOM) (Kohonen 1990) and are usually deployed to portion or stratify the objects into various groups (clusters) based on similarities and differences. Clustering algorithms are applied to identify the novel patterns in gut microbiota research (Shi et al. 2022), i.e., discovering human enterotypes of microbiota (Arumugam et al. 2011; Karlsson et al. 2013) and groups of

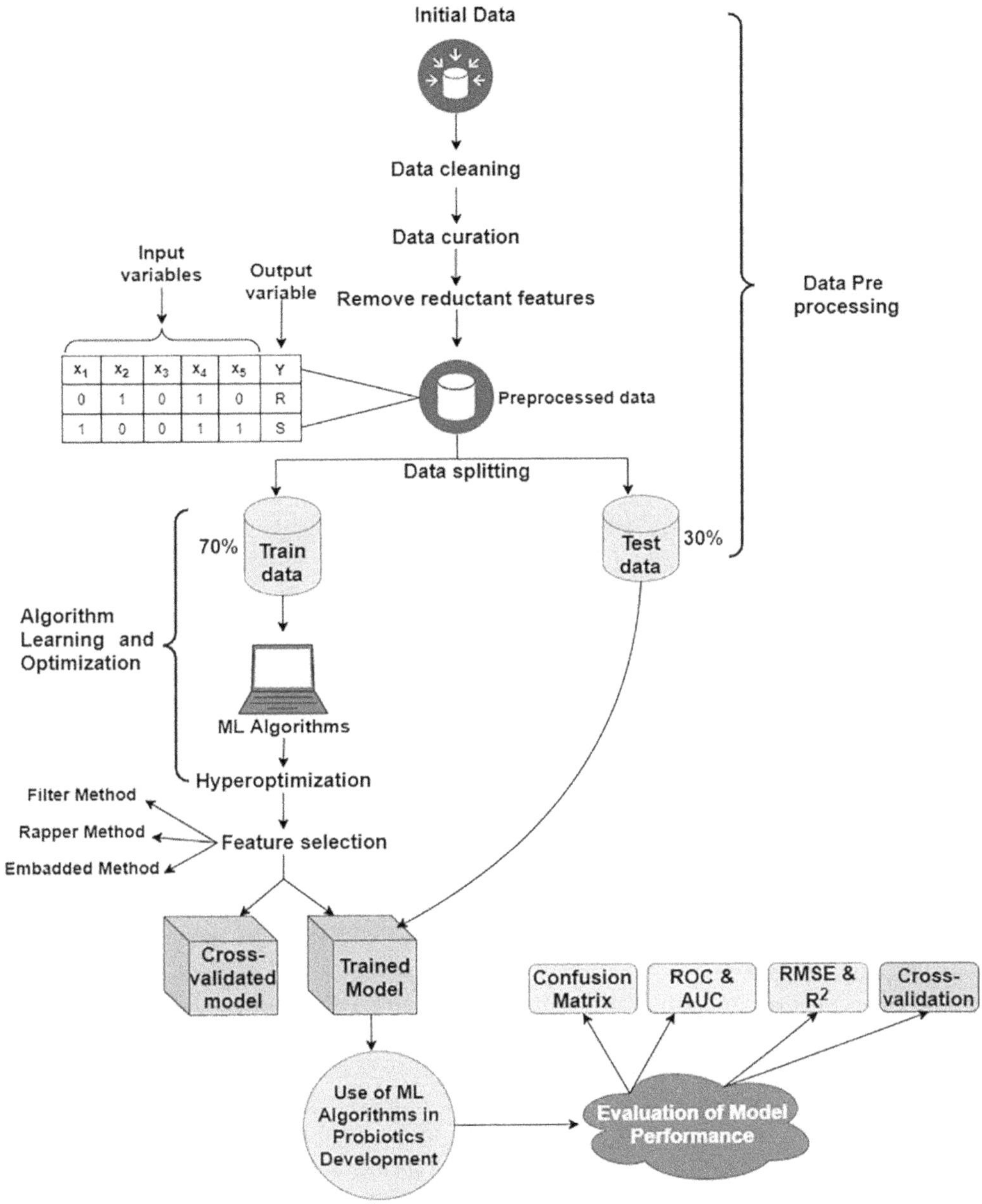

FIGURE 1.3 The proposed study model about the working of supervised learning algorithms.

co-abundance genes (Nielsen et al. 2014). On the other hand, the dimension reduction techniques include principal-component analysis (Ma and Dai 2011), discriminant analysis, and principal-coordinate analysis (Zuur, Ieno, and Smith 2007). These are extensively used to visualize and extract the principal variables from the highly dimensioned omics data.

Probiotics have gained considerable popularity; however, literature on probiotic-specific metabolism during the fermentation process is scarce. A recent study by Sun et al. (2023) utilized ultra-performance liquid chromatography with q-exactive orbitrap mass spectrometry-based metabolomics to monitor alterations in the milk metabolomes during fermentation by two probiotic strains, *Bifidobacterium adolescentis* B8589 and *Lacticaseibacillus paracasei* PC-01. The PCA and OPLS-DA algorithms were employed to analyze the variation in milk metabolome over time. Significant variations in the probiotic-fermented milk metabolome were observed between 0h and 36h of fermentation, with less pronounced variations between the interim period (36h and 60h) and the ripening stage (60h and 72h). Several time point-specific differential metabolites were identified, predominantly associated with organic, amino, and fatty acids. Nine of these identified metabolites were associated with the Krebs cycle, fatty acid, and glutamate metabolism. Toward the end of fermentation, there was an increase in the quantity of pyruvic, capric, and γ-aminobutyric acid (GABA), potentially enhancing the functional properties and nutritional quality of the probiotic fermented milk. This time-based metabolomics study provided a comprehensive analysis of probiotic-specific fermentative changes in milk, offering comprehensive insights into probiotic metabolism within a milk matrix and elucidating the potential favorable mechanisms of probiotic-fermented milk.

Another study by Mazlumi et al. (2022) used unsupervised PCA to evaluate the probiotic characteristics of lactic acid bacteria (LAB) isolates from the intestinal tract of saltwater fish. For this purpose, LAB isolates (n=19), including *Lactiplantibacillus pentosus*, *Pediococcus pentosaceus*, *Lactobacillus acidophilus*, *Levilactobacillus brevis*, *Pediococcus acidilactici*, and *Lactiplantibacillus plantarum* were differentiated by biochemical and morphological characteristics of colonies on Man, Rogosa, and Sharpe broth. All isolates demonstrated resilience under simulated gastrointestinal tract conditions. Auto-aggregation exhibited a range from 1.3±0.5% to 82.6±1.4%, and hydrophobicity with toluene varied from 3.7±1.6% to 69.4±1.3%, while hydrophobicity with xylene spanned from 2.2±1.6% to 56.4±2.1%. Sensitivity and resistance to clinical antibiotics varied among *lactobacilli*, *pediococci*, *enterococci*, and *lactococci* isolates. The non-neutralized cell-free supernatant of isolates F 15 and F 12 exhibited antimicrobial properties against each of the eight assessed enteric pathogens. Cluster analysis (CA), heat-map analysis (HMA), and principal component analysis (PCA) algorithms revealed the significance of isolates F 15, F 12, F7, and F 3 as potential biocontrol mediators in the fishery sector. Results of this study substantially contributed to comprehending the probiotic significance of LAB in saltwater fish, facilitating their evaluation of probiotic properties for potential utilization as biocontrol mediators in the fishery.

Zhang et al. (2022) utilized PCA and HMA to compare probiotic properties of *Lactobacillus* isolated from human GIT and traditional fermented food products. In this study, the probiotic characteristics of *Lactobacillus* strains (n=40) were identified from the GIT of local fermented foods and the human longevity population in China. The determined characteristics included aggregation, adhesion rate, antimicrobial properties, antioxidant ability, hydrophobicity, and survival rates in simulated bile salt and gastric acid. Statistical analyses, including t-tests and PCA, were employed to assess the differences between human and nonhuman strains. Significant variations were observed in the survival rate at 0.3% bile salt, adhesion ability, and antioxidant ability of the fermentation broth (P-value <0.05). PCA results indicated that the first principal component (PC1) score of strains from human sources was considerably greater than that of nonhuman sources (P-value <0.01). Probiotic *Lactobacillus* strains were further selected based on unsupervised CA, HMA, and K-means algorithms. Four strains, CS 128, CS 39, CS 01, and CS 1301, alongside *Lactobacillus rhamnosus* GG (LGG), formed cluster I. Notably, all four human-derived strains were part of this selection. This suggested that *Lactobacillus* strains from human sources exhibit superior probiotic characteristics and functional prospects compared to those from nonhuman sources. The combined use of PCA, CA, and HMA proved to be a visually and effectively insightful approach for the analysis and selection of *Lactobacillus* strains.

Romero et al. (2021) analyzed microorganisms derived from Mexican milk-kefir grains in terms of their probiotic abilities, in vitro fermentation ability with prebiotics (lactulose, citrus-pectin, and inulin), and their ability to produce GABA (Atique et al. 2023).

Additionally, the similarity of microorganisms in terms of the observed properties was identified using CA. Identification of yeast and bacteria from the milk-kefir was done by MALDI-TOF mass spectrometry (MS), which revealed microorganisms belonging to the genus *Leuconostoc*, *Lactococcus*, *Lactobacillus*, and *Kluyveromyces*. The probiotic properties were evaluated through resistance to in vitro gastrointestinal digestion, aggregation abilities, antibiotic susceptibility, and antimicrobial activity. The performance of these isolates compared favorably with commercial probiotics, with most maintaining a concentration >6 log CFU/mL after the intestinal phase. Specific isolates of *Lactobacillus* (BIOTEC-014 and BIOTE-C15), *Kluyveromyces* (BIOTEC-009 and BIOTEC-010), and *Leuconostoc* (BIOTEC-011 and BIOTEC-012) exhibited high fermentability in media supplemented with commercial prebiotics. The ability to produce GABA as compared to commercially available prebiotics was moderate for *K. lactis* BIOTEC-009; *L. lactis* BIOTEC-006, BIOTEC-007, and BIOTEC-008; *L. pseudomesenteroides* BIOTEC-012; and *L. kefiri* BIOTEC-014. The Ward's hierarchical clustering algorithm revealed 2–5 clusters of microorganisms on the basis of linkage distance. The *L. lactis* group was found to be most different from all the other microorganisms involved in the study and was considered to be good starter culture. This multivariate analysis suggested the potential of kefir-isolates as promising cultures for functional dairy food applications.

Nami et al. (2020) characterized LAB isolates for their probiotic properties, as genotypic and phenotypic characterization of the strain of probiotic is crucial for the utilization of nourishment. For this purpose, LAB (n=63) from dairy sources that were previously identified through pyrosequence-based 16S-rRNA profiling were collected and refined, and various probiotic properties were evaluated. This study also evaluated the efficiency of an unsupervised K-means clustering algorithm and HMA in selecting probiotic strains by assessing their phenotypic properties. Findings of this study indicated that ShP 1, YP 9, YP 8, and TP 15 isolates exhibited the maximum acid and bile tolerance, along with cell surface hydrophobicity. Additionally, CP 12, CP 3, TP 17, TP 16, TP 15, ShP 1, YP 9, YP 8, and CuP 3 demonstrated the highest bile salt hydrolase activity. Notably, YP 8, ShP 1, and YP 9 displayed inhibitory effects against *Yersinia* and *Listeria*, respectively. Effective clustering and HMA facilitated the selection and differentiation of nine *Lactobacillus* isolates (YP 8, YP 9, CP 3, TP 15–17, ShP 1, CP 12, and CuP 3) as promising candidate probiotic strains. This research introduced a novel approach to the selection and characterization of bacterial isolates on the basis of their probiotic properties, paving the way for potential use in future functional foods.

1.2.3 Semi-supervised Learning

Semi-supervised learning is positioned at the intersection between supervised and unsupervised learning (Camacho et al. 2018). It entails the utilization of a partially labeled dataset. The ML algorithm uses unsupervised learning to assign labels to the unlabeled data by making inferences from the labeled data. Subsequently, supervised techniques can be used to address specific inquiries pertaining to the labeled information. In practical applications, semi-supervised learning proves to be advantageous, given that real-world datasets often suffer from incomplete labeling due to suboptimal formatting during their inception (Namkung 2020a). The manual process of labeling data can be exceedingly time-consuming, prone to errors, and costly, particularly when dealing with large and intricate datasets. The automated labeling steps in semi-supervised learning offer a prompt and impartial approach to formatting.

1.2.4 Reinforcement Learning

It is target-oriented learning that addresses predefined problems through a self-teaching and iterative approach (Neftci and Averbeck 2019). An algorithm executes actions to solve the problem, with the level of reward experienced in the system increasing as an action gets closer to a solution. Conversely, if an action leads to an outcome that deviates from a solution, the system faces punishment. Over time, the algorithm learns which actions are rewarding, thus paving the way toward problem solution. However, to construct an optimal path to problem solution, reinforcement learning algorithms encounter a dilemma: they must explore new actions while simultaneously striving to maximize rewards. This trade-off between exploration and exploitation is a well-known challenge in reinforcement learning (Sledge and Príncipe 2017). Reinforcement learning includes Q-learning, deep Q-learning, and dtate-action-reward-state-action (SARSA).

1.3 WORKING OF MACHINE LEARNING ALGORITHM

The fundamental steps include feature engineering, training and hyper-parametrization, evaluation of performance, and model testing on the real-world data.

1.3.1 Feature Engineering

Feature engineering is a crucial phase that precedes data preparation for ML models. This involves adding new features (variables) or improving the old ones so that algorithms work better (Saeys, Inza, and Larrañaga 2007). Feature engineering tries to provide informative data to an algorithm for predicting/classifying with better accuracy. The following are the features of engineering techniques commonly used in the probiotic domain.

1.3.1.1 Genomic Feature Engineering

The genomic data is very important in determining the genetic makeup of probiotic strains. One of the key steps in the genomics feature engineering process is extracting meaningful data from the DNA sequence. This includes, for instance, determining functional genes, regulatory items, or probiotic-related gene markers. Raw genomic data may also be transformed into informative features that can be used by ML models using techniques such as k-mer counting or sequence motif analysis (Uddin et al. 2018).

1.3.1.2 Metabolomic Feature Engineering

The fermentation process of microbial community profiles is captured in small molecule data called metabolomics. The feature engineering of metabolomic data includes treating raw spectra or chromatograms to have peak heights, areas, or ratios that are meaningful (Loganathan and Priya Doss C 2022). Other ways of generating higher-level features include the use of metabolic pathway analysis for determining the metabolic activity of probiotic strains during fermentation.

1.3.1.3 Temporal Feature Engineering

Microbial changes are dynamic in fermentation processes. Temporal feature engineering refers to capturing the temporal changes in microbial population and metabolite concentration. Growth rates, fermentation kinetics, or changes in diversity indices are some examples of time-series features that give information on how the fermentation develops and may even be used to predict the ultimate product qualities (Loganathan and Priya Doss C 2022; Uddin et al. 2018).

1.3.1.4 Spatial Feature Engineering

Spatial features are related to the spread of microorganisms in fermentation tanks or the clustering of microcolonies in microscopes. For example, they can use spatial statistics such as Moran's I or Geary's C (Loganathan and Priya Doss C 2022) to measure spatial autocorrelation and understand whether microbes clump together or disperse. Spatial features can be extracted from microscopy images by employing image processing techniques such as texture analyses, leading to a better understanding of the microbial behavior.

1.3.1.5 Textural and Sensory Feature Extraction

Feature engineering can make use of sensory data, such as descriptive sensory analysis or consumer panels. Word embeddings and sentiment analysis, which are among many NLP techniques, can also be applied to extracting features from textual descriptions provided by sensory panels or consumers (Hany, Rashwan, and Abdelmotilib 2023). These features can sense the flavored, texturized, and overall sensory profiles associated with probiotic products, allowing tailor-made manufactured products to meet consumers' preferences.

1.3.2 Feature Engineering Process

It is further subdivided into:

- Data-preprocessing
- Extraction and selection features

1.3.2.1 Data Preprocessing

In ML, data preprocessing refers to a series of steps and techniques applied to raw data prior to use in building the algorithms (Brownlee 2020). It transforms the raw data into a more refined form to extract meaningful insights (Kotsiantis, Kanellopoulos, and Pintelas 2006; Zelaya 2019). Preprocessing of data consists of the following steps.

1.3.2.1.1 Data Cleaning

It is the most critical and fundamental step in ML. Data cleaning includes the deletion of repeating values, imputation of missing values, and streamlining the variables (Chu et al. 2016). The deletion of missing values results in data loss, so based on the nature of the missing values, an optimum imputing algorithm must be selected. Missing values can be missing at random, not at random, and complete at random (Alabadla et al. 2022; Jerez et al. 2010).

1.3.2.1.2 Data Splitting

In ML, the whole data can be used to train the model, but it is a common practice to split the input data into train, test, and validation dataset. The main goal is to evaluate the model performance on the unseen test data (Uçar et al. 2020). This also facilitates assessing the generalized prediction ability of the model and to overcome the issues of overfitting and underfitting.

1.3.3 Feature Selection and Extraction

Feature extraction aims to create a feature vector that represents a reduced number of variables derived from the raw measured data (Brownlee 2020). This feature vector should contain sufficient relevant information from the original data, which in turn can facilitate subsequent training processes. However, it is highly unlikely that many features within the dataset would still lack informative value and be irrelevant for the construction of a predictive model. An example of this is when constructing a model with an extensive number of variables such as genes, proteins, or metabolites, which demands substantial computational power and memory and is prone to overfitting (Ditzler et al. 2015). Hence, the selection of features becomes crucial in order to obtain an optimal and non-redundant subset of the initial features. This subset plays a critical role in enabling fast model training, improving performance, and even enhancing model interpretation.

1.3.3.1 Feature Selection Techniques

Feature selection is broadly classified into filter, wrapper, and embedded methods (Chandrashekar and Sahin 2014). Filter methods are the statistical measurements that rank the features on the basis of their correlation with the output variables. Highly correlated variables are less important and are removed from the training data. The commonly used statistical measures in filter methods include Chi-square, correlation coefficient, mutual information, and ANOVA. After feature ranking, the most important ones are selected by setting a threshold value. In wrapper methods, subsets of features are selected by training model on specific features subset and evaluating performance on validation data. This procedure is repeated for different features subsets, and the best-performing features are opted for. The main advantage of wrapper methods is that these techniques consider the feature interactions that could be vital for some models (Domingos 2012). In the embedded method, feature selection is integrated into the ML process. Examples of embedded methods include L1 regularization and DTs. L1 regularization reduces the value of feature coefficients, causing some coefficients to be set to zero, effectively eliminating those features from the model. DTs can be utilized to choose the features by assessing the Gini values that divide tree nodes. It's important to note that feature selection is not always necessary or advantageous. In situations, utilizing all features can result in improved model performance, especially when dealing with a manageable number of features. Furthermore, certain models, like networks, have the ability to learn relevant features automatically from raw data, thereby reducing the significance of feature selection (Saeys, Inza, and Larrañaga 2007). Feature selection methods rely on data nature and the problem to be solved, and it is usually a hit-and-trail method to select an optimum feature extraction method.

1.3.4 Model Training and Hyper-parametrization

ML algorithms are trained on training data, and it usually comprises a major portion of input data (most commonly 60–80%). Train data is applied to fit the model parameters and to find out the relationship between independent variables and outcome variables. During the training phase, model performance is evaluated on validation data, usually 20% of the original data. It is used to fine-tune the hyperparameters of models to improve performance (Wei et al. 2019). Hyperparameters are model configurations not learned during training but rather fixed before training. Examples include numbers of trees, the number of variables tried at each split, and the maximum tree depth in the RF Model (Sun et al. 2020). After the model training and optimization, its accuracy is measured on the unseen test data, 20% of input data. Testing set is crucial to determine the final performance of the model, whether it can be applied in real-world scenarios.

1.3.5 Selecting a Sophisticated ML Algorithm

Developing a robust model, especially when high-quality training data is not available, relies greatly on choosing a sophisticated and appropriate algorithm. Although the model is built on data, the selection of an appropriate algorithm can substantially influence its predictive ability. Hence, choosing the proper algorithm involves considering numerous aspects, such as the nature of input data, learning and prediction type, desired labels, and the entirety of data, including data quality and quantity (Namkung 2020b).

The selection of the appropriate ML algorithm often relies on the input data. The input/feeding data can be in various forms, such as numeric, categorical, images, or text. The CNN could be appropriate for image classification tasks since it is developed to acquire information from two-dimensional input data, e.g., images. Likewise, an RF algorithm may be suitable for handling structured data like numeric or categorical data.

The other vital aspect in the selection of an ML algorithm involves learning and prediction type. There exist different algorithms that are meant for various purposes, including classification, clustering, regression, and dimensionality reduction (Ghannam and Techtmann 2021). Therefore, one must understand the kind of task to be done and the desired output format for choosing the right algorithm. For example, classifying data is related to making a prediction of a categorical variable, while regression is about predicting a continuous variable (Ghazanfar et al. 2023).

The targeted variables are another important factor for an algorithm selection between the learning and prediction types. There are algorithms built to accommodate multi-label or multi-class classification, while there are others designed for binary classification. Thus, proper selection of an algorithm presupposes a clear understanding of the target variable and the required output format.

Other considerations in algorithm selection include data completeness, quantity, and quality. The performance of the algorithm depends on the provided data quantity and quality. While some algorithms require large amounts of data to yield accurate results, others can be more reactive to the quality rather than the quantity of the data used (McCoubrey et al. 2021). Therefore, one needs to thoroughly understand data coupled with specific matters of concern to select a suitable algorithm.

1.3.6 Evaluation of ML Algorithm Performance

It is a crucial step to construct a reliable and efficient model. ML algorithms are trained on datasets, and the fundamental aspects are to make predictions similar to real-world scenarios by evaluating performance on validation data. Confusion matrix, receiver-operation characteristic (ROC), area-under curve (AUC), root means square error (RMSE), and coefficient of determination (R^2) are commonly used metrics to evaluate performance of classification or regression algorithms (Namkung 2020a). In a binary classification problem, metrics such as sensitivity, specificity, F-score, and Matthews correlation coefficient (MCC) are commonly used to evaluate a model's performance.

1.3.7 Confusion Matrix

A confusion matrix is an important and useful tool for evaluating binary classification model performance. It is a table that compares the predicted and actual values in terms of true positives (TP), false positives (FP), true negatives, and false negatives (FN) (Marom, Rokach, and Shmilovici 2010) (Figure 1.4).

Sensitivity is the ability to correctly identify the positive instances out of all the positive instances.

$$Sensitivity = \frac{TP}{TP + FN}$$

Specificity is the ability of a model to correctly identify the true negative instances out of all the instances.

$$Specificity = \frac{TN}{TN + FN}$$

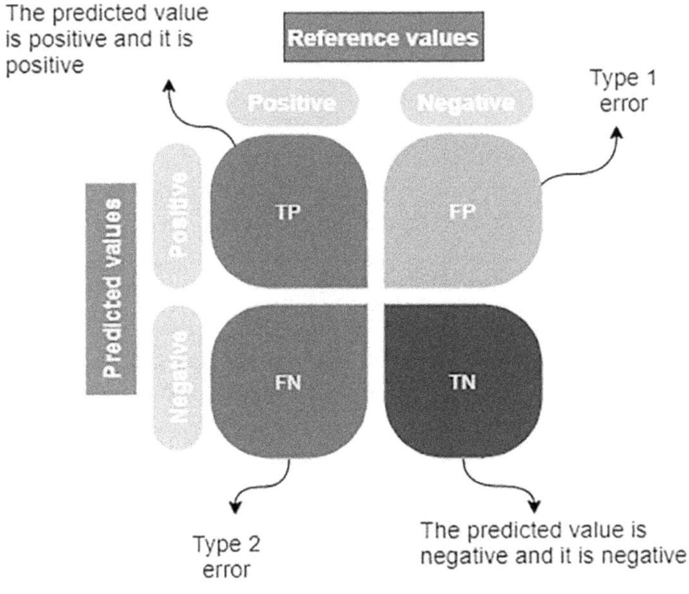

FIGURE 1.4 Confusion matrix to evaluate the performance of binary classification algorithms.

A high specificity score means that the model is effective in identifying most of the negative instances in the data.

The F-score is a weighted average of precision and recall, which are both important metrics for evaluating classification models. Precision of a model is calculated by dividing the TP to sum of TP+FP. The recall is measured by dividing the TP to sum of TP+FN. F-score is calculated by the harmonic mean of both precision and accuracy. Its values range from 0 to 1, and higher values indicate better performance. The F-score is a useful metric when the data is imbalanced, i.e., when the number of positive and negative instances is significantly different.

$$Precision = \frac{TP}{TP + FP}$$

$$Recall = \frac{TP}{TP + FN}$$

$$F\ score = \frac{TP}{TP + 1/2(FP + FN)}$$

The MCC is a metric that considers all four values in the confusion matrix to evaluate a model's performance. It ranges from –1 to +1.

Disagreement b/w actual & predicted.	Random prediction	Perfect prediction
–1	0	+1

MCC score values.

MCC is especially useful when the data is imbalanced or when there are differences in the class distributions.

$$MCC = \frac{TP \times TN - FP \times FN}{\sqrt{(TP + FP)(TP + FN)(TN + FP)(TN + FN)}}$$

1.3.8 ROC-AUC

ROC is used to measure the model performance by plotting true-positive rate (TPR); sensitivity versus true-negative rate (TNR); 1-specificity at different threshold values of the classification model as depicted in Figure 1.5.

The ROC curve of a better-performance classifier must be closer to the top left corner. AUC values range from 0 to 1 and are used to compare various model performances (Sokolova, Japkowicz, and Szpakowicz 2006).

1.3.9 Cross-Validation

The performance of an ML classifier can also be measured by cross-validation (CV). CV involves splitting the data into several folds or subsets, with onefold serving as the validation set while the remaining folds are used for training the model. The procedure is undertaken several times, whereby each of the folds constitutes the validation subset (Figure 1.6). Finally, the results from each model validation step are averaged to provide more accurate estimates of the model's performance (Wong 2015).

1.3.10 MSE & RMSE

MSE and RMSE are usually taken as the most important assessment criteria for regression models. MSE is the average squared difference between the predicted and the actual values, and RMSE is the square root of MSE. Therefore, RMSE presents a more interpretable measure of the error.

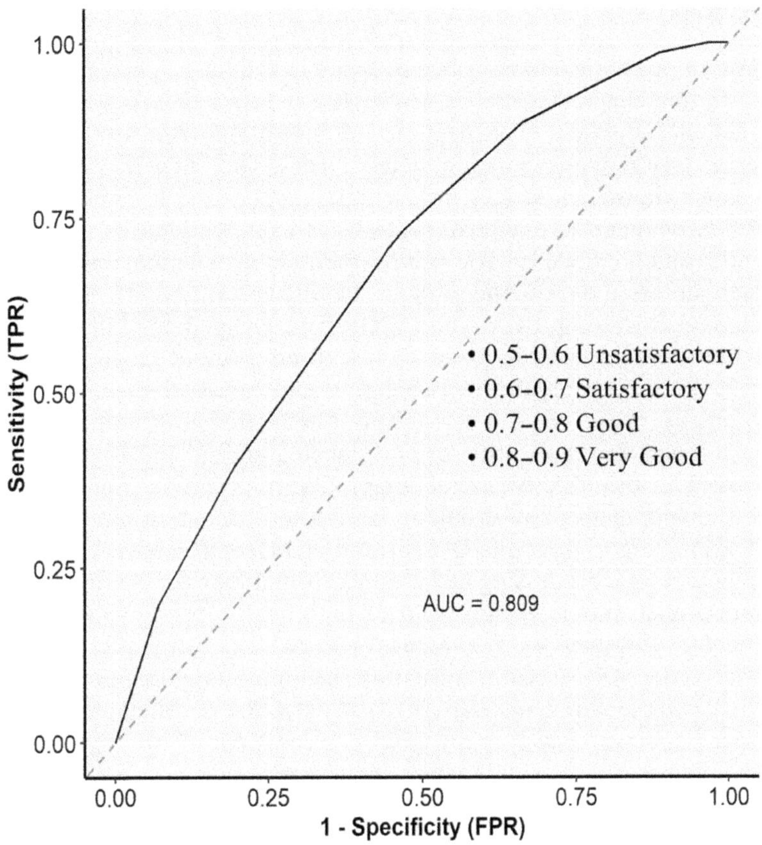

FIGURE 1.5 AUC-ROC to calculate the prediction performance.

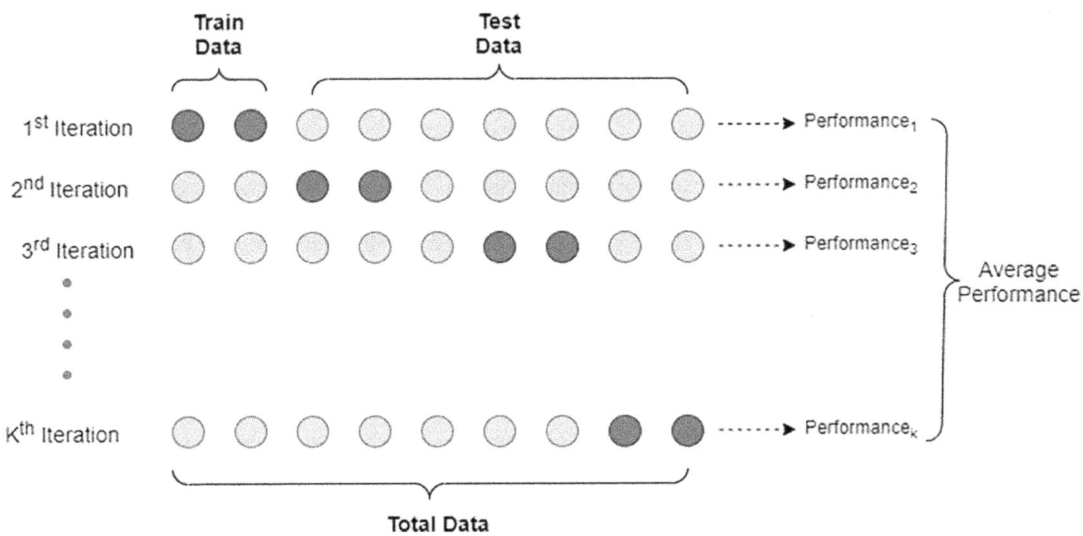

FIGURE 1.6 K-fold cross-validation to split the train data into various sub-folds.

1.4 CONCLUSION

The growing data in probiotics domain has presented pivotal insights into the application of ML algorithms. In this chapter, we discussed different facets of ML applications in probiotics, including their introduction, classification, the algorithms used in developing probiotics, feature engineering, and model evaluation metrics. The comprehensive review of ML usage in probiotics has shown significant strides in leveraging supervised learning, such as RF, SVM, and NN. The ability of supervised algorithms to learn from the input-labeled data has enabled to correctly identify and predict probiotic strains, their mode of action, and interaction, resulted in the development of tailored probiotic products having specific health benefits. Unsupervised algorithms, including clustering, PCA, and t-SNE, have enabled us to understand and explore the hidden complex data patterns in probiotics, leading to the discovery of novel probiotics and revealing probiotic-microorganism association and behavior. Semi-supervised learning techniques like label propagation and self-training have proven to be useful in label-limited data situations. Through the implication of both labeled and unlabeled data, semi-supervised learning has allowed for the effective extraction of vital information from insufficient datasets, enhancing overall comprehension of probiotic interactions. In addition, the recent adoption of reinforcement learning algorithms like Q-Learning and Deep Q-networks provides new channels for improving probiotic processing and product formulation, integrating adaptive decision-making frameworks that maximize the yield of high-quality fermented dairy products. The integration of Feature Engineering Techniques with ML algorithms like PCA and RFE has been instrumental in deriving useful information from probiotic data to allow the determination of key variables or features and patterns that drive the development of potent probiotic solutions. Another important aspect is the careful assessment of ML algorithm performance, which inherently emphasizes the significance of rigorous model assessment and validation for ML-inspired probiotic progress. In brief, the application of ML to probiotics and fermented dairy products has the potential to revolutionize the field. These recent developments in ML approaches and their use specifically in probiotic studies have opened an opportunity to use a precision approach when developing probiotics, customized health measures, and innovating products. Moving forward with ML to shape the future of probiotics and fermented dairy products, it has become mandatory to stress on interdisciplinary collaboration, data quality, and ethical concerns to ensure maximum impact of ML in probiotics and fermented dairy foods. The synergy between ML and probiotics continues to evolve, exploring a novel frontier at the nexus of human health and wellness.

REFERENCES

Alabadla, Mustafa, Fatimah Sidi, Iskandar Ishak, Hamidah Ibrahim, Lilly Suriani Affendey, Zafienas Che Ani, Marzanah A Jabar, Umar Ali Bukar, Navin Kumar Devaraj, and Ahmad Sobri Muda. 2022. "Systematic review of using machine learning in imputing missing values." *IEEE Access* 10: 44483–502.

Arumugam, M., J. Raes, E. Pelletier, D. Le Paslier, T. Yamada, D. R. Mende, G. R. Fernandes, J. Tap, T. Bruls, J. M. Batto, M. Bertalan, N. Borruel, F. Casellas, L. Fernandez, L. Gautier, T. Hansen, M. Hattori, T. Hayashi, M. Kleerebezem, K. Kurokawa, M. Leclerc, F. Levenez, C. Manichanh, H. B. Nielsen, T. Nielsen, N. Pons, J. Poulain, J. Qin, T. Sicheritz-Ponten, S. Tims, D. Torrents, E. Ugarte, E. G. Zoetendal, J. Wang, F. Guarner, O. Pedersen, W. M. de Vos, S. Brunak, J. Doré, M. Antolín, F. Artiguenave, H. M. Blottiere, M. Almeida, C. Brechot, C. Cara, C. Chervaux, A. Cultrone, C. Delorme, G. Denariaz, R. Dervyn, K. U. Foerstner, C. Friss, M. van de Guchte, E. Guedon, F. Haimet, W. Huber, J. van Hylckama-Vlieg, A. Jamet, C. Juste, G. Kaci, J. Knol, O. Lakhdari, S. Layec, K. Le Roux, E. Maguin, A. Mérieux, R. Melo Minardi, C. M'Rini, J. Muller, R. Oozeer, J. Parkhill, P. Renault, M. Rescigno, N. Sanchez, S. Sunagawa, A. Torrejon, K. Turner, G. Vandemeulebrouck, E. Varela, Y. Winogradsky, G. Zeller, J. Weissenbach, S. D. Ehrlich, and P. Bork. 2011. "Enterotypes of the human gut microbiome." *Nature* 473 (7346): 174–80. https://doi.org/10.1038/nature09944.

Atique, U., M. Altaf, D. Sinha, S. Ghazanfar, M. A. Haque, and S. Chowdhury. 2023. "The Role of probiotics and prebiotics in gut modulation." In *The Gut Microbiota in Health and Disease*, edited by Nimmy Srivastava, Salam A. Ibrahim, Jayeeta Chattopadhyay, and Mohamed H. Arbab, 205–16. John Wiley & Sons.

Azad, H., M. Y. Akbar, J. Sarfraz, W. Haider, and S Ghazanfar. 2025. "Simulation studies to identify high-affinity probiotic peptides for inhibiting PAK1 gastric cancer protein: A comparative approach." *Computational Biology and Chemistry* 115: 10834.

Bergamini, Carlo M., Nicoletta Bianchi, Valerio Giaccone, Paolo Catellani, Leonardo Alberghini, Alessandra Stella, Stefano Biffani, Sachithra Kalhari Yaddehige, Tania Bobbo, and Cristian Taccioli. 2022. "Machine learning algorithms highlight tRNA information content and Chargaff's second parity rule score as important features in discriminating probiotics from non-probiotics." *Biology* 11 (7): 1024.

Breiman, Leo. 2001. "Random forests." *Machine Learning* 45: 5–32.

Brownlee, Jason. 2020. *Data Preparation for Machine Learning: Data Cleaning, Feature Selection, and Data Transforms in Python*. Machine Learning Mastery.

Camacho, D. M., K. M. Collins, R. K. Powers, J. C. Costello, and J. J Collins. 2018. "Next-generation machine learning for biological networks." *Cell* 173 (7): 1581–92. https://doi.org/10.1016/j.cell.2018.05.015.

Chandrashekar, Girish, and Ferat Sahin. 2014. "A survey on feature selection methods." *Computers & Electrical Engineering* 40 (1): 16–28. https://doi.org/10.1016/j.compeleceng.2013.11.024.

Chu, Xu, Ihab F. Ilyas, Sanjay Krishnan, and Jiannan Wang. 2016. "Data Cleaning: Overview and Emerging Challenges." Proceedings of the 2016 International Conference on Management of Data, San Francisco, California, USA.

Cortes, Corinna, and Vladimir Vapnik. 1995. "Support-vector networks." *Machine Learning* 20: 273–97.

Ditzler, G., J. C. Morrison, Y. Lan, and G. L. Rosen. 2015. "Fizzy: Feature subset selection for metagenomics." *BMC Bioinformatics* 16: 358. https://doi.org/10.1186/s12859-015-0793-8.

Domingos, Pedro. 2012. "A few useful things to know about machine learning." *Commun. ACM* 55: 78–87. https://doi.org/10.1145/2347736.2347755.

Friedman, Nir, Dan Geiger, and Moises Goldszmidt. 1997. "Bayesian network classifiers." *Machine Learning* 29: 131–63.

Ghahramani, Zoubin. 2004. "Unsupervised Learning." In *Advanced Lectures on Machine Learning: ML Summer Schools 2003, Canberra, Australia, February 2–14, 2003, Tübingen, Germany, August 4–16, 2003, Revised Lectures*, edited by Olivier Bousquet, Ulrike von Luxburg and Gunnar Rätsch, 72–112. Berlin, Heidelberg: Springer Berlin Heidelberg.

Ghannam, Ryan B., and Stephen M. Techtmann. 2021. "Machine learning applications in microbial ecology, human microbiome studies, and environmental monitoring." *Computational and Structural Biotechnology Journal* 19: 1092–107. https://doi.org/10.1016/j.csbj.2021.01.028.

Ghazanfar, S., R. Abid, A. Khan, M. Y. Akbar, I. Asif, R. Nadeem, and A Ansari. (2023). "Machine Learning to Explore the Psychrophilic Soil Microbiome." In *Soil Microbiome of the Cold Habitats*, 213–22. CRC Press.

Ghyselinck, Jonas, Lynn Verstrepen, Frédéric Moens, Pieter Van Den Abbeele, Arnout Bruggeman, Jawal Said, Barry Smith, Lynne Ann Barker, Caroline Jordan, and Valentina Leta. 2021. "Influence of probiotic bacteria on gut microbiota composition and gut wall function in an in-vitro model in patients with Parkinson's disease." *International Journal of Pharmaceutics: X* 3: 100087.

Hany, Maha, Shaheera Rashwan, and Neveen M Abdelmotilib. 2023. "Amachine learning method for prediction of yogurt quality and consumers preferences using sensory attributes and image processing techniques." *Machine Learning and Applications: An International Journal* 10 (1): 1–7.

Hartigan, John A, and Manchek A Wong. 1979. "Algorithm AS 136: A k-means clustering algorithm." *Journal of the Royal Statistical Society. Series c (applied Statistics)* 28 (1): 100–8.

Jerez, José M, Ignacio Molina, Pedro J García-Laencina, Emilio Alba, Nuria Ribelles, Miguel Martín, and Leonardo Franco. 2010. "Missing data imputation using statistical and machine learning methods in a real breast cancer problem." *Artificial Intelligence in Medicine* 50 (2): 105–15.

Kamarudin, M. B., C. S. Ong, and S. C. Tan. 2022. "Quantum-Inspired Differential Evolution Algorithm in Probiotics Marker Genes Selection." 2022 10th International Conference on Information and Communication Technology (ICoICT), 2–3 Aug. 2022.

Karlsson, Fredrik H, Valentina Tremaroli, Intawat Nookaew, Göran Bergström, Carl Johan Behre, Björn Fagerberg, Jens Nielsen, and Fredrik Bäckhed. 2013. "Gut metagenome in European women with normal, impaired and diabetic glucose control." *Nature* 498 (7452): 99–103.

Kohonen, T. 1990. "The self-organizing map." *Proceedings of the IEEE* 78 (9): 1464–80. https://doi.org/10.1109/5.58325.

Kotsiantis, Sotiris B, Dimitris Kanellopoulos, and Panagiotis E Pintelas. 2006. "Data preprocessing for supervised leaning." *International Journal of Computer Science* 1 (2): 111–7.

Lam, Ka-Lung, Wai-Yin Cheng, Yuting Su, Xiaojie Li, Xiyang Wu, Ka-Hing Wong, Hoi-Shan Kwan, and Peter Chi-Keung Cheung. 2020. "Use of random forest analysis to quantify the importance of the structural characteristics of beta-glucans for prebiotic development." *Food Hydrocolloids* 108: 106001. https://doi.org/10.1016/j.foodhyd.2020.106001.

Loganathan, Tamizhini, and Priya Doss C. George. 2022. "The influence of machine learning technologies in gut microbiome research and cancer studies – a review." *Life Sciences* 311: 121118. https://doi.org/10.1016/j.lfs.2022.121118.

Long, Y., M. Wu, C. K. Kwoh, J. Luo, and X. Li. 2020. "Predicting human microbe-drug associations via graph convolutional network with conditional random field." *Bioinformatics* 36 (19): 4918–27. https://doi.org/10.1093/bioinformatics/btaa598.

Ma, Shuangge, and Ying Dai. 2011. "Principal component analysis based methods in bioinformatics studies." *Briefings in Bioinformatics* 12 (6): 714–22.

Manandhar, Ishan, Ahmad Alimadadi, Sachin Aryal, Patricia B Munroe, Bina Joe, and Xi Cheng. 2021. "Gut microbiome-based supervised machine learning for clinical diagnosis of inflammatory bowel diseases." *American Journal of Physiology-Gastrointestinal and Liver Physiology* 320 (3): G328–37.

Marom, Nadav David, Lior Rokach, and Armin Shmilovici. 2010. "Using the confusion matrix for improving ensemble classifiers." 2010 IEEE 26-Th Convention of Electrical and Electronics Engineers in Israel.

Martinez-Guryn, Kristina, Vanessa Leone, and Eugene B Chang. 2019. "Regional diversity of the gastrointestinal microbiome." *Cell Host & Microbe* 26 (3): 314–24.

Mazlumi, Atefeh, Bahman Panahi, Mohammad Amin Hejazi, and Yousef Nami. 2022. "Probiotic potential characterization and clustering using unsupervised algorithms of lactic acid bacteria from saltwater fish samples." *Scientific Reports* 12 (1): 11952. https://doi.org/10.1038/s41598-022-16322-z.

McCoubrey, Laura E., Moe Elbadawi, Mine Orlu, Simon Gaisford, and Abdul W. Basit. 2021. "Harnessing machine learning for development of microbiome therapeutics." *Gut Microbes* 13 (1): 1872323. https://doi.org/10.1080/19490976.2021.1872323.

Moens, Frédéric, Pieter Van den Abbeele, Abdul W Basit, Cornelius Dodoo, Rangan Chatterjee, Barry Smith, and Simon Gaisford. 2019. "A four-strain probiotic exerts positive immunomodulatory effects by enhancing colonic butyrate production in vitro." *International Journal of Pharmaceutics* 555: 1–10.

Nami, Yousef, Bahman Panahi, Hossein Mohammadzadeh Jalaly, Reza Vaseghi Bakhshayesh, and Mohammad Amin Hejazi. 2020. "Application of unsupervised clustering algorithm and heat-map analysis for selection of lactic acid bacteria isolated from dairy samples based on desired probiotic properties." *LWT* 118: 108839. https://doi.org/10.1016/j.lwt.2019.108839.

Namkung, J. 2020a. "Machine learning methods for microbiome studies." *Journal of Microbiology* 58 (3): 206–16. https://doi.org/10.1007/s12275-020-0066-8.

Namkung, Junghyun. 2020b. "Machine learning methods for microbiome studies." *Journal of Microbiology* 58: 206–16.

Neftci, Emre O, and Bruno B Averbeck. 2019. "Reinforcement learning in artificial and biological systems." *Nature Machine Intelligence* 1 (3): 133–43.

Nielsen, H. B., M. Almeida, A. S. Juncker, S. Rasmussen, J. Li, S. Sunagawa, D. R. Plichta, L. Gautier, A. G. Pedersen, E. Le Chatelier, E. Pelletier, I. Bonde, T. Nielsen, C. Manichanh, M. Arumugam, J. M. Batto, M. B. Quintanilha Dos Santos, N. Blom, N. Borruel, K. S. Burgdorf, F. Boumezbeur, F. Casellas, J. Doré, P. Dworzynski, F. Guarner, T. Hansen, F. Hildebrand, R. S. Kaas, S. Kennedy, K. Kristiansen, J. R. Kultima, P. Léonard, F. Levenez, O. Lund, B. Moumen, D. Le Paslier, N. Pons, O. Pedersen, E. Prifti, J. Qin, J. Raes, S. Sørensen, J. Tap, S. Tims, D. W. Ussery, T. Yamada, P. Renault, T. Sicheritz-Ponten, P. Bork, J. Wang, S. Brunak, and S. D. Ehrlich. 2014. "Identification and assembly of genomes and genetic elements in complex metagenomic samples without using reference genomes." *Nature Biotechnology* 32 (8): 822–8. https://doi.org/10.1038/nbt.2939.

Noor, F., M. Junaid, A. H. Almalki, M. Almaghrabi, S. Ghazanfar, and M Tahir ul Qamar. 2024. "Deep learning pipeline for accelerating virtual screening in drug discovery." *Scientific Reports* 14 (1): 28321.

Oliveira, Francislon S, John Brestelli, Shon Cade, Jie Zheng, John Iodice, Steve Fischer, Cristina Aurrecoechea, Jessica C Kissinger, Brian P Brunk, and Christian J Stoeckert Jr. 2018. "MicrobiomeDB: A systems biology platform for integrating, mining and analyzing microbiome experiments." *Nucleic Acids Research* 46 (D1): D684–91.

Romero, Alejandra Hurtado, Mariano Del Toro-Barbosa, Misael Sebastián Gradilla-Hernández, Luis Eduardo Garcia-Amezquita, and Tomás García-Cayuela. 2021. "Probiotic properties, prebiotic fermentability, and GABA-producing capacity of microorganisms isolated from Mexican milk kefir grains: A clustering evaluation for functional dairy food applications." *Foods* 10 (10): 2275.

Saeys, Y., I. Inza, and P. Larrañaga. 2007. "A review of feature selection techniques in bioinformatics." *Bioinformatics* 23 (19): 2507–17. https://doi.org/10.1093/bioinformatics/btm344.

Sen, Pratap Chandra, Mahimarnab Hajra, and Mitadru Ghosh. 2019. "Supervised classification algorithms in machine learning: A survey and review." *Advances in Intelligent Systems and Computing*. doi: 10.1007/978-981-13-7403-6_11.

Shi, Yushu, Liangliang Zhang, Christine B Peterson, Kim-Anh Do, and Robert R Jenq. 2022. "Performance determinants of unsupervised clustering methods for microbiome data." *Microbiome* 10 (1): 1–12.

Sledge, Isaac J, and José C Príncipe. 2017. "Balancing Exploration and Exploitation in Reinforcement Learning Using a Value of Information Criterion." 2017 IEEE International Conference on Acoustics, Speech and Signal Processing (ICASSP).

Sokolova, Marina, Nathalie Japkowicz, and Stan Szpakowicz. 2006. "Beyond accuracy, f-score and ROC: A family of discriminant measures for performance evaluation." *Australasian Joint Conference on Artificial Intelligence*. doi: 10.1007/11941439_114.

Sun, S., Z. Cao, H. Zhu, and J. Zhao. 2020. "A survey of optimization methods from a machine learning perspective." *IEEE Transactions on Cybernetics* 50 (8): 3668–81. https://doi.org/10.1109/TCYB.2019.2950779.

Sun, Yaru, Shuai Guo, Ting Wu, Jingwen Zhang, Lai-Yu Kwok, Zhihong Sun, Heping Zhang, and Jicheng Wang. 2023. "Untargeted mass spectrometry-based metabolomics approach unveils biochemical changes in compound probiotic fermented milk during fermentation." *Npj Science of Food* 7 (1): 21. https://doi.org/10.1038/s41538-023-00197-z.

The Human Microbiome Project Consortium. 2012. Structure, function and diversity of the healthy human microbiome. *Nature* 486 (7402): 207–14. https://doi.org/10.1038/nature11234.

Uçar, Muhammed Kürşad, Majid Nour, Hatem Sindi, and Kemal Polat. 2020. "The effect of training and testing process on machine learning in biomedical datasets." *Mathematical Problems in Engineering*. https://doi.org/10.1155/2020/2836236.

Uddin, Muhammad Fahim, Jeongkyu Lee, Syed Rizvi, and Samir Hamada. 2018. "Proposing enhanced feature engineering and a selection model for machine learning processes." *Applied Sciences* 8 (4): 646.

Veiga, Patrick, Jotham Suez, Muriel Derrien, and Eran Elinav. 2020. "Moving from probiotics to precision probiotics." *Nature Microbiology* 5 (7): 878–80.

Vujkovic-Cvijin, I., J. Sklar, L. Jiang, L. Natarajan, R. Knight, and Y. Belkaid. 2020. "Host variables confound gut microbiota studies of human disease." *Nature* 587 (7834): 448–54. https://doi.org/10.1038/s41586-020-2881-9.

Wei, Jing, Xuan Chu, Xiang-Yu Sun, Kun Xu, Hui-Xiong Deng, Jigen Chen, Zhongming Wei, and Ming Lei. 2019. "Machine learning in materials science." *InfoMat* 1 (3): 338–58.

Westfall, Susan, Francesca Carracci, Molly Estill, Danyue Zhao, Qing-li Wu, Li Shen, James Simon, and Giulio Maria Pasinetti. 2021. "Optimization of probiotic therapeutics using machine learning in an artificial human gastrointestinal tract." *Scientific Reports* 11 (1): 1067. https://doi.org/10.1038/s41598-020-79947-y.

Wong, Tzu-Tsung. 2015. "Performance evaluation of classification algorithms by k-fold and leave-one-out cross validation." *Pattern Recognition* 48 (9): 2839–46.

Yu, Y, S Dunaway, J Champer, J Kim, and A Alikhan. 2020. "Changing our microbiome: Probiotics in dermatology." *British Journal of Dermatology* 182 (1): 39–46.

Zelaya, C. V. Gonzalez. 2019. "Towards Explaining the Effects of Data Preprocessing on Machine Learning." 2019 IEEE 35th International Conference on Data Engineering (ICDE), 8–11 April 2019.

Zhang, Longfei, Hengxian Qu, Xiaoxiao Liu, Qiming Li, Yang Liu, Wenqiong Wang, Dawei Chen, Lixia Xiao, and Ruixia Gu. 2022. "Comparison and selection of probiotic *Lactobacillus* from human intestinal tract and traditional fermented food in vitro via PCA, unsupervised clustering algorithm, and heat-map analysis." *Food Science & Nutrition* 10 (12): 4247–57. https://doi.org/10.1002/fsn3.3018.

Zhou, Y, ZZ Xu, Y He, Y Yang, L Liu, Q Lin, Y Nie, M Li, F Zhi, and S Liu. 2018. "Gut microbiota offers universal biomarkers across ethnicity in inflammatory bowel disease diagnosis and infliximab response prediction." *mSystems* 3: e00188-17.

Zhu, Jie, Liu Tian, Peishan Chen, Mo Han, Liju Song, Xin Tong, Xiaohuan Sun, Fangming Yang, Zhipeng Lin, and Xing Liu. 2022. "Over 50,000 metagenomically assembled draft genomes for the human oral microbiome reveal new taxa." *Genomics, Proteomics & Bioinformatics* 20 (2): 246–59.

Zuur, Alain F, Elena N Ieno, and Graham M Smith. 2007. *Analysing Ecological Data*. Vol. 680. Springer.

1 (Part II) Machine Learning Perspective

Precision Probiotics and Fermented Dairy Foods

Muhammad Hussain Ghazali, Abdullah Ahmed Butt, Muhammad Tahir, Rameesha Azhar, Faizan Saleem, and Mehwish Rafiq

1.1 INTRODUCTION

In recent years, probiotic foods have attracted extra interest due to their potential health benefits. Fortification or enrichment of functional ingredients can be accomplished, masking an extensive spectrum of products consisting of probiotics, prebiotics, and symbiotics. As customer awareness of fitness and nutrients has extended, ongoing efforts have brought about the continuing discovery of sorts with healing houses derived from traditional fermented ingredients. Those useful meals fall into two number-one classes: dairy-based and non-dairy-based. Similarly, to meet dietary desires, those foods may additionally exert an expansion of physiological consequences, doubtlessly gambling on an unfavorable or beneficial position in certain illnesses (Koletzko et al. 1998). The call for probiotic dairy merchandise keeps pushing upward, usually attributed to their fitness benefits and their capability to relieve the soreness of lactose intolerance, but research on pediatric probiotics suggests sizable variability, preventing it from attaining conclusive findings. Mainly, probiotics are lactic acid bacteria (LAB), famous for *Lactobacillus*. Other genera also contain probiotic lines (Lee et al. 2018). LAB, often known as probiotics, produce antimicrobial substances that influence the composition of the oral microflora. Nevertheless, there is a lack of comprehensive information on their exact functions in the oral cavity. Specific strains of LAB have beneficial compounds like ACE inhibition, EPS synthesis, etc. (James and Wang 2019).

Probiotic therapy is gaining interest, particularly in oral medicine and dentistry, in the context of growing antimicrobial drug resistance (Meurman 2005). Consistently incorporating probiotic-rich foods into your diet is a sure-fire approach to ensuring that these beneficial microorganisms successfully enter your intestinal tract (Ghazanfar et al. 2019). It is worth noting that dairy-based probiotic products, such as ice cream, act as effective carriers of probiotics (Homayouni Rad 2008; Ravula and Shah 1998). Health benefits are linked with dairy products. Prebiotics, which can be incorporated into various foods, actively promote health functions. The trend of these probiotic variants is becoming more and more popular. These products, dairy or non-dairy, such as probiotic ice cream, cheese, organic yogurt, and kefir, are included in the list (Kanwal et al. 2021).

Restoration of intestinal microflora, mainly *Lactobacillus* and *Bifidobacterium* strains, is integrated into dairy products, whether they are fermented or not. Specific attributes are required in

DOI: 10.1201/9781003399797-2

products to fulfill standards. Various dairy products have been specially developed for this purpose. Fermentation gives by-products, which improve the quality of food. Importantly, LAB have a greater tolerance to acidity and organic acids compared to many pathogens and spoilage microorganisms (Abdelnour et al. 2023).

1.2 RECENT TRENDS AND MACHINE LEARNING

In recent years, the landscape of probiotic and fermented dairy food product development has witnessed a transformative shift, largely driven by advancements in machine learning (ML). This dynamic intersection of cutting-edge technology and food science has ushered in a new era of precision and efficiency. More companies use ML tools to enhance their entire development procedures, starting from choosing suitable strains to improving production quality. Research proves that ML technology enhances how bacteria grow during fermentation to create better nutritional products that last longer while finding superior microbe strains. Companies use measured results to drive new product development. Advanced ML will enhance manufacturing processes, which can reduce performance problems while creating better, human-friendly, healthy, and customizable probiotics and fermented dairy products.

The collection of microorganisms within humans influences both our immune response and total well-being (McDonald et al. 2018; Quinn and Erb 2020). Studies become simpler to see how bacteria affect both host and microbial relationships while aiding research on chronic disease effects. Scientists investigate both fungal and protozoal eukaryotic microorganisms by using 18S rRNA marker gene sequencing (Elekwachi et al. 2017; Yarza et al. 2017). Through research on host-microbiome interactions, our knowledge of microbiome characteristics has grown. 18S rRNA testing determined many details about these interactions (Ashburner et al. 2000) and pathways (Kanehisa et al. 2004; Scholz et al. 2016). To properly use biological information, scientists need advanced analysis systems. Scientists developed taxonomy-based feature selection as a useful method to forecast host characteristics. Modern science uses this technique better to link medical conditions with microbiome findings to create precise medical treatments for every patient (Naiel et al. 2023).

Scientists repeatedly use ML tools to enhance knowledge of human bacterial systems. Modern studies have primarily helped scientists better understand health relationships to microbiota (Bonder et al. 2012; Cai and Sun 2011; Vervier et al. 2016). This study has strong potential for enhancing our knowledge of the function of the microbiome in disease and health (Niazifar et al. 2024).

1.3 TRADITIONAL AND FUNCTIONAL FOODS BASED ON MACHINE LEARNING AND IMPORTANT BRANDS IN THE MARKET

The trend in food consumption habits has changed radically, with consumers having more interest in functional foods (Turkmen, Akal, and Özer 2019).

Traditional foods are made up of diverse ingredients. Kimchi, Kefir, and Fermented Grain Bowls are a few examples, all of which have been researched for their probiotic characteristics. In vivo studies have shown that LABs from these foods produce varying health outcomes. These include the management of conditions such as obesity and type 2 diabetes, as well as reducing plasma cholesterol to improve hypocholesterolemia and immunomodulation.

Additionally, several LAB strains have high levels of chemicals linked to anti-inflammatory, anti-microbial, antioxidant, angiotensin-converting enzyme inhibition, exopolysaccharide, and bacteriocin synthesis (Table 1.1).

TABLE 1.1
Industries with Manufactured Products with Potential Probiotics (Granato et al. 2010; Mishra et al. 2018)

Food Item	Industry	Producer	Name of Probiotics
Fermented foods	Batavito	Batavo	*Lb. casei*
	Bob Sponja	Batavo	*Lb. casei*
Yogurt	Activia (creamy yoghurt)	Danone Co., Ltd	*B. lactis*
	ProViva	Skanemejerier	*Lb. plantarum*
Cheese	Kraft	Kraft Heinz Co	*B. lactis and Lb. rhamnosus*
	Saputo	Saputo Inc	

1.4 VARIOUS LABs BASED ON MACHINE LEARNING

Organizations use ML methods to find LAB strains suitable for dairy product production. The process needs to analyze complex databases with microbial samples and genetic stats. Data collection involves multiple LAB strain samples with metabolic profiles and genetic information. Researchers look for necessary traits from the data, such as metabolic pathways or genetic markers, to start building their distinction models.

The machine-learning model grabs knowledge about data relationships from a training set, which is later tested on a separate set called a test set. The ML system picks an algorithm based on the data type it handles and selects among decision trees, support vector machines, and neural networks. By undergoing training, the model finds natural patterns that indicate particular LAB strains in milk products. Monitoring performance on test data tells us how accurately the model works in new data situations.

The trained model is ready to determine LAB strains in fresh dairy product samples. It uses the patterns and features it has learned while training to make accurate identifications. Continuous improvement is achieved through iterative updates and refinements to the model as more data becomes available or as the understanding of LAB strains evolves. This application of ML not only enhances the precision of LAB strain identification but also contributes to advancements in the development of high-quality dairy products. Researchers in the field often refer to recent publications in microbiology, genomics, and food science to stay abreast of the latest methodologies and innovations in LAB strain identification using ML (Table 1.2).

1.5 MACHINE LEARNING ALGORITHMS USED

The accumulation of billions of bacterial genomes that reside on and within the human body is referred to as the "human microbiome". The previous microbiota, which occurs in nearly every part of its host, is made up of bacteria, viruses, fungi, archaea, and protozoa. Bacteria alone express almost 100 times as many different genes as humans. As a result, it is generally accepted that microbiota strongly impacts human metabolism and health (DeSalle and Perkins 2015).

ML algorithms were gradually used for purposes other than winning the games at the turn of the millennium. The pace of ML research took off, with computers being able to identify faces from photos, anticipate consumer behavior, talk like native speakers, lip-read, and detect cancers more precisely than radiologists, to mention just a few applications (McCoubrey et al. 2021). These days, ML technology is being used globally in practically every industry. Nevertheless, ML capabilities are developing quickly; every day, new research publications solving seemingly insurmountable problems are published. In particular, the healthcare industry has benefited from ML; algorithms are now able to detect illnesses, create medications, suggest customized treatment regimens, and forecast future health outcomes. A future of identical probabilistic and

TABLE 1.2
Attributes of LABs Isolated from Different Fermented Foods Based on Machine Learning (Kariyawasam, Lee, and Paik 2021)

Name	Product	Attribute	Origin	Source
Lb. plantarum KLAB21	Kimchi	Anti-mutagenic activity	Korea	Park and Rhee (2001)
Lb. plantarum SY11 and *Lb. plantarum* SY12	Kimchi	Anti-allergic activity	Korea	Lee et al. (2016)
Lb. plantarum AAS3	Hentak (fermented fish)	Hemolytic, DNase, and gelatinase Activities	India	Aarti and Khusro (2019)
Lb. paraplantarum MTCC 9483	Gundruck (fermented leafy green vegetable)	Anti-inflammatory activity	Nepal	Devi, Kurrey, and Halami (2018)
Lb. plantarum JLK0142	Fermented tofu (fermented soy milk)	Immunomodulatory activity	China	Wang et al. (2018)
Lb. s193 and s292	Funazushi (fermented soybean)	Anti-inflammatory activity	Japan	Okada et al. (2018)
W. hellenica BCC 7293	Nham (fermented pork sausage)	Bacteriocin production	Thailand	Woraprayote et al. (2015)

evidence-based medicine is promised by ML, even though it will take a while for broad clinical implementation (Chen et al. 2021).

Artificial intelligence (AI) algorithms that are also known as algorithms for ML can be used to distinguish between subsets of data or can be used to forecast future information through learning from historical data (Ciaburro and Iannace 2021).

These days, an extensive range of scientific fields, such as medicine, pharmacology, financing, and the arts, use ML algorithms. The genomes of bacteria that live in the gut or have infected it are being used in several recently developed nutrition sciences programs to automatically acquire critical information about foodborne outbreak verification, antibiotic resistance prediction, potential infection sources, and risk assessment (Bergamini et al. 2022).

Work on preventative agents and therapies via ML can start once it is established which components of the microbiome produce disease. Many diseases have been linked to dysbiosis at different body sites. For example, variations in the total number of Firmicutes and Bacteroidetes in the intestinal tract have been linked directly with several diseases, including ulcerative colitis, overweight and obesity, and motor neuron disorders (Belizário, Faintuch, and Garay-Malpartida 2018).

The etiological elements are believed to be a rise in intestinal permeability and inflammation, decreased synthesis of shorter-chain fatty acids (SCFAs), and elevated levels of lipoprotein lipase. Several techniques based on ML have been used to identify microbiome characteristics that are common in different illness stages, such as logistic regression, artificial neural networks, and support vector machines. Based on microbiome profiling, Gupta et al. have created a classification ML algorithm that can forecast an individual's health condition (Gupta et al. 2020). Protein-protein interactions between humans and microorganisms have been accurately predicted elsewhere (Lian et al. 2019). In order to predict postprandial glycemic response, Zeevi et al. developed an ML algorithm that integrated the blood parameters, eating patterns, anthropometrics, activity level, and gut microbiome profile of 800 participants. Such research demonstrates the possibility of precision microbiome medicine. Scientists have provided a thorough description of the potential for ML-led microbiome precision therapy in oncology (Zeevi et al. 2015).

In the publication, they demonstrate a novel approach that uses four ML algorithms to discover novel probiotic microorganisms and pinpoint the important genetic features that set probiotic microorganisms for human consumption apart from non-probiotics. The lack of adequate and predictive models and the tiny number of these organisms that have been found thus far are two of the primary obstacles facing research on human gastrointestinal probiotics. Therefore, we think that the field's study will be substantially enhanced by the development of bioinformatics approaches, particularly ML techniques (Gupta et al. 2021). In actuality, the high cost and advanced technologies that accompany numerous experiments in use today substantially limit their practical use. In light of this, techniques like ML and AI have emerged as critical mathematical tools for finding trends and relationships in huge datasets that can be verified by more conventional methods of analysis (Vora et al. 2023).

These RNA molecules serve as an interface between the ordered sequence of amino acids that are part of the encoded proteins and the genetic language represented by messenger RNA (mRNA), assisting in moving specific amino acids to the building of polypeptide chains on ribosomes. But tRNAs likely do more for the cell than only carry amino acids. The study of eukaryotes may benefit from some of the characteristics found in bacteria, such as those found in yeast, like yeast *Saccharomyces cerevisiae* (*Sc*), or in various areas of human and animal medical practice, such as the gut microbiome, as new behaviors are becoming increasingly recognized and understood (Chatterjee and Yadav 2019).

Thus, breakthroughs in tRNA-sequencing and filtering methods are certain to have an integral part shortly and enhance our comprehension of the biological investigation of bacteria's RNA in connection to the transcriptional and translational management of the host. These developments may offer fresh perspectives on the characteristics of these particles, which were previously believed to be solely transported amino acids but now seem to be essential for cellular control (de Crécy-Lagard and Jaroch 2021). Furthermore, the future sequencing of novel bacterial genomes will be of great relevance for food and fundamental research purposes, not only because of their inherent scientific value but also because the inclusion of these genomes in the training dataset will improve the performance of our ML algorithms. Furthermore, by using the approaches outlined for multiple species, both prokaryotic and eukaryotic, data concerning the progress of infectious organisms and also those that have gained biological characteristics that are useful for surviving with the host will be obtained (Lupolova, Lycett, and Gally 2019; Marcos-Zambrano et al. 2021).

The gut microbiome has come to be recognized as a critical component in a wide range of illnesses, including neurological conditions like Major Depressive Disorder and Alzheimer's disease, as well as irritable bowel syndrome and colon cancer (Doroszkiewicz, Groblewska, and Mroczko 2021). An important barrier in the creation of therapeutic probiotics is the lack of humanized GI models that can be modified in controlled environments to study the effects of beneficial bacteria on gut microbe populations and their metabolome. Because the gut microbiota plays a significant role in the digestion and absorption of vitamins, minerals, dietary polyphenols, all nutraceuticals, and nutritional components, estimating its complexity is a difficult task (Sorbara and Pamer 2022). While several in vitro models of the intestinal tract in humans have been produced so far, their application is severely limited by their size and/or complexity. The work developed and proved the ability of the ABIOME, a novel and state-of-the-art model of a human GI tract, to consider the varied characteristics of the human digestive system microbiota. After that, it was mixed with an algorithm for ML to produce probiotic formulations with significant potential for therapy that were optimized (Westfall et al. 2021).

1.5.1 Limitations

Key challenges that impede ML in microbiome research are GI constraints owing to pH, cellular makeup, size, individuality, and complexities. The database should be easily accessible and could provide data related to drug-related interactions with probiotics, like drug metabolism, plasma level, and pharmacokinetic details for probiotic-concerned drugs. Clear and accurate information exchange between research institutions and commercial cooperation (Schuhmacher et al. 2020). For

ML scientists, it is necessary to have a pharmaceutics background because without these anomalies, data are unable to be identified. ML algorithms are more difficult to be perceived by pharmaceutical scientists. This requires huge investment for both ML and pharmaceutical scientists to have training on correlated models (McCoubrey et al. 2021).

1.6 CONCLUSION

In today's business environment, innovation is an urgent necessity, especially in the vital process of creating new functional foods. This effort is of great importance from farm to fork. The manufacturing of innovative food products is a complex task that requires reconciling consumer preferences for attractive and healthy products. As a result, the process of creating and marketing functional food products is complicated, expensive, and uncertain.

Applying ML techniques is a contemporary procedure to improve probiotic knowledge and production system comprehension. Research teams use ML methods to process large databases containing information about microbial genetics, genomics, and clinical measures for studying probiotic strains and their predicted activities and health evaluations. Research teams use ML approaches to discover notable findings that usual patterns would struggle to reveal from colossal databases. The implementation of ML helps scientists understand probiotic mechanisms better while speeding up the process of discovering valuable new prospects.

REFERENCES

Aarti, Chirom, and Ameer Khusro. 2019. "Functional and technological properties of exopolysaccharide producing autochthonous *Lactobacillus plantarum* strain AAS3 from dry fish based fermented food." *LWT* 114: 108387.

Abdelnour, S. A., S. Ghazanfar, M. Abdel-Hamid, H. M. Abdel-Latif, Z. Zhang, and M. A Naiel. 2023. "Therapeutic uses and applications of bovine lactoferrin in aquatic animal medicine: An overview." *Veterinary Research Communications* 47 (3): 1015–29.

Ashburner, Michael, Catherine A. Ball, Judith A. Blake, David Botstein, Heather Butler, J. Michael Cherry, Allan P. Davis, Kara Dolinski, Selina S. Dwight, and Janan T. Eppig. 2000. "Gene ontology: Tool for the unification of biology." *Nature Genetics* 25 (1): 25–9.

Belizário, José E., Joel Faintuch, and Miguel Garay-Malpartida. 2018. "Gut microbiome dysbiosis and immunometabolism: New frontiers for treatment of metabolic diseases." *Mediators of Inflammation* 2018: 2037838.

Bergamini, Carlo M., Nicoletta Bianchi, Valerio Giaccone, Paolo Catellani, Leonardo Alberghini, Alessandra Stella, Stefano Biffani, Sachithra Kalhari Yaddehige, Tania Bobbo, and Cristian Taccioli. 2022. "Machine learning algorithms highlight tRNA information content and Chargaff's second parity rule score as important features in discriminating probiotics from non-probiotics." *Biology* 11 (7): 1024.

Bonder, Marc J., Sanne Abeln, Egija Zaura, and Bernd W Brandt. 2012. "Comparing clustering and preprocessing in taxonomy analysis." *Bioinformatics* 28 (22): 2891–7.

Cai, Yunpeng, and Yijun Sun. 2011. "ESPRIT-tree: Hierarchical clustering analysis of millions of 16S rRNA pyrosequences in quasilinear computational time." *Nucleic Acids Research* 39 (14): e95–e95.

Chatterjee, Sankar, and Surya Yadav. 2019. "The origin of prebiotic information system in the peptide/RNA world: A simulation model of the evolution of translation and the genetic code." *Life* 9 (1): 25.

Chen, Irene Y, Emma Pierson, Sherri Rose, Shalmali Joshi, Kadija Ferryman, and Marzyeh Ghassemi. 2021. "Ethical machine learning in healthcare." *Annual Review of Biomedical Data Science* 4: 123–44.

Ciaburro, Giuseppe, and Gino Iannace. 2021. "Machine learning-based algorithms to knowledge extraction from time series data: A review." *Data* 6 (6): 55.

de Crécy-Lagard, Valérie, and Marshall Jaroch. 2021. "Functions of bacterial tRNA modifications: From ubiquity to diversity." *Trends in Microbiology* 29 (1): 41–53.

DeSalle, Rob, and Susan L Perkins. 2015. *Welcome to the Microbiome: Getting to Know the Trillions of Bacteria and Other Microbes in, on, and Around You*. Yale University Press.

Devi, Sundru Manjulata, Nawneet K. Kurrey, and Prakash M. Halami. 2018. "In vitro anti-inflammatory activity among probiotic *Lactobacillus* species isolated from fermented foods." *Journal of Functional Foods* 47: 19–27.

Doroszkiewicz, Julia, Magdalena Groblewska, and Barbara Mroczko. 2021. "The role of gut microbiota and gut–brain interplay in selected diseases of the central nervous system." *International Journal of Molecular Sciences* 22 (18): 10028.

Elekwachi, Chijioke O., Zuo Wang, Xiaofeng Wu, Alaa Rabee, and Robert J. Forster. 2017. "Total rRNA-seq analysis gives insight into bacterial, fungal, protozoal and archaeal communities in the rumen using an optimized RNA isolation method." *Frontiers in Microbiology* 8: 1814.

Ghazanfar, S., A. Riaz, M. N. Tahir, S. Maqbool, G. M. Ali, F. Tariq, and I. Arif. (2019). Probiotic Supplement Improves the Health Status and Lactation Performance in Dairy Animals. In *Lactation in Farm Animals-Biology, Physiological Basis, Nutritional Requirements, and Modelization*. IntechOpen.

Granato, Daniel, Gabriel F. Branco, Adriano Gomes Cruz, José de Assis Fonseca Faria, and Nagendra P. Shah. 2010. "Probiotic dairy products as functional foods." *Comprehensive Reviews in Food Science and Food Safety* 9 (5): 455–70.

Gupta, Rohan, Devesh Srivastava, Mehar Sahu, Swati Tiwari, Rashmi K. Ambasta, and Pravir Kumar. 2021. "Artificial intelligence to deep learning: Machine intelligence approach for drug discovery." *Molecular Diversity* 25: 1315–60.

Gupta, Vinod K, Minsuk Kim, Utpal Bakshi, Kevin Y. Cunningham, John M. Davis, Konstantinos N. Lazaridis, Heidi Nelson, Nicholas Chia, and Jaeyun Sung. 2020. "A predictive index for health status using species-level gut microbiome profiling." *Nature Communications* 11 (1): 4635.

Homayouni Rad, A. 2008. "Therapeutical effects of functional probiotic, prebiotic and synbiotic foods." *Tabriz: Tabriz University of Medical Sciences*: 17–22.

James, Armachius, and Yousheng Wang. 2019. "Characterization, health benefits and applications of fruits and vegetable probiotics." *CyTA-Journal of Food* 17 (1): 770–80.

Kanehisa, Minoru, Susumu Goto, Shuichi Kawashima, Yasushi Okuno, and Masahiro Hattori. 2004. "The KEGG resource for deciphering the genome." *Nucleic Acids Research* 32 (suppl_1): D277–80.

Kanwal, H., A. Di Cerbo, F. Zulfiqar, C. Sabia, A. Nawaz, F. M. Siddiqui, … S Ghazanfar. 2021. "Probiotic characterization and population diversity analysis of gut-associated pediococcus acidilactici for its potential use in the dairy industry." *Applied Sciences* 11 (20): 9586.

Kariyawasam, Kariyawasam Majuwana Gamage Menaka Menike, Na-Kyoung Lee, and Hyun-Dong Paik. 2021. "Fermented dairy products as delivery vehicles of novel probiotic strains isolated from traditional fermented Asian foods." *Journal of Food Science and Technology* 58: 2467–78.

Koletzko, Berthold, Peter J. Aggett, J. G. Bindels, P. Bung, P. Ferre, Angel Gil, Michael J. Lentze, Marcel Roberfroid, and S. Strobel. 1998. "Growth, development and differentiation: A functional food science approach." *British Journal of Nutrition* 80 (S1): S5–45.

Lee, Eunjung, So-Ra Jung, So-Young Lee, Na-Kyoung Lee, Hyun-Dong Paik, and Seong-Il Lim. 2018. "*Lactobacillus plantarum* strain Ln4 attenuates diet-induced obesity, insulin resistance, and changes in hepatic mRNA levels associated with glucose and lipid metabolism." *Nutrients* 10 (5):643.

Lee, N-K, RKC Jeewanthi, E-H Park, and H-D Paik. 2016. "Physicochemical and antioxidant properties of Cheddar-type cheese fortified with *Inula britannica* extract." *Journal of Dairy Science* 99 (1): 83–8.

Lian, Xianyi, Shiping Yang, Hong Li, Chen Fu, and Ziding Zhang. 2019. "Machine-learning-based predictor of human–bacteria protein–protein interactions by incorporating comprehensive host-network properties." *Journal of Proteome Research* 18 (5): 2195–205.

Lupolova, Nadejda, Samantha J Lycett, and David L Gally. 2019. "A guide to machine learning for bacterial host attribution using genome sequence data." *Microbial Genomics* 5 (12): e000317.

Marcos-Zambrano, Laura Judith, Kanita Karaduzovic-Hadziabdic, Tatjana Loncar Turukalo, Piotr Przymus, Vladimir Trajkovik, Oliver Aasmets, Magali Berland, Aleksandra Gruca, Jasminka Hasic, and Karel Hron. 2021. "Applications of machine learning in human microbiome studies: A review on feature selection, biomarker identification, disease prediction and treatment." *Frontiers in Microbiology* 12: 313.

McCoubrey, Laura E., Moe Elbadawi, Mine Orlu, Simon Gaisford, and Abdul W. Basit. 2021. "Harnessing machine learning for development of microbiome therapeutics." *Gut Microbes* 13 (1): 1872323.

McDonald, Daniel, Embriette Hyde, Justine W. Debelius, James T. Morton, Antonio Gonzalez, Gail Ackermann, Alexander A. Aksenov, Bahar Behsaz, Caitriona Brennan, and Yingfeng Chen. 2018. "American Gut: An open platform for citizen science microbiome research." *Msystems* 3 (3):10.1128/msystems. 00031-18.

Meurman, Jukka H. 2005. "Probiotics: Do they have a role in oral medicine and dentistry." *European Journal of Oral Sciences* 113 (3): 188–96.

Mishra, Swati S., Prafulla K. Behera, Biswabandita Kar, and Ramesh C. Ray. 2018. "Advances in probiotics, prebiotics and nutraceuticals." In *Innovations in Technologies for Fermented Food and Beverage Industries*, edited by Sandeep Kumar Panda and Prathapkumar Halady Shetty, 121–41. International Publishing AG.

Naiel, M. A., S. S. Negm, S. Ghazanfar, M. Shukry, and S. A Abdelnour. 2023. "The risk assessment of high-fat diet in farmed fish and its mitigation approaches: A review." *Journal of Animal Physiology and Animal Nutrition* 107 (3): 948–69.

Niazifar, M., M. Besharati, M. Jabbar, S. Ghazanfar, M. Asad, V. Palangi, … M Lackner. 2024. "Slow-release non-protein nitrogen sources in animal nutrition: A review." *Heliyon* 10:e33752.

Okada, Yoshikiyo, Yoshikazu Tsuzuki, Takajo Takeshi, Hirotaka Furuhashi, Masaaki Higashiyama, Chikako Watanabe, Kazuhiko Shirakabe, Chie Kurihara, Shunsuke Komoto, and Kengo Tomita. 2018. "Novel probiotics isolated from a Japanese traditional fermented food, funazushi, attenuates DSS-induced colitis by increasing the induction of high integrin αv/β8-expressing dendritic cells." *Journal of Gastroenterology* 53:407–18.

Park, Heui-Dong, and Chang-Ho Rhee. 2001. "Antimutagenic activity of *Lactobacillus plantarum* KLAB21 isolated from Kimchi Korean fermented vegetables." *Biotechnology Letters* 23: 1583–9.

Quinn, Thomas P, and Ionas Erb. 2020. "Interpretable log contrasts for the classification of health biomarkers: A new approach to balance selection." *Msystems* 5 (2): e00230-19.

Ravula, RR, and NP Shah. 1998. "Viability of probiotic bacteria in fermented frozen dairy desserts." *Food Australia* 50:136–139.

Scholz, Matthias, Doyle V. Ward, Edoardo Pasolli, Thomas Tolio, Moreno Zolfo, Francesco Asnicar, Duy Tin Truong, Adrian Tett, Ardythe L. Morrow, and Nicola Segata. 2016. "Strain-level microbial epidemiology and population genomics from shotgun metagenomics." *Nature Methods* 13 (5):435–8.

Schuhmacher, Alexander, Alexander Gatto, Markus Hinder, Michael Kuss, and Oliver Gassmann. 2020. "The upside of being a digital pharma player." *Drug Discovery Today* 25 (9): 1569–74.

Sorbara, Matthew T., and Eric G. Pamer. 2022. "Microbiome-based therapeutics." *Nature Reviews Microbiology* 20 (6): 365–80.

Turkmen, Nazli, Ceren Akal, and Barbaros Özer. 2019. "Probiotic dairy-based beverages: A review." *Journal of Functional Foods* 53: 62–75.

Vervier, Kévin, Pierre Mahé, Maud Tournoud, Jean-Baptiste Veyrieras, and Jean-Philippe Vert. 2016. "Large-scale machine learning for metagenomics sequence classification." *Bioinformatics* 32 (7): 1023–32.

Vora, Lalitkumar K., Amol D. Gholap, Keshava Jetha, Raghu Raj Singh Thakur, Hetvi K. Solanki, and Vivek P. Chavda. 2023. "Artificial intelligence in pharmaceutical technology and drug delivery design." *Pharmaceutics* 15 (7): 1916.

Wang, Ji, Tong Wu, Xiaobin Fang, Weihong Min, and Zhennai Yang. 2018. "Characterization and immunomodulatory activity of an exopolysaccharide produced by *Lactobacillus plantarum* JLK0142 isolated from fermented dairy tofu." *International Journal of Biological Macromolecules* 115: 985–93.

Westfall, Susan, Francesca Carracci, Molly Estill, Danyue Zhao, Qing-li Wu, Li Shen, James Simon, and Giulio Maria Pasinetti. 2021. "Optimization of probiotic therapeutics using machine learning in an artificial human gastrointestinal tract." *Scientific Reports* 11 (1): 1067.

Woraprayote, Weerapong, Laphaslada Pumpuang, Amonlaya Tosukhowong, Sittiruk Roytrakul, Rodney Honrada Perez, Takeshi Zendo, Kenji Sonomoto, Soottawat Benjakul, and Wonnop Visessanguan. 2015. "Two putatively novel bacteriocins active against Gram-negative food borne pathogens produced by *Weissella hellenica* BCC 7293." *Food Control* 55:176–84.

Yarza, Pablo, Pelin Yilmaz, Katrin Panzer, Frank Oliver Glöckner, and Marlis Reich. 2017. "A phylogenetic framework for the kingdom fungi based on 18S rRNA Gene sequences." *Marine Genomics* 36: 33–9.

Zeevi, David, Tal Korem, Niv Zmora, David Israeli, Daphna Rothschild, Adina Weinberger, Orly Ben-Yacov, Dar Lador, Tali Avnit-Sagi, and Maya Lotan-Pompan. 2015. "Personalized nutrition by prediction of glycemic responses." *Cell* 163 (5): 1079–94.

2 Role of Metagenomics for Searching of Novel Probiotic Strains

Mohammad Javad Fakkari Afjadi,
Shakira Ghazanfar, and Leila Haghshenas

2.1 PROBIOTICS

2.1.1 The Science of Studying Probiotic Bacteria

Probiotics are organisms that are beneficial to the human body. The term probiotics comes from the Greek word meaning "life". "Probiotics" was first used in 1965 to describe metabolites that promote the growth of other microorganisms. Thus, these metabolites have the opposite function of antibiotics. Additionally, the term probiotic was used in 1971 to describe tissue extracts that stimulate the growth of microorganisms. As a concept, the term refers to organisms and substances that contribute to the intestinal microbial balance (Fuller 2012). Probiotics occur naturally in the human body and have potential therapeutic benefits (Fuller 1989). Probiotics have many health benefits and are inexpensive to manufacture, transport, and store, making them an attractive alternative to traditional treatments. There is evidence that probiotics have many beneficial effects on the host's mucosal defense system. By producing bactericidal substances, it blocks the effects of pathogens and competes with pathogens and toxins for adherence to the intestinal epithelium. Probiotics support intestinal epithelial homeostasis by supporting intestinal epithelial cell survival, enhancing barrier function, and stimulating protective responses (Akmal et al. 2022). The most important effect of probiotics on human health is the regulation of the immune system. They have been shown to modulate inflammatory signaling pathways (Vanderpool, Yan, and Brent 2008). The use of genetically engineered probiotic strains as therapeutics, especially to relieve pain, fight disease, and prevent infections, is a viable alternative to conventional methods (Sleator 2010). *Lactobacillus*, *Bifidobacterium*, and *Saccharomyces* can be mentioned among the common probiotics found in animals and humans. There is a wide variety of probiotics available, but *Lactobacillus* spp. is most commonly used as a commensal microorganism. Therefore, these bacterial species are suitable for probiotic research (Lehri, Seddon, and Karlyshev 2015).

2.1.2 Efficacy of Probiotic Bacteria

To date, over 1,000 complete bacterial genome sequences have been annotated in GenBank. The clinical efficacy of probiotics is largely related to their adhesion to the intestinal mucosa. Multiple filamentous projections protruding from the cell surface contribute to adhesion. Poor reproducibility of probiotics can result in species with genome instability and loss of population genetic information (Khan et al. 2021). Therefore, some species that are active in vitro may not adhere to the mucus layer of the gastrointestinal tract and thus be ineffective in the clinical setting (Vitetta et al. 2017). The ideal probiotic strain is one that is genetically stable and lacks the plasmid transfer mechanism. In addition, probiotics must have the right technical properties to withstand and survive the conditions encountered during food manufacturing, processing, storage, and transportation. Moreover, they must maintain a satisfactory level of viability at the time of consumption (Bajaj, Claes, and Lebeer 2021).

 DOI: 10.1201/9781003399797-3

2.1.3 Mechanism of Action of Probiotics

Diverse and different effects are exerted by host probiotic bacteria. Different organisms can influence the intestinal lamina environment, epithelial and mucosal barrier function, and the mucus immune system. Microbial substances act on epithelial cells and various immune cells such as dendritic cells, monocytes/macrophages, B cells, T cells, T cells with regulatory functions, and NK cells (Ng et al. 2008). Changes in the gut flora can occur as a result of stress, illness, or antibiotic treatment, leading to diarrhea and anorexia. Inflammatory, immune, neurological, and endocrine conditions are often associated with or caused by an overgrowth of harmful bacteria. One option for standardizing health status is the induction of beneficial bacteria (Khan et al. 2024). One possible solution is to add viable bacterial cells to the host. Probiotics can increase the number of beneficial bacteria in the gut and even eliminate pathogenic bacteria entirely. Additionally, these bacteria can also release several enzymes that aid in the digestion of feed. It is important to realize that there are a variety of microbes that can act as probiotics. *Lactobacillus* and *Bifidobacterium* are the bacteria most commonly used in probiotic supplements. The use of probiotics also extends to the use of non-pathogenic species of the genera *Saccharomyces, Streptococcus,* and *Lactococcus.* The effectiveness of probiotics depends on their high adhesion and colonization in the gut, which strengthens the host's immune system. However, no host-specific pathogenic, toxic, mutagenic, or carcinogenic responses of probiotic metabolites are expected (Hemaiswarya et al. 2013). The mechanism of action of probiotics is illustrated in Figure 2.1.

2.1.4 Use of Probiotics for the Treatment

Probiotics have the potential not only to improve our health and control pathogenic infections but also to help treat and manage the disease. How can it help in treatment and management? A more important issue is understanding the behavior of the disease and its causative agents.

Below are some of the roles probiotics play in keeping us healthy and in treating and managing diseases.

1. Reduces the destructive effects of antibiotics and restores any kind of loss of beneficial microflora.
2. Treatment of diarrheal diseases.

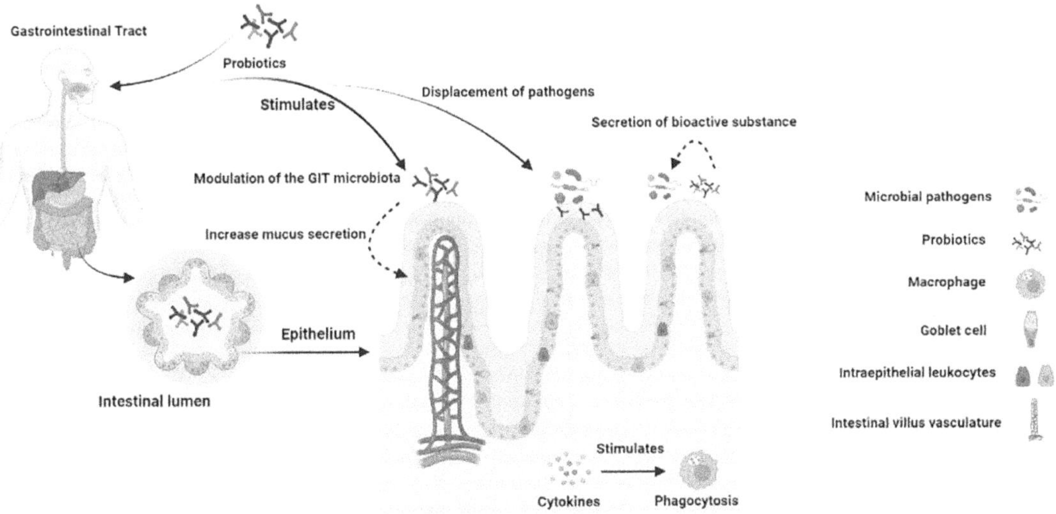

FIGURE 2.1 Mechanisms of action of probiotics.

3. Improves digestive health.
4. Strengthening the immune system.
5. Reduces symptoms of lactose intolerance and reduces the incidence of allergies in susceptible individuals.
6. Reduces the risk of some cancers.
7. Regulation of serum cholesterol levels.
8. Probiotics can also improve the function of the intestinal mucosa, including the synthesis and secretion of antimicrobial peptides (Amara and Shibl 2015).

In addition, results from evidence-based analyses of human studies and animal models suggest that probiotics may be clinically effective for intestinal diseases, including infectious diarrhea, antibiotic-associated diarrhea, allergic disease, necrotizing enterocolitis, ulcerative colitis, irritable bowel syndrome, and intestinal diseases such as allergies (Yan and Polk 2011).

Probiotics work in different ways to promote their effects. Here are some of the main mechanisms of action of probiotics on the host (Table 2.1).

Probiotics can affect many components of epithelial barrier function by decreasing enterocyte apoptosis and increasing mucin production. *Lactobacillus rhamnosus* GG prevented cytokine-induced apoptosis in an intestinal epithelial cell model by inhibiting tumor necrosis factor (TNF). *Lactobacillus* species have been shown to increase mucin expression in human intestinal epithelial cells in vitro, thereby blocking invasion and attachment of pathogenic *Escherichia coli. Lactobacillus rhamnosus* GG has been shown to prevent inflammation and programmed cell death of intestinal epithelial cells, exert mitogenic effects, and promote mucosal regeneration (Kanwal et al. 2021). Probiotics intervene to combat pathogens and prevent epithelial invasion by stimulating host cells to produce peptides or by directly releasing peptides. Defensins (hBD proteins) and cathelicidin are antimicrobial peptides constitutively expressed by intestinal epithelial cells and exhibit antimicrobial activity against a variety of bacteria, fungi, and some viruses. Certain strains of probiotics, such as *E. coli* strain DSM 17252 G2, and certain species of lactobacilli have been shown to express certain defensins. *Lactobacillus casei* has been shown to increase the levels of total and pathogen-specific secreted IgA upon infection in mice by stimulating class transition from B cells to IgA. Specific antibodies against *L. casei* were not produced, indicating that the intestinal immune system did not respond to this beneficial bacterium. In neonatal rabbits pretreated with *L. casei*, the incidence of subsequent hemorrhagic *enteric E. coli* (EHEC) infection was reduced due

TABLE 2.1
Different Mechanisms of Probiotic Action on Human Intestinal Cells

Bacteria	Disease/Affected Organ	Mechanism/Explanation
***Lactobacillus rhamnosus* GG**	Atopy Intestine	Decreasing apoptosis of epithelial cells via reducing TNF-α production
***Lactobacillus* sp**	Intestine	Increasing mucin production via overexpression of MUC 2
***E coli strain* DSM 17252S2**	Intestine	Cathelcidins (CATHLs) by producing butyrate
Salmonella tyhimurium **VSL#3 probiotics**	Intestine	Blocking pro-inflammatory molecules by attenuating IL-8 secretion or blocking the degradation of the counter-regulatory factor IκB
L. casei	Gastroenteritis	Increasing mucosal immunity by increasing IgA production
L. acidophilus	Intestine	Blocks the communication between pathogenic bacteria by secreting molecules that blocks quorum sensing signaling

Abbreviations: CATHLs: Cathelcidins; IL-8: Interleukin 8; IκB: Inhibitor of kappa B; MUC 2: Mucin 2; and TNF-α: Tumor necrosis factor alpha.

to increased mucosal levels of anti-EHEC and anti-Shiga toxin IgA antibodies compared with the rabbits that were witnessed. *L. casei* downregulates the transcription of several genes encoding inflammatory agents, such as cytokines and chemokines, and adhesion molecules by *Shigella flexneri*. This results in anti-inflammatory effects that appear to be mediated by inhibition of the NF-κB pathway, particularly through stabilization of I-κBα. Bacteria communicate with each other and with the environment through chemical signaling molecules called autoinducer, a phenomenon known as quorum sensing. The use of this cell-to-cell signaling mechanism facilitates the control of key traits of gut microbes, enabling them to successfully colonize the host or initiate infection. Medellin Pena et al., demonstrated that *Lactobacillus acidophilus* secretes molecules that inhibit quorum-sensing signaling or interact directly with bacterial transcription of the *E. coli* O157 gene involved in colonization to counteract bacterial virulence (Gogineni, Morrow, and Malesker 2013).

2.2 PROBIOTICS IN BIOENGINEERING

2.2.1 Development of Probiotics and Use for Therapy

Evidence suggests that probiotics have an important role in the treatment and prevention of infectious diseases. Currently, infectious diseases are often managed through the use of antibiotics. However, inappropriate antibiotic use can have consequences at the patient level, such as drug-specific side effects, and at the public health level, such as the selection of multiresistant bacteria. Therefore, the search for new alternative antibacterial therapies, with particular interest in therapies based on natural products, is urgently needed (Silva et al. 2020). Probiotics confer many health benefits through their biological mechanisms in the host entity. The "Human Microbiome Project" has shown that the number of other bacterial cells in the human body is about 10 times greater than the number of human cells. These bacteria contain beneficial microbes and are necessary for human health. The exact mechanism of action is not completely understood, but probiotics produce postbiotics such as short-chain fatty acids, enzymes, and lactic acid and secrete antimicrobial peptides that destroy strains of pathogenic bacteria. Probiotics have also been combined with nutrients to compete with pathogens, thereby inhibiting/preventing the adhesion of pathogenic bacteria in the colon lumen. This improves mucus production, which helps strengthen the intestinal epithelial barrier to stimulate the immune system. Interestingly, some probiotics assist disease treatment by producing specific anticarcinogenic metabolites and antioxidants. Supplementation with endogenous probiotics can help replenish the gut microbiota after antibiotic therapy. Using probiotics in combination with antibiotics or after antibiotic treatment can prevent antibiotic-associated diarrhea. Researchers have also found that probiotic-specific mechanisms that affect neuronal signaling in the central nervous system may promote therapeutic effects on neuropsychiatric disorders and stress-related health. In addition to improving digestive health, probiotics have also been linked to brain function to help treat irritable bowel syndrome. Probiotics help prevent vaginal and urinary tract infections (yeast/bacterial) in women, prevent pancreatitis, and improve pancreatic health. Contributes to respiratory tract health, inhibits tumor development, and modulates immune system responses. Probiotics are used to treat metabolic diseases such as diabetes, nonalcoholic fatty liver disease, cardiovascular disease, cancer, oral Candida infections, and periodontitis. Different strains have unique health outcomes. It is clear that identification and selection of probiotic types and their strains are therefore essential for individual applications (Yoha et al. 2022).

2.2.2 Treatment and Prevention of Infectious Diseases by Probiotics

Probiotics play an important role in controlling infectious agents through their action on the epithelium, production of antimicrobial compounds, and elimination of competition. Probiotic supplementation can help reduce the risk of infections and the consumption of antibiotics and help reduce or slow the development of multidrug-resistant bacteria. A serious problem is that antibiotic therapy usually fails due to antibiotic resistance. As a result, infections can lead to serious illness if left untreated. A growing body of research shows promising results for probiotics of various strains used in the prevention or treatment of acute and recurrent infections. However, more standardized clinical studies are needed (Li et al. 2022). The gut microbiota is known to play a role in enhancing resistance to extra-gut pathogens, inhibiting the production of antipathogens, and enhancing immune function. Research on probiotics has shown that they can modulate certain influencing factors of the immune system, such as lymphocytes, antibodies, and natural killer cells. The ability of probiotics to reduce common infections is one of their general functions being studied. A study on the efficacy of probiotics in preventing acute upper respiratory tract infections (URTIs) showed that probiotics reduced the incidence of URTIs. Studies have also shown that probiotics reduce the duration and severity of flu symptoms in children. A community-based clinical trial in the United States found that consumption of probiotic milk reduced the frequency of acute diarrhea by 24% and URTIs by 18%, but unlike other trials, there was no effect on lower respiratory tract infections. A community-based pediatric clinical trial conducted in precarious living conditions in India compared the consumption of milk containing the bacteria *L. casei* Shirota with the consumption of a nutritious drink in 3,758 children from 1 to 5 years old. The incidence of acute diarrhea decreased to 14% in the probiotic group. Another study reported that probiotics significantly reduced the duration of URTI disease and the severity of URTI symptoms but did not affect disease prevalence. Overall, these studies suggest that probiotics, when used in unhealthy individuals, can rebalance the microbiome and help alleviate some infectious diseases (Rostami et al. 2018).

2.2.3 Anticancer Therapies and Tumor Targeting Using Designer Probiotics

Bacteria have unique abilities that make them ideal anticancer agents. Recently, the mechanism of action and antitumor effects of bacteria on tumor cells have been studied. Bacteria exhibit intrinsic antitumor activity because they express chemotactic receptors that direct their chemotaxis to molecular signals in the tumor microenvironment. They are also equipped with flagella, which facilitate tissue penetration. They can also be engineered to sense and respond to the tumor microenvironment, leading to innate and adaptive antitumor immune responses (Yu et al. 2020). The advent of synthetic biology has enabled commensal microbes and probiotics to take on new therapeutic roles. Over the years, scientists have developed probiotics to help fight diseases such as colitis, diabetes, and HIV infection. This section reviews recent developments in synthetic therapeutic probiotics. With multidrug-resistant pathogens and the depletion of antibiotic options, there is an urgent need to use other methods to combat these deadly microbes. One such approach is the use of engineered probiotics, which have several advantages, including increased specificity, regulatory release of antimicrobial agents, and reduced risk of developing antibiotic resistance. Conventional cancer therapies often do not lead to complete remission of cancer. The idea of using bacteria to treat cancer dates back to the 19th century. Over the years, microbiologists have reported a variety of anaerobic agents that preferentially proliferate in solid tumors, thereby encouraging the use of these microbes as vectors of anticancer agents. Among the various anaerobic strains tested, strains of the genera *Clostridium, Bifidobacterium, Salmonella,* and *Escherichia* are most frequently genetically modified. Salmonella is an optional anaerobe commonly developed for cancer treatment. Therapeutic applications of attenuated Salmonella are diverse, expressing apoptotic genes such as Fas-associated death domain (FADD), and cytokines such as interferon-gamma (IFN-g) to manipulate Salmonella as a vehicle for drug delivery. This includes direct killing. Salmonella is also designed to target difficult-to-access or poor-prognosis tumors such as glioma and neuroblastoma. Aside from *Salmonella, Escherichia coli* Nissle (EcN) has also been demonstrated to function as a cancer diagnostic tool. The modified EcN was developed to selectively colonize liver tumors in mice. EcN also releases small molecules in the urine, indicative of liver metastases. *E. coli* can also be engineered to detect glucose and sugar ribose in solid tumor cell masses. Given recent advances in synthetic probiotic therapy, as described here, it is widely expected that a new generation of therapy will become available in the near future (Chua et al. 2017).

2.2.4 Limitations of the Use of Bacterial Probiotics

Despite the many health benefits and benefits associated with probiotics and probiotic food products, there are risks associated with probiotic therapy. These risks are mainly related to the safety of vulnerable groups such as immunocompromised people (pregnant women, infants, and the elderly) and critically ill or hospitalized patients (Jankovic et al. 2010). Either way, the main challenge is to understand how they work to reveal the health benefits of each strain of beneficial bacteria. Clinical studies are therefore urgently needed to better understand the interface between bacteria, host cells, mucus, and the immune defense system and to develop effective interventions. These studies should include molecular examination of the gut microbiota (not just stool) and the long-term effects (5–10 years) of probiotic microorganisms. Such as problems, especially enterococci, that may contain transmissible antibiotic resistance determinants and the *Bacillus* cereus group that is known to produce enterotoxins and emetic toxins (Anadón, Martínez-Larrañaga, and Aranzazu Martínez 2006). Probiotics can cause four types of side effects: Systemic infections, adverse metabolic activity, overstimulation of the immune system in susceptible individuals, and gene transfer. Currently, the U.S. Food and Drug Administration (FDA) has not approved any probiotic claims linking probiotics to a reduced risk of disease. Another important issue for probiotic survival in dairy products is oxygen toxicity. The high oxygen concentration in the products reduces the viability of these anaerobic bacteria. Maintaining the viability of probiotic bacteria is a major commercial and technological challenge in the application of probiotic cultures in functional foods. Many active bacteria are killed during the production, storage, or transportation of the finished product, as well as during defecation. Therefore, most of them are destroyed before benefiting consumers (Ayichew et al. 2017). For bacterial probiotics to play their intended role in animal production, they must meet certain microbiological criteria. Bacteria must be: (1) non-pathogenic, (2) host specificity, (3) resistant to the digestive environment (such as pH), and (4) genetic stability. Failure to meet some of these criteria may explain why no effect was observed (Kathirvelan).

2.3 METAGENOMICS

2.3.1 Metagenomics Analysis Study

Metagenomics is a community-wide analysis of genomic DNA, which is different from genomics, which is the analysis of the genomic DNA of an organism or a cell. In fact, the most appropriate translation of meta in Greek is "beyond", so the term literally means "beyond the study of a single genome". The term was first published in 1998 in a study of soil bacteria by random cloning of environmental DNA (Gilbert and Dupont 2011). To date, applications of metagenomics include infectious disease diagnosis in various types of syndromes and specimens, microbiome analysis in diseased and healthy conditions, characterization of human host responses to transcriptomic infection and tumor-associated disease, and identification of the virus and its genomic integration sites (Chiu and Miller 2019). Metagenomics allows the study of the genomes of uncultured microorganisms. Faster, cheaper sequencing technologies and the ability to sequence uncultured bacteria directly from their habitats are changing the way we view the microbial world (Wooley, Godzik, and Friedberg 2010). One of the most notable events in the field of microbial ecology in the last decade has been the emergence and development of metagenomics. The field began with the cloning of environmental DNA, followed by screening for functional expression, and was quickly complemented by direct random sequencing of environmental DNA. The first project not only demonstrated the principle of the metagenomic approach but also revealed various functional genes in the microbial kingdom around us. The rapid and significant reduction in the cost of next-generation sequencing (NGS) has greatly accelerated the development of sequence-based metagenomics. Indeed, metagenomic sequence datasets have skyrocketed in recent years (Thomas, Gilbert, and Meyer 2012). The first metagenomic studies by analyzing the genetic content of environmental samples enabled the identification of metabolic signatures, the characterization of organisms, and the discovery of new antibiotics and enzymes. Metagenome research today covers a wide range of research areas, including research into the marine environment, plant and agricultural biotechnology, human genetics, and the diagnosis of human disease. The range of metagenomic applications can range from environmental microbiome research to standard clinical diagnostics, such as patient palliative care, public health, and industry. Metagenomics originally relied on cloning to analyze double-stranded DNA genomes, but today high-throughput sequencing techniques can be applied to any type of genome, including DNA and single-stranded RNA. Basically, the metagenomic analysis consists of three main steps: (1) sample preparation, (2) high-throughput sequencing, and (3) bioinformatics analysis (Mokili, Rohwer, and Dutilh 2012), which will be discussed in the next sections.

2.3.2 Functional Metagenomics

Metagenomics is a relatively modern but rapidly growing field of environmental biology that aims to gather knowledge about the genomes of environmental bacteria and entire microbial communities. A number of "functional metagenomics" approaches have been deployed in recent years to enable high-resolution genomic analysis of uncultivated bacteria, linking them to environmental functions. Specifically, the recent emergence of new (next-generation) sequencing technologies, which will lead to higher sequence production and significantly reduced sequencing prices, will define a new era of metagenomics. The term "functional metagenomics", in a broad sense, is intended to interpret the association between the identification of bacteria or communities detected by metagenomics and their respective environmental functions. The power of metagenomics is to harness the immense metabolic potential of uncultivated microbes that represent most of Earth's microbes, including entirely new microorganisms and novel metabolic pathways. Recent advances in next-generation (ultra-high-throughput) sequencing technology, which greatly reduce the cost of DNA sequencing, are paving the way for a new era of metagenomics (Chistoserdova 2009). Only a small fraction of microorganisms on Earth can be grown on standard laboratory media, a problem that is a significant limitation of genomics and biotechnology studies. To overcome this problem, different methodological approaches have been developed to approach the genetic potential of microorganisms present in the environment. These procedures help to analyze all microbial genomes, known as metagenomes, in a specific medium, independent of the culture medium. One such approach is functional metagenomics, which isolates functional genes from the cloned metagenome in the library. These genes can be expressed in cultured microorganisms. One of the most important points in building metagenomic libraries is choosing an appropriate DNA extraction method. Therefore, this method must ensure sufficient yield and suitable average insert size. The isolated DNA is fragmented using appropriate restriction enzymes or cut mechanically. The resulting fragments are ligated into linearized cloning vectors such as bacterial plasmids or artificial chromosomes (BACs). The resulting metagenomics library is presented as a microbial host, especially the well-known model bacterium *E. coli*, which is susceptible to genetic manipulation, and many genetic tools have been developed. By appropriate screening assays, a single gene or a set of genes expressing a specific enzyme function can be identified, and their products can be further analyzed. The benefits of using functional metagenomics are: (1) the whole genes can be isolated, (2) genes identified in the discovery process have a function, and (3) new genes can be isolated, as this method does not depend on known gene sequence information. Accordingly, functional screening of metagenomic libraries enables the discovery of new and known classes of genes involved in various cellular functions (Mirete, Morgante, and González-Pastor 2016).

2.3.3 Clinical Metagenomics

Clinical metagenomics (CMg) is the discipline that deals with the sequencing of all nucleic acids found in a patient sample for the purpose of retrieving clinically relevant microbial information. From a diagnostic standpoint, NGS allows rapid identification of suspected pathogens and prediction of their antibiotic resistance properties to optimize targeted treatment patterns. Since the introduction of metagenomics nearly a decade ago, numerous reports have described successful applications in an increasingly diverse range of biological samples, including respiratory secretions, cerebrospinal fluid, feces, blood, and tissue. The tremendous advances made in recent years in sequencing and computing technologies have made CMg a promising tool in clinical microbiology laboratories. Infectious manifestations and symptoms are not usually unique to any particular pathogen and may lead to a general differential diagnosis. The diagnosis is further complicated because viral and bacterial infections are difficult to distinguish. In this regard, CMg has the potential to revolutionize diagnostics. However, further exploration is needed to investigate the ability of CMg to detect emerging pathogens or novel genetic variants, such as those observed during the SARS outbreak (Forbes et al. 2018). There are many literature reviews detailing the technical methods of CMg for the diagnosis of infectious diseases. Importantly, while CMg uses NGS technologies, the use of these technologies for CMg is fundamentally different from other uses (such as whole-genome sequencing). CMg is a diagnostic approach with the goal of providing viable outcomes for patient care. In summary, the steps of CMg include sample collection, sample processing, nucleic acid extraction, library preparation, sequencing, bioinformatics analysis, and interpretation and communication of results. Each step is a variable that can be changed, and the method of taking each step can have a significant impact on the final result. It is important to understand the different steps involved in a specific CMg approach, including the critical step in data interpretation; changes in these steps can have a positive or negative impact on the quality, usefulness, and cost of the end result (Gaston 2023).

2.3.4 Diagnosis of Infectious Diseases

Infectious diseases continue to be the leading cause of morbidity and mortality in all patient populations worldwide. Accurate diagnosis can be difficult due to the variety of pathogens that cause clinically indistinguishable diseases. The NGS method allows for the universal detection of pathogens regardless of the type of microorganism (viruses, bacteria, fungi, and parasites) and can even be applied to detect new organisms (Simner, Miller, and Carroll 2018). Most studies have shown that mNGS has a similar sensitivity to certain PCR tests and can detect more potential pathogens than conventional methods. The diagnosis of infectious diseases is based either on direct detection of microorganisms in patient samples using culture, antigens, or molecular markers or on indirect detection based on the host and antibody responses to infection. mNGS uses the identification of nucleic acid sequences (DNA and/or RNA) as biomarkers to detect bacterial pathogens in patient samples. Culturing is the preferred approach in most cases of infection for recovery and isolation for antimicrobial susceptibility testing, but it is often the case that pathogens that have not been cultured or have been suppressed by the previous administration of antibiotics are not generated by culture. This is especially true for viral infections, but it can also apply to bacterial, fungal, and parasitic infections. PCR testing for specific agents is the standard method for detecting many difficult-to-cultivate microorganisms; however, the number of tests required could be sufficient to cover rare and emerging pathogens (Miller and Chiu 2021).

2.4 METAGENOMICS IN THE DISCOVERY OF NEW GENES

2.4.1 Strategy for New Gene Identification

Bacteria grown in pure culture are the starting point for the discovery of many currently used antibacterial agents. Metagenomics, which uses culture-independent methods to access the common genomes of naturally occurring bacterial populations, provides a way to discover antimicrobial products produced by populations of bacteria. Large numbers of bacteria are known to be present in the medium but are still resistant to the culture. New small-molecule antibiotics and novel active antibacterial proteins have been identified using metagenomics (Banik and Brady 2010). In a metagenomic detection method, the target genes constitute a small percentage of the total nucleic acids. Therefore, sample pre-enrichment provides an interesting means of improving screening success rates. Target gene detection can be greatly enhanced by applying one of many enrichment options, ranging from whole-cell enrichment to the selection and enrichment of target genes and genomes. Methods for the extraction of nucleic acids from many communities have been developed. The two main strategies for metagenomic DNA recovery are cell recovery and direct lysis. Machine threshing has been shown to recover more varieties than chemical treatment. Either way, chemical lysis is a gentler method that recovers higher molecular weight DNA. Chemical lysis can also select specific taxes by taking advantage of their unique biochemical properties. These studies laid the foundation for the ultimate discovery of metagenomic genes. Recently, with the advent of high-throughput automated sequencing and powerful sequencing algorithms, these projects are now technically feasible, albeit financially ambitious (Cowan et al. 2005).

2.4.2 Next-generation Sequencing Technology

The recent development of nucleic acid sequencing technology, called next-generation sequencing (NGS), has created a true revolution, opening up new possibilities for research and diagnostic applications by increasing the speed and throughput of data generation. The first commercial NGS system was developed by 454 Life Sciences and introduced in 2005. Current NGS methods use a three-step sequencing process: library preparation, DNA collection and enrichment, and sequencing/detection. During the library preparation, DNA (or cDNA) fragments of appropriate length are prepared, either by breaking long molecules or by preparing for the synthesis of short molecules. During the DNA capture and enrichment steps, these short molecules are labeled primers and are used to physically capture and separate each short fragment for immobilization on a solid substrate. These steps are performed simultaneously on a myriad of cloned amplified fragments (Capobianchi, Giombini, and Rozera 2013). Once the sequencing is complete, the raw data from the batch must be analyzed in several steps. A general data analysis procedure for NGS data that includes data preprocessing to remove adapter sequences and low-quality reads, mapping data to base or copy genomes, a new record of the compiled sequence read, and parsed sequence analysis includes a variety of bioinformatics assessments, including genetic variants requiring the detection of SNPs, detection of novel genes or regulatory elements, and evaluation of transcript expression levels. Although whole genome and whole exon sequencing are possible, often targeted sequencing of specific genes or genomic regions is recommended when a suspected disease or condition has been identified. Targeted sequencing is inexpensive and provides significant coverage of genomic regions of interest, reducing sequencing costs and time (Grada and Weinbrecht 2013).

2.5 METAGENOMICS IN PROBIOTICS RESEARCH

2.5.1 The Role of Metagenomics in Microbiome and Probiotics Research

A wide range of sequencing methods have been developed to perform metagenomic studies on the human gut microbiome. This has greatly improved our understanding of gut microbiota composition and activity, allowing us to identify microbial abnormalities associated with various diseases. Probiotic agents are commonly used to modulate the intestinal flora. Most recently, the development of quick and inexpensive DNA sequencing techniques has made it possible for these modern techniques to be applied to the study of microbial communities directly in their environment without requiring the culture of species and isolation to characterize the microbiome of the human intestine. In the same way, the sequenced genomes of probiotics and intestinal bacteria help to understand the role of the microbiota and probiotics in influencing human health (Gueimonde and Collado 2012). The ultimate goal of global microbiome research is to explore associations between bacterial species, communities, genes, and genomes of the human gut and various clinical conditions, such as obesity and colitis. Pathogens as well as probiotics are invaders of the general microbiome, which may originate in the gastrointestinal microbiome or be introduced to or selected from dietary interventions and pharmacology, such as probiotics or antibiotics. Epithelial cells have been regarded as the major physical barrier against pathogens, forming selective cellular barriers, but are now recognized as the major interface with the microbiota that initiates innate and adaptive immune responses. Several studies suggest that "the production of microbial metabolites following amino acid conversion leads to the production of low molecular weight immunoregulatory signals that help the bacteria inhibit mucosal immunity and improved nutrition". Many beneficial microorganisms participate in fermentation and produce metabolic products such as indole and acetate, which can protect against *E. coli* infection. Small molecules derived from microorganisms can perform signaling functions. Another class of soluble small molecules is reactive oxygen species (ROS). Recent studies have demonstrated that bacteria rapidly stimulate the production of ROS in intestinal epithelial cells (Bienenstock et al. 2013). Furthermore, the human gut microbiome contains an estimated 3.3 million microbial genes. The presence of this highly diverse gene suggests that a large impact of gut microbiota in the human body is to be expected. Metagenomics studies provide only insight into the genetic potential of the gut microbiome and fail to demonstrate its true functional contribution to the maintenance of health and well-being. Overall, more samples, more extensive sequencing, longer sequence readings, and further comparative analysis are required (Gerritsen et al. 2011).

2.5.2 Targeting the Human Microbiome with Probiotics

The concept of the human microbiome, which includes the notion that humans are highly sensitive to the microbial communities that live in and on our bodies, is an indisputable step forward in scientific research and has identified evidence that these microbiomes play a key role in human health and disease. Deciphering the function and composition of our microbiome, the common genome of the resident microbial community is a challenge that has been explored by scientists in a series of initiatives, such as the Human Microbiome Project (HMP) and the Integrated Human Microbiome Project (IHMP). The results of these projects have provided valuable data on the function of the human microbiome in different pathways (e.g., nose, mouth, skin, gastrointestinal tract, and urogenital tract). In this sense, gut microorganisms have been shown to perform several basic functions, such as synthesizing essential amino acids and vitamins and processing cellulosic material, which plays important roles in many aspects of human health (Alves et al. 2018). These diverse and complex gene collections encode a wide range of biochemical and physiological functions that may benefit the host as well as neighboring microbes. Most of the information collected to date has been synthesized using Sanger sequencing platforms (dideoxy). NGS systems introduced collection opportunities and deep sequencing strategies for microbial identification using single genetic targets or whole-genome approaches. The large intestine has a very complex microbial population, and the relative proportions of different bacteria can vary in different parts of the large intestine. Metagenomic studies have highlighted differences between the bacterial populations associated with the mucosa of the colon and fecal bacterial populations in humans and non-human primates (Petrosino et al. 2009). 16S ribosomal RNA (16S rRNA) gene sequencing and shotgun metagenomic analysis have revealed great diversity in the human microbiome. The 16S rRNA gene-based approach uses targeted NGS based on PCR amplification. In this case, we use conserved universal primers for 16S rRNA gene amplification, followed by sequencing of all 16S rRNA genes amplified from biomaterials and a reference sequence to identify the microbial species through advanced bioinformatics. The shotgun metagenomic approach, on the other hand, shears microbial DNA isolated from biomaterials, sequences all fragments via NGS, and identifies microbial species using reference libraries and complex bioinformatics approaches. From a simple point of view, metagenomics is the study of all the nucleic acids of a given biological material, such as human samples, environmental samples, etc. In the context of the microbiome, metagenomics refers to the study of microbial nucleic acids obtained from complex biological and environmental samples. These metagenomics studies involve sequencing a variety of nucleic acids from trillions of microorganisms, including bacteria, archaea, protists, viruses, and fungi. This can be done using two main approaches: (1) amplify the target genes by PCR and identify these amplicons, or (2) sequence by shotgun, fragment the genomic material into small fragments, sequence each fragment, and identify and sequence known nucleotide sequences, both of which require extensive and complex bioinformatics (Kanangat and Skaljic 2021).

2.5.3 According to the Science of Metagenomics, What Events Are Ahead?

Using NGS, metagenomics has the potential to detect all microorganisms in a sample, whether it is a known pathogen or an emerging pathogen. Anyway, this is not always an accurate method of indicating the spread of the infection, due to the lack of a definitive diagnosis of the disease. The most reliable strategy is laboratory surveillance, which allows several methods to be implemented to detect and confirm the presence of pathogens. Instead of using targeted analyses to identify a specific pathogen, metagenomics has the potential to completely overhaul this process, detecting all microorganisms with a single sequencing technique, and no culture is required. In addition, genomic information that can be collected through metagenomic NGS can be used for purposes other than outbreak investigation, including virulence gene identification and prediction of antibiotic resistance (Sawyer, Free, and Martin 2021). Despite the successes described above, the application of metagenomics to laboratory diagnosis is still in its infancy. However, the discovery of a new pathogen or an unusual subset of microorganisms in a specimen is only the first step in determining its role in disease. For other complex diseases with multiple microbial etiologies, metagenomics offers more targeted quantitative analysis in broader cohorts to distinguish between disease-causing microbes and non-pathogenic commensals, which can be used as a basis for. Despite these precautions, new discoveries must begin with new metagenomics. It can be used as a unique comprehensive screening test for potential, known, and emerging pathogens and can also assess an individual's microbiome status. Other targeted diagnostic tests can then be used to better understand the clinical disease and identify treatment options. As sequencing becomes cheaper and faster, it will be possible to characterize the human microbiome to study changes over time. Although metagenomics could one day be used to screen clinical samples, it is now better positioned as a complementary technique to be used alongside culture and other conventional methods. As whole genome sequencing technologies increase in speed and simplicity, however, metagenomic approaches are expected to be applied more frequently in public health emergencies, and routine pipelines have the potential to evolve through ongoing collaboration between researchers and clinicians. Such measures will be essential to expanding our public health tools, enabling us to quickly detect and manage new and emerging infectious diseases (Miller et al. 2013).

2.6 CONCLUSION

Microorganisms offer interesting opportunities to prevent diseases that cannot be treated by conventional medicines. Despite inconsistent and counterintuitive recognition for recombinant microbes and their metabolites, designer probiotics are envisaged as emerging living therapeutics for human and veterinary applications. Designer probiotics with antimicrobial, anti-inflammatory, and immunomodulatory properties could be used in a combined therapy to combat infectious and metabolic diseases. It is important to research whether designer probiotics can avoid the side effects of antibiotic therapy. In conclusion, recombinant probiotics offer powerful and futuristic alternative therapies for treating infectious and chronic diseases. First human observations, researchers, and the various applications of probiotics in their various forms highlight how much these amazing microorganisms can contribute to promoting our health, protecting us, and ensuring the treatment and management of diseases. It's crucial that probiotics come in natural forms and perform safe activities, which is perhaps the most crucial aspect. The application of metagenomics offers great potential to reveal the mechanisms and relationships between the human gut microbiome and disease. Not only can metagenomics identify the diversity of the human gut microbiome, but it can also reveal new genes and microbial pathways and reveal functional gut microbiome dysbiosis. However, metagenomics also has limitations and needs to be improved. The rapid development and application of metagenomics make it possible to identify new microbial diagnostic markers that will enable earlier diagnosis and new therapeutic approaches.

REFERENCES

Akmal, U., I. Ghori, A. M. Elasbali, B. Alharbi, A. Farid, A. S. Alamri, … S Ghazanfar. 2022. "Probiotic and antioxidant potential of the *Lactobacillus* spp. Isolated from artisanal fermented pickles." *Fermentation* 8 (7): 328.

Alves, Luana de Fátima, Cauã Antunes Westmann, Gabriel Lencioni Lovate, Guilherme Marcelino Viana de Siqueira, Tiago Cabral Borelli, and María-Eugenia Guazzaroni. 2018. "Metagenomic approaches for understanding new concepts in microbial science." *International Journal of Genomics* 2018: 2312987. doi: 10.1155/2018/2312987.

Amara, A. A., and A. Shibl. 2015. "Role of probiotics in health improvement, infection control and disease treatment and management." *Saudi Pharmaceutical Journal* 23 (2): 107–14.

Anadón, A., M. R. Martínez-Larrañaga, and M. Aranzazu Martínez. 2006. "Probiotics for animal nutrition in the European union. Regulation and safety assessment." *Regulatory Toxicology and Pharmacology* 45 (1): 91–5. doi: 10.1016/j.yrtph.2006.02.004.

Ayichew, T., A. Belete, T. Alebachew, H. Tsehaye, H. Berhanu, and A. Minwuyelet. 2017. "Bacterial probiotics their importances and limitations: A review." *Journal of Nutrition and Health Sciences* 4 (2): 202.

Bajaj, Bijender Kumar, Ingmar J. J. Claes, and Sarah Lebeer. 2021. "Functional mechanisms of probiotics." *Journal of Microbiology, Biotechnology and Food Sciences* 2021: 321–7.

Banik, Jacob J., and Sean F. Brady. 2010. "Recent application of metagenomic approaches toward the discovery of antimicrobials and other bioactive small molecules." *Current Opinion in Microbiology* 13 (5): 603–9.

Bienenstock, John, Glenn Gibson, Todd R. Klaenhammer, W. Allan Walker, and Andrew S. Neish. 2013. "New insights into probiotic mechanisms: A harvest from functional and metagenomic studies." *Gut Microbes* 4 (2): 94–100.

Capobianchi, M. R., E. Giombini, and G. Rozera. 2013. "Next-generation sequencing technology in clinical virology." *Clinical Microbiology and Infection* 19 (1): 15–22.

Chistoserdova, Ludmila. 2009. "Functional metagenomics: Recent advances and future challenges." *Biotechnology and Genetic Engineering Reviews* 26 (1): 335–52.

Chiu, Charles Y., and Steven A. Miller. 2019. "Clinical metagenomics." *Nature Reviews Genetics* 20 (6): 341–55.

Chua, Koon Jiew, Wee Chiew Kwok, Nikhil Aggarwal, Tao Sun, and Matthew Wook Chang. 2017. "Designer probiotics for the prevention and treatment of human diseases." *Current Opinion in Chemical Biology* 40: 8–16.

Cowan, Don, Quinton Meyer, William Stafford, Samson Muyanga, Rory Cameron, and Pia Wittwer. 2005. "Metagenomic gene discovery: Past, present and future." *TRENDS in Biotechnology* 23 (6): 321–9.

Forbes, Jessica D., Natalie C. Knox, Christy-Lynn Peterson, and Aleisha R. Reimer. 2018. "Highlighting clinical metagenomics for enhanced diagnostic decision-making: A step towards wider implementation." *Computational and Structural Biotechnology Journal* 16: 108–20.

Fuller, R. 1989. "Probiotic in man and animals." *Journal of Applied Bacteriology* 66: 131–9.

Fuller, Ray. 2012/1992. "Probiotics: The Scientific Basis."
Gaston, D. C. 2023. "Clinical metagenomics for infectious diseases: Progress toward operational value." *Journal of Clinical Microbiology* 61 (2): e0126722. doi: 10.1128/jcm.01267-22.
Gerritsen, Jacoline, Hauke Smidt, Ger T. Rijkers, and Willem M. de Vos. 2011. "Intestinal microbiota in human health and disease: The impact of probiotics." *Genes & Nutrition* 6: 209–40.
Gilbert, Jack A., and Christopher L. Dupont. 2011. "Microbial metagenomics: Beyond the genome." *Annual Review of Marine Science* 3: 347–71.
Gogineni, Vijaya K., Lee E. Morrow, and M. A. Malesker. 2013. "Probiotics: Mechanisms of action and clinical applications." *Journal of Probiotics & Health* 1 (1): 1–11.
Grada, Ayman, and Kate Weinbrecht. 2013. "Next-generation sequencing: Methodology and application." *The Journal of Investigative Dermatology* 133 (8): e11.
Gueimonde, M., and M. C. Collado. 2012. "Metagenomics and probiotics." *Clinical Microbiology and Infection* 18: 32–4.
Hemaiswarya, Shanmugam, R. Raja, R. Ravikumar, and S. Isabel. 2013. "Mechanism of action of probiotics." *Brazilian Archives of Biology and Technology* 56: 113–9.
Jankovic, I., W. Sybesma, P. Phothirath, E. Ananta, and A. Mercenier. 2010. "Application of probiotics in food products–challenges and new approaches." *Current Opinion in Biotechnology* 21 (2): 175–81. doi: 10.1016/j.copbio.2010.03.009.
Kanangat, Sivadasan, and Ina Skaljic. 2021. "Microbiome analysis, the immune response and transplantation in the era of next generation sequencing." *Human Immunology* 82 (11): 883–901.
Kanwal, H., A. Di Cerbo, F. Zulfiqar, C. Sabia, A. Nawaz, F. M. Siddiqui, … S. Ghazanfar. 2021. "Probiotic characterization and population diversity analysis of gut-associated pediococcus acidilactici for its potential use in the dairy industry." *Applied Sciences* 11 (20): 9586.
Kathirvelan, C. "Role of Probiotics in Designer Milk Production."
Khan, A. N., H. Yasmin, S. Ghazanfar, M. N. Hassan, R. Keyani, I. Khan, and A. Ahmad. 2021. "Antagonistic, anti-oxidant, anti-inflammatory and anti-diabetic probiotic potential of *Lactobacillus agilis* isolated from the rhizosphere of the medicinal plants." *Saudi Journal of Biological Sciences* 28 (11): 6069–76.
Khan, A. N., S. Ghazanfar, M. N. Hassan, A. Ahmad, N. Khan, S. Khalid, and H. Yasmin. 2024. "Probiotic potential of *Lactobacillus agilis* against oxidative, inflammatory and diabetic stresses." *Journal of King Saud University-Science* 36 (5): 103144.
Lehri, B., A. M. Seddon, and A. V. Karlyshev. 2015. "*Lactobacillus fermentum* 3872 genome sequencing reveals plasmid and chromosomal genes potentially involved in a probiotic activity." *FEMS Microbiology Letters* 362 (11):fnv068. doi: 10.1093/femsle/fnv068.
Li, X., Q. Wang, X. Hu, and W. Liu. 2022. "Current status of probiotics as supplements in the prevention and treatment of infectious diseases." *Frontiers in Cellular and Infection Microbiology* 12: 789063. doi: 10.3389/fcimb.2022.789063.
Miller, Ruth R., Vincent Montoya, Jennifer L. Gardy, David M. Patrick, and Patrick Tang. 2013. "Metagenomics for pathogen detection in public health." *Genome Medicine* 5: 1–14.
Miller, Steve, and Charles Chiu. 2021. "The role of metagenomics and next-generation sequencing in infectious disease diagnosis." *Clinical Chemistry* 68 (1): 115–24. doi: 10.1093/clinchem/hvab173.
Mirete, Salvador, Veronica Morgante, and José Eduardo González-Pastor. 2016. "Functional metagenomics of extreme environments." *Current Opinion in Biotechnology* 38: 143–9.
Mokili, John L., Forest Rohwer, and Bas E. Dutilh. 2012. "Metagenomics and future perspectives in virus discovery." *Current Opinion in Virology* 2 (1): 63–77.
Ng, S. C., A. L. Hart, M. A. Kamm, A. J. Stagg, and S. C. Knight. 2008. "Mechanisms of action of probiotics: Recent advances." *Inflammatory Bowel Diseases* 15 (2): 300–10. doi: 10.1002/ibd.20602 Inflammatory Bowel Diseases.
Petrosino, Joseph F., Sarah Highlander, Ruth Ann Luna, Richard A. Gibbs, and James Versalovic. 2009. "Metagenomic pyrosequencing and microbial identification." *Clinical Chemistry* 55 (5): 856–66.
Rostami, Farzaneh Mohammadzadeh, Hadis Mousavi, Morteza Rabi Nezhad Mousavi, and Milad Shahsafi. 2018. "Efficacy of probiotics in prevention and treatment of infectious diseases." *Clinical Microbiology Newsletter* 40 (12): 97–103.
Sawyer, Abigail, Tristan Free, and Joseph Martin. 2021. Metagenomics: Preventing Future Pandemics. Future Science, 1–4.
Silva, Diego Romário, Janaína de Cássia Orlandi Sardi, Nayla de Souza Pitangui, Sindy Magri Roque, Andréa Cristina Barbosa da Silva, and Pedro Luiz Rosalen. 2020. "Probiotics as an alternative antimicrobial therapy: Current reality and future directions." *Journal of Functional Foods* 73: 104080.

Simner, P. J., S. Miller, and K. C. Carroll. 2018. "Understanding the promises and hurdles of metagenomic next-generation sequencing as a diagnostic tool for infectious diseases." *Clinical Infectious Diseases* 66 (5): 778–88. doi: 10.1093/cid/cix881.

Sleator, Roy D. 2010. "Probiotic therapy-recruiting old friends to fight new foes." *Journal of Gut Pathogens* 2 (1): 1–4.

Thomas, Torsten, Jack Gilbert, and Folker Meyer. 2012. "Metagenomics – A guide from sampling to data analysis." *Microbial Informatics and Experimentation* 2 (1): 12.

Vanderpool, Charles, Fang Yan, and D Brent. 2008. "Mechanisms of probiotic action: Implications for therapeutic applications in inflammatory bowel diseases." *Inflammatory Bowel Diseases* 14 (11): 1585–96.

Vitetta, Luis, Samantha Coulson, Michael Thomsen, Tony Nguyen, and Sean Hall. 2017. "Probiotics, D–Lactic acidosis, oxidative stress and strain specificity." *Gut Microbes* 8 (4): 311–22. doi: 10.1080/19490976.2017.1279379.

Wooley, John C., Adam Godzik, and Iddo Friedberg. 2010. "A primer on metagenomics." *PLoS Computational Biology* 6 (2): e1000667.

Yan, Fang, and D. B. Polk. 2011. "Probiotics and immune health." *Current Opinion in Gastroenterology* 27 (6): 496.

Yoha, K. S., Sundus Nida, Sayantani Dutta, J. A. Moses, and C. Anandharamakrishnan. 2022. "Targeted delivery of probiotics: Perspectives on research and commercialization." *Probiotics and Antimicrobial Proteins* 14 (1): 15–48. doi: 10.1007/s12602-021-09791-7.

Yu, X., C. Lin, J. Yu, Q. Qi, and Q. Wang. 2020. "Bioengineered *Escherichia coli* Nissle 1917 for tumour-targeting therapy." *Microbial Biotechnology* 13 (3): 629–36. doi: 10.1111/1751-7915.13523.

3 Probiotic Genomics and Cancer

Leila Haghshenas, Maryam Bakhtiyari, Shakira Ghazanfar, and Ali Ajami

3.1 INTRODUCTION

Symbiotic bacteria in the human gut are named intestinal flora in general, and their number reaches more than 100 trillion. Interestingly, the number of these bacteria is more than the number of host cells. As these bacteria are essential and even vital for human health, nowadays the stability of normal flora via the use of probiotics has universal application (Lu et al. 2021). The use of probiotics dates back to ancient times when the ancient Chinese used human feces to treat food poisoning and infections. In 1965, it took a professional form of use. When Lilly and Stillwell used materials produced by bacteria to grow other bacteria. With the passage of time, this use got a new name, and in 1989, living microbial components was a term that was only used for nutritional health. Today, however, this use of bacteria has a wider meaning and is not only related to nutritional health (Śliżewska et al. 2021). But the use of probiotics in the field of cancer goes back to Goldin and Gorbach's research when they found that lactobacilli-enriched foods reduce the incidence of colorectal cancer (CRC) by 37% (Wieërs et al. 2019). Since then, many research studies have been conducted in this field, and many successful results have been obtained. In this chapter, we intend to examine the relationship between probiotics and cancer and also to express a more specialized concept of this use by presenting the genomics information.

3.2 EFFECTS OF PROBIOTICS ON HOST

Like other biological and chemical interactions, interactions of probiotics and the host have some consequents. These consequents are divided into nonspecific and specific effects.

3.2.1 Nonspecific Physiological Effects

The regulation of gastrointestinal flora: probiotics are able to preserve a healthy balance between intestinal flora. In the study of fecal samples, it was established that supplementing with probiotics in healthy adults might lead to a higher concentration of particular strains of bacteria and thus increase the overall number, diversity, or composition of gut flora (Lu et al. 2021). As the research progresses, it has been proven that regulating intestinal flora can help to enhance immunity, treat metabolic disorders, and even treat mental illness. Probiotics play an important role in regulating the intestinal flora as a class of beneficial microbes. In mice, Baglihar et al. reported major changes in intestinal flora that were associated with a variety of diseases following feeding on a fat-rich diet. The injection of the probiotics into obese animals has revealed a reduction in intestinal flora firmicutes and an increased incidence of *Actinobacteria*. Treatment with probiotics can reverse intestinal flora dysbiosis and treat inflammatory responses in mice. The critical element for the treatment of inflammatory diseases is intestinal flora. Explanation: it is likely that the gut flora regulation by probiotics

DOI: 10.1201/9781003399797-4

may in fact be effective at combating related inflammatory disorders. The researchers found a significant increase in short-chain fatty acids (SCFAs) concentrations from the gut flora of probiotic-treated mice compared to control mice, as well as beneficial bacteria such as *Oscillibacter* and *Prevotella*. SCFAs, like important metabolites of intestinal microflora, are very valuable in the field of research. The SCFA has been shown to improve immunity and serve as a chemical signal of brain–gut communication. The quantity of SCFAs producing microorganisms in the gut can be increased by using probiotics. This can be beneficial to our immune system, as well as maintaining a healthy lifestyle. In addition, the regulatory effect of probiotics on intestinal flora has been confirmed by a large number of studies, and this is indisputable. Studies by Oh et al. have shown that the decrease in diversity of intestinal flora, caused by use of antibiotics, may be improved with probiotics. Another study showed an increase of probiotics, including *Bifidobacterium* and *Lactobacillus*, which contributed to improving intestinal function after consumption of yogurt by patients with inflammatory bowel disease (IBD). Probiotics seem to play an important role in regulating gut flora and have a very promising potential for the treatment of several diseases over time (Wang et al. 2021). Stabilize the intestinal epithelial cell (IEC) barrier: to prevent bacterial colonization in the epithelium, probiotics regulate the cytoskeletal structure with a view to stabilizing the mucosal barrier and promoting mucin secretion. They can induce expression and distribution of tight junction proteins by sealing the top epithelium and endothelium, an increase in epithelial permeability and damage to the epithelial structure are prevented. Probiotics can also reduce the inflammation response and excessive pathogenic apoptosis by returning normal transepithelial resistance caused by lipopolysaccharide (LPS). Furthermore, certain strains of probiotics are able to influence the polarization of T helper 17 (Th17) cells and specifically activate production of IL-17a, which causes type 3 innate lymphocytes (ILC3s) to produce IL-12. The crucial immunological defense cytokines that play a major role in the maintenance of intestinal homeostasis and promote healing and tissue regeneration are IL-22. Animal studies have shown that, due to defects in the secretion of defensin and damaged epithelial tight junctions, mice lacking these cytokines are prone to experimental colitis (Lu et al. 2021). Hou et al. demonstrated that *Limosilactobacillus reuteri* D8 stimulates lamina propria lymphocytes to secrete IL-22 via the aryl hydrocarbon receptor and then activates the phosphorylation of signal transducer and activator of transcription 3 (STAT3) to promote the proliferation of IECs and increase the growth of intestinal organs, thus recovering the intestinal mucosal barrier, which restores the structural damage to the intestinal epithelium caused by TNF treatment. In a mouse model of IBD, continuous administration of *Bacillus subtilis* during the remission period can maintain the integrity of the intestinal barrier by regulating the proliferation of IECs and alleviating IBD (Gou et al. 2022). Inhibition of pathogens: there are, in particular, two different mechanisms for inhibiting the pathogen. One's a member of the Physical Defense System. The pathogen's infection is caused by colonization on the surface of the gastrointestinal mucosa, leading to tissue damage. When the wall of the gut is completely occupied by probiotics, there isn't room for pathogens, and they can also inhibit their adhesion to a pathogenic bacterium through increased nutrient intake. These mechanisms are related to the antagonistic properties of probiotics that could reduce microenvironment pH by producing SCFAs, as described below.

3.2.2 Specific Physiological Effects

In the same way as live probiotics, specific molecules produced by probiotics may contribute to the improvement of host health by promoting specific physiological functions; it is not yet possible to fully elucidate the precise mechanisms.

In order to achieve an immunological modulation without the potential risks of living microbes, such as those caused by immune-deficient patients, postbiotic supernatants collected from bacterial cultures could be used. It is well known that the effect of probiotics can be mediated by their metabolites, such as SCFA, in particular propionate, acetate, and butyrate, which may exercise anti-inflammatory effects. Bifidobacilli, lactobacilli, and some commensal bacteria are involved in the production of SCFAs. The postbiotics exert their effects by binding to a specific receptor for IECs. This results in inhibition of NFkB pathway, Treg cell suppression, and Pro-inflammatory Cytokine Production by Neutrophils and Macrophages; therefore, inflammation is blocked, and an antioxidative effect has been observed. Through the induction of differentiation of Treg cells, butyrate may have a positive effect on reducing inflammation in the gut. For instance, *Lactobacillus acidophilus* CRL 1014, recently studied in the Simulator of Human Microbial Ecosystem reactor, has been shown to produce SCFAs, while BB. Longum SP07/03 and BB. Only propionate and acetate can be produced with bifidum MF 20/5. In addition, interaction with tryptophan derived from food and indolic acid derivatives produced by probiotics or intestinal bacteria expressing specific receptors for immune cells can produce anti-inflammatory effects. In gut homeostasis, these molecules play a significant role; for instance, indole-3 ionic acid is the main factor promoting IPA production through anti-inflammatory activities and decreasing TNFα release. Both in vitro and in vivo murine studies have shown how *Clostridium sporogenes* can convert tryptophan into IPA, resulting in a protective effect against dextran sulfate sodium-induced acute ulcerative colitis (UC). In another study, the whole bunch of supernatants from probiotic cultures was collected to check for sets of regulatory molecules that might stimulate immune cells. In patients with infection-related diarrhea, necrotizing enterocolitis, and IBD, *L. reuteri* 17938 is commonly used as an adjuvant. *L. reuteri*'s supernatant has been mixed with peripheral blood mononuclear cells (PBMCs), inducing CCR7 on DC membranes and the production of FOXP3 and IL-10 in Treg cells. De Marco et al. found that *L. acidophilus* and *Lactobacillus casei* supernatant may reduce the TNF-α release and stimulate IL-10 secretion. In addition, *L. caseinatants* may inhibit the activation of LPS1, which could explain its positive effect in IBD, by inhibiting the activation of IL-1b (Cristofori et al. 2021).

3.3 THE EFFECTS OF HOST ON PROBIOTICS

There are reports of differences in host physiology caused by the same strain. Unlike medicinal products, the effectiveness of probiotics varies widely between individuals. The variability of effects is the result of different ages, physical conditions, gut microbial composition, colonization permission, and dietary habits in the host. The development of intestinal flora plays an important role in the balanced development of a child's immune system, especially for infants and younger children who have not yet fully developed their immunity during the first month after birth. In addition to the non-cytotoxicness of *Bifidobacteria* in breast milk, this bacterium has good immune-stimulatory ability, and there is no evidence for a beneficial effect on infants' health by supplementing with probiotics. In an observational study, the incidence of mucosa-related diseases was higher in early childhood, although supplementation increased the response of infants to sIgA. The benefits of exposure to probiotics at an early age are not restricted in comparison with healthy adults but may also be associated with a higher incidence of infections during adulthood. In cancer patients, after undergoing treatments, such as chemotherapy, radiotherapy, or surgical eradication, underlying medical conditions, such as cachexia combined with treatment-related side effects, and the microenvironment are more complicated and can directly lead to intestinal mucosal barrier destruction and immune system dysfunction.

As a result, beneficial prokaryotes in the colon are not stimulated by these changes. A decrease in the number of probiotics has been observed in individuals with CRC. Zmora, N. et al. have established that the dominant role of local intestinal microbes in probiotics development is also played by host flora, and its usefulness depends upon support from gut flora. These results suggest that the barrier of colonization will have a major influence on the therapeutic effects, even though the use of probiotics is beneficial. In particular populations, for example, cancer patients, it is important to understand the effects of probiotics. The intestinal microecology is composed of flora and prebiotics and enteral nutrition, which are complementary to each other (Lu et al. 2021).

3.4 USE OF PROBIOTICS IN CHEMOPREVENTION OF CANCER

A large number of studies describing possible modes of action have demonstrated that the use of probiotic medicinal products is associated with a reduction in cancer risk. In the human gut, polymerized polyphenols such as ellagitannins and proanthocyanidins have very limited absorption and are catabolized by probiotic bacteria, including *Bifidobacterium* and *Lactobacillus*, which are naturally present in the human microbiome (Azad et al. 2025). By reducing colon exposure to carcinogens or toxins, bacillus is also contributing to the reduction of cancer risk. Probiotics reduce the levels of β-glucuronidase, nitroreductase, and azoreductase enzymes by colonic cells to discourage the conversion of procarcinogens into carcinogens. In addition, the uptake of nitrites and other bile salts, which are substrates for these enzymes, is also able to be used by probiotics. In addition, the bacteria are capable of removing toxins, including mycotoxins such as aflatoxin B1, which have been shown to be beneficial for colon and liver cancer. The primary cause of cancer is oxidative stress generated by reactive oxygen species (ROS). Adequate availability of antioxidants and antioxident enzymes is also important for neutralizing ROS before further damage to cell structure, in particular DNA.

Antioxidants like glutathione, vitamin E, and C are created by the bacteria. In addition, ROS neutralization is enhanced by enzymes such as superoxide dismutase, glutathione peroxidase, and catalases generated by probiotics. Probiotics also exhibit cancer prevention mechanisms at a higher level of interaction with the host at the cellular level. A number of studies have suggested that probiotics can stimulate the production of apoptosis, necrosis, inhibition of proliferation, and migration of colonic cancer cells. By increasing BAX translocation, releasing cytochrome C, and degrading caspases 3 and 9, the bacterium *Lactobacillus rhamnosus GG* or *Bacillus lactis Bb12* has been shown to induce apoptosis in Caco2 cells. *Saccharomyces boulardii* limits the proliferation and induces apoptosis in human colonic cancer cells, HT29, SW480, and HTC-116, by inactivating EGFR-Mek-Erk pathway signaling. In addition, by reducing the cell potential degradation of collagen matrix in tissues, L. acidophilus and *L. casei* may also limit Caco-2-cell migration and invasiveness. A possible intervention for effective prevention of colon cancer could be the potential to use probiotics in host immune modulation. Wistar rats induced for colonic cancer using 1, 2-dimethylhydrazine (DMH), demonstrated a lesser number of aberrant crypt foci (AFC) and cancer incidence when administered with Enterococcus faecium bacteria. Improved immune response expressed by elevated production of IL-4, TNF-α, and IFN-γ (confirmed by histological experiment) protected the rats from DMH-induced colonic cancer. A significant contribution to reducing cancer risks is contributed by the role of probiotic regulation in limiting pathogenic invasion. Pathogenic microbes cause tumorigenesis in 15–20% of cancer cases. A typical feature of such cases is dysbiosis and changes to the natural microbiome (Thilakarathna et al. 2018). The communication between properia lymphocytes and IECs is shown in Figure 3.1.

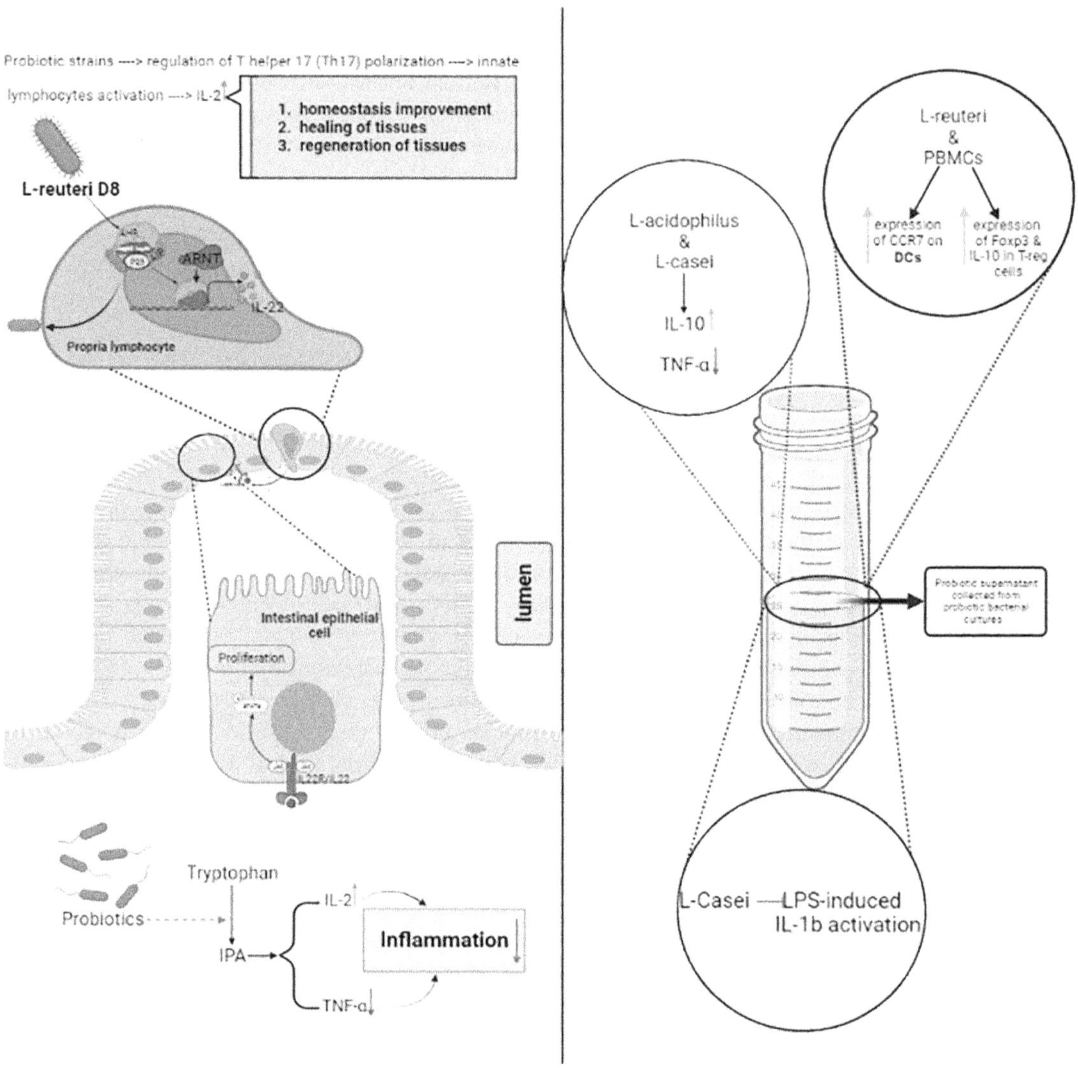

FIGURE 3.1 Communication of properia lymphocytes and intestinal epithelial cells.

3.5 MECHANISM OF PROBIOTICS ACTION IN CANCER PREVENTION AND THERAPY

3.5.1 The Metabolic Activities of Intestinal Microbiota, and Bacterial Genomics

Dysbiosis is a change in the microbial composition that leads to an interruption of physiological balance within IECs. Dietary changes, antibiotic therapy, and IBD are the causes of biological disorders. The mucosa barrier may be compromised by a prevalence of dysbiosis. As a result, the microbiome can influence cancer through multiple mechanisms, such as creating oncogenic metabolites that involve pro-inflammatory pathways and suppressing immune system function. The regulation of the growth and reproduction of intestinal flora is aided by microbial first metabolites, for example, amino acids, nucleotides, polysaccharides, lipids, or vitamins. These metabolites are the same for each cell type of bacteria. The process of synthesizing primary metabolites is continuous, and all disorders can have an effect on normal microbial activity. The specificity and function of the flora are determined by microbial secondary metabolites, such as alkaloids, phenols, antibiotics, or pigments. Gastrointestinal tumors have different in vivo metabolisms, including amino acid metabolism linked to liver cancer cells and glucose metabolism related to gastric cancer. The metabolism of lipids is associated with liver cancer. Studies have shown that the microbes in the gut are degrading and absorbing nutrients, which produce SCFAs. It is vital to maintain the correct composition of the microbiota in order to prevent pathogens and contributes to a host's own immunity.

3.5.1.1 Carbohydrates

There are many complex chemical reactions occurring inside the human body, called carbohydrate metabolism. As the main pathway of carbohydrate metabolism, the citric acid cycle is the final metabolic pathway and the center of the three primary nutrients, carbohydrates, lipids, and amino acids. In the case of intestinal flora and colon cancer, carbohydrates play a major role in metabolism. The intestine is made up of anaerobic and aerobic bacteria. CRC develops in a very direct relationship with superoxides, oxidation radicals, and oxygen molecules. Glucose and lactose are broken up by a variety of bacteria in order to synthesize acid. Maintaining an acid-base balance and regulating the osmotic pressure are important ways of controlling intestinal microbial populations.

3.5.1.2 Short-Chain Fatty Acids

The SCFA is a signaling molecule that mediates the interaction between diet, gut microbiota, and host. In the body, they play a vital role in immune, metabolic, and endocrine systems. Maltodex inhibits growth of cancer cells, induces an apoptosis, upregulates MUC2 gene expression, and differentiates malignant cells during the production of SCFA and increased butyrate due to wheat grain fermentation. *Bifidobacterium bifidum* inhibits the growth of cancer cells in CRC by arresting the cell cycle within the G0/G1 phase. Moreover, it increases the activity of alkaline phosphatase, a unique marker whose levels are reduced in malignant cells.

3.5.1.3 Tryptophan

Most of the tryptophan comes from food with various biological functions. Tryptophan metabolism in humans is mainly based in the small intestines and the central nervous system, making it a major regulator of inflammation and immunity. Its metabolites are endogenous ligands of AHR, regulating the intestinal immune function. It is an important factor for maintaining homeostasis of the mammalian gastrointestinal immunological system. The most recent findings have shown that the consumption of high tryptophan by diet supplemented with *L. reuteri* may be associated with an increased number of T cells. Predictive experiments on host-microbe interactions have revealed that TNF-α and IFN-γ are related to specific microbial metabolic pathways, including palmitic acid and tryptophan-degrading alcohols. A study has shown that inflammation is suppressed by tryptophan and its metabolites, for example Kynurenic acid. A specific tryptophan concentration can have a significant inhibitory effect on gene expression of ionizing factors. Therefore, it slows down the progression of illness and is associated with a positive effect on health. Studies have shown that tumor cells are capable of using large amounts of tryptophan as an immunosuppressant for the synthesis of a tumor microenvironment, thereby promoting proliferation. Microorganisms may regulate tryptophan metabolism and thus control the regulation of immunity in the gut.

3.5.1.4 The Production of Compounds with Anti-Carcinogenic Characteristics, Such as SCFAs and Conjugated Linoleic Acid

SCFAs consist of organic acids, including acetic acid, propionic acid, Butyric acid, valec and caprenic acid in the aliphatic chain composed of 1–16 carbon atoms. The main and final product of the metabolism of light living bacteria in the large intestine is SCFA.

SCFAs are catalysts of energy in colonocytes and support an acidic environment, which promotes anticancer cells' acidosis and apoptosis, thereby inhibiting the formation of high levels of primary bile acids. Butyric acid helps regulate the balance between cell proliferation, division, and apoptosis of colonocytes.

The colonocytes break up about 70–90% of butyrate. Compared to CRC patients, this acid is found in higher amounts in the feces of healthy people. In addition, it is assumed that a reduction in the 1 µg/L butyrate concentration in stools increases the risk of colon cancer by 84.2%. However, the risk of developing adenoma increases by 71.3% when the concentration of acetic acid is reduced by 1 g/L. The bacteria that make up the intestinal microbiota are all natural sources of SCFAs. However, in order to inhibit the development of CRC, the amount produced may not be sufficient. Thus, an increase in SCFA intake can be expected from the consumption of probiotics. The presence of SCFAs in the lumen of the large intestine can inhibit the growth of pathogens, for example, *Salmonella typhimurium*, in the presence of propionate and butyrate; inhibit the expression of invasive genes encoding *Salmonella SPI-1* pathogen islands; and prevent attack on cells in tissue culture in vitro. Because of its ability to improve the intestinal barrier by increasing mucus production and proliferation of beneficial cells, butyric acid may play an important role in preventing CRC. Butyrate has the ability to decrease the production of inflammatory cytokines by inhibiting the activation of NF-κB, regulating the activity of proteins involved in apoptosis (Bcl-2, Bak, and caspases 3 and 7), increasing the immunogenicity of tumor cells, suppressing cyclooxygenase (COX)-2 activity, increasing the activity of the antioxidant enzyme glutathione S-transferase (GST), and inhibiting the deacetylation of histones. These effects may have an effect on the suppression or regulation of certain genes that are involved in controlling cell cycle proliferation, differentiation, and apoptosis. The isomers of linoleic acid (LA), in which two carbon pairs are connected to each other and formed double bonds, are conjugated linoleic acids (CLAs). Csi-9, trans-11, cis-10, and cis-12 are the most frequently discovered CLA isomers in nature. CLAs have been shown to affect cellular response to cell growth factors and the expression of genes related to apoptosis process, Bcl-2, caspase 3, and 9. Secondly, CLAs are capable of two ways of reducing the eicosanoids produced by colonocytes. CLAs cause the elimination of arachidonic acid from cell membranes, and this is followed by interference with activity in cyclooxygenase or lipoxygenase enzymes to synthesize eicosanoids. According to Ewaschuk et al., some species of probiotic bacteria, such as *Lactobacillus (acidophilus, casei, delbrueckii, and plantarum), Bifidobacterium (breve, infantis, and longym), Propionibacterium freudenreichii, and Streptococcus salivaris subsp. Thermophilus* is able to produce CLAs from LA. In fact, this fatty acid is produced by a bacterium in the proximal ileum and can be consumed or mixed with colonocytes within the lumen of the intestine to confer beneficial effects (Śliżewska et al. 2021).

3.6 INHIBITION OF CELL PROLIFERATION AND INDUCTION OF APOPTOSIS IN CANCER CELLS

Interest is growing in the use of bacteria that are able to regulate cell proliferation and apoptosis, because cancer cells will be eliminated at a slower rate, which means they won't damage neighboring cells or cause inflammation. The apoptosis signaling pathways can be activated by probiotic bacteria (e.g., lactic acid bacteria, or LAB) through a mitochondria-dependent (intrinsic) and a death receptor–dependent, mitochondria-independent (extrinsic) pathway. The antiproliferative activity of *L. acidophilus* and *L.* against LS513 gastric cell lines through cellular apoptosis has been evaluated by Baldwin et al. By inhibiting NFB- and mTOR-mediated apoptosis, Hwang et al. have shown that probiotic activation in gastric cancer cells KATO3 is induced. The SCFA, notably butyrate, which is able to induce epigenetic changes, interrupts the cell cycle and stimulates the expression of proapoptotic genes, has been associated with an increased incidence of apoptosis among cancer cells induced by probiotics. For instance, in colonic crypts, a relationship exists between the amount of SCFA in feces and cell proliferation. Probiotic *Lactobacillus spp.* induces selective genotoxic, cytotoxic, and proapoptotic effects on leukemia and colon cancer cell lines and as anti-inflammatory effects on macrophage cells at molecular level (Śliżewska et al. 2021).

3.7 BINDING AND DEGRADATION OF THE CARCINOGENIC COMPOUNDS PRESENT IN THE INTESTINAL LUMEN

Some probiotic bacteria may be bound to the cell walls by carcinogenic compounds. This is also linked to the presence of anionic exchange between carcinogenic substances and peptidoglycans found on cell walls in some strains of bacteria. Therefore, both carcinogenic substances and bacteria are eliminated through feces. In studies, the *Bifidobacterium longum*, *L. acidophilus*, and Strepto have been found. Coccus salivarius strains could bind and cause the release in feces of heterocyclic amines and mutagens such as 2-amino-3,4-dimethylimidazo [4,5-f] quinoline (MeIQ), 2-amino-3-methyl-3H-imidazo [4,5-f] quinoline (MHIQ), 5-phenyl-2-amino-l-methylimidazo [4,5-f] pyridine (PhMIP), and 3-amino-1-methyl 5H pyrido[4,3-b] indole acetate (TrpP2).

In vitro, Rowland and Grasso studied the effects of intestinal microbial species known as genera *Lactobacillus, Bifidobacterium,* and *Streptococcus* on dimethylnitrosamine by showing that these substances were most strongly degraded by bacteria in this genus. As a result, the amine was converted to its precursor's dimethylamine and other volatile metabolites. In contrast, the ability of *L. casei* to remove mutagenic heterocyclic amines has been shown by Morotomi and Mutai. They've studied the effects of live and heat-inactivated bacteria, and only live bacteria have shown this ability. This may indicate that live bacteria produce metabolites or catalyze reactions that lead to amine detoxification. According to the literature, mycotoxins that may have carcinogenic characteristics are also capable of being excreted by probiotics (Śliżewska et al. 2021).

3.8 PROBIOTICS' ROLE IN THE TREATMENT OF ANTITUMOR COMPLICATIONS

By biological action, probiotics derived from *Bifidobacterium* and *Lactobacillus* are capable of efficiently suppressing the development of pathogenic bacteria. The environment of the gastrointestinal tract may be improved by supplementing it with probiotics, thereby increasing intestinal mucosal barrier function and reducing incidences of diarrhea. Recent studies have shown that innate immunity is also related to the improvement in antitumor side effects of probiotics. For example, by stimulating intracellular signal recognition receptors NOD2, bacterial cell walls' acyl dipeptides prevent damage to the mucosal system caused by antibiotic chemotherapeutics. In general, probiotics may have a beneficial effect by improving diarrhea caused by radio-chemotherapy or surgery and rarely cause side effects. Besides restoring the mucosal barrier in the gut, probiotics are also effective at reducing damage to the oral mucosa caused by chemotherapy. Probiotics may also be helpful in the treatment of systemic inflammation, such as graft versus host disease (GVHD). The primary causes for GVHD, in which the intestines are one of the most severely affected organs and play a key role in its severity, are donor-derived T cells, pro-inflammatory cytokines, or LPS. The feasibility and efficacy of HSCT are significantly limited by the occurrence of GVHD. An intact intestinal wall has a major role to play in the development of GVHD, and LPS may enter the bloodstream through damage to the mucosal barrier capable of causing GVHD (Lu et al. 2021).

3.9 PROBIOTICS TO BE USED FOR A VARIETY OF CANCERS

So far, we talked about theories and mechanisms of probiotic actions. In this section, we want to talk about cancer cases specifically.

3.9.1 Probiotics Using in Colon Cancer

A number of studies have shown that regular consumption of probiotics may improve the quantitative and qualitative profile of the intestinal microbiota, thereby reducing the incidence of chronic inflammation and the production of carcinogenic compounds during intestinal dysbiosis. In a study by Liu et al., the diversity and microbial richness of CRCs undergoing colorectomy were increased in patients when they had been regularly injected with *L. plantarum, L. acidophilus,* or *B. longum* at high doses for 16 days. The patient's gastrointestinal microbiota composition resembled that of healthy individuals in this study. As mentioned, some intestinal enzymes, such as β-glucosidase, β-glucuronidase, nitrate reductase, azoreductase, and 7-α-dehydroxylase, which are able to convert aromatic hydrocarbons and amines into active carcinogens by synthesizing aglycones, phenols, cresols, ammonia, and N-nitroso compounds, can have cytotoxic and genotoxic activities, thereby contributing to the development of colon cancer. The human study conducted by Hatakka et al. has shown that the consumption of some strains of antibiotic bacteria may have reduced these enzyme activities as well as prevented colon cancer. Galdeano et al., for their part, highlighted the importance of a dosage around 109 colony-forming units (CFU), day longevity, and an intestinal duration between 48 and 72 h as optimal characteristics to induce host immune stimulation. Human microbiota manipulation, as well as use of selective probiotics, has been identified as a promising therapeutic option for CRC.

Treatment with a mixture of probiotics (*L. plantarum, L. acidophilus, and B. longum*) in individuals with CRC improved the outcome and increased the amount of cell junction proteins (claudin, occludin, and JAM-1) as well as their distribution throughout the colonic epithelium. The pro-apoptotic activity induced by the consumption of probiotics, especially by increasing TNF-α production, is another factor well investigated in human cancer. Wan and colleagues have established that an increased expression of caspase 3 is responsible for the ability to cause apoptosis in tumor cells via bacterium *L. delbrueckii* (Al Mohaini et al. 2022).

Recently, Kotzampassi et al. observed a significant decrease in all major postoperative complications in patients undergoing surgery for CRC after treatment with a probiotic formulation versus a placebo (28.6% vs. 48.8%, $p = 0.010$, OR 0.42). These studies are too early to clearly demonstrate the effectiveness of probiotics in preventing and treating colon cancer, but they may be an indication that further research into this fascinating area is warranted.

Recently, it has been shown that the immune response to cancer therapy and its susceptibility to adverse impacts can be influenced by microbiota, in particular those originating in the gut (Al Mohaini et al. 2022). Evidence of the ability of the microbiota to influence chemotherapy, radiotherapy, and immunotherapies with a particular focus on microbial composition has begun to emerge (Drago 2019).

3.9.2 Probiotic Using in Breast Cancer

3.9.2.1 Animal Models

There are several animal studies that demonstrate beneficial effects of probiotics against BC. From two weeks prior to the transplantation of BC tumors, oral administration of *L. acidophilus* in 0.5 mL suspension containing 2.7×10^8 CFU/mL was administered every day for a period of 30 days with three-day intervals. The results showed a significant increase in the survival time among the *L. acidophilus* group compared to that of the controls, demonstrating that this treatment can promote the immune responses via stimulation of the production of pro-inflammatory cytokines such as IFN-γ and inhibition of the production of anti-inflammatory cytokines such as IL-4 and IL-10. Sixty-five additional animal studies confirmed that oral administration of *L. acidophilus* displays anticancer activity in mice bearing breast tumors. In another study, 0.5 mL of this suspension containing 2.7×10^8 CFU/mL *L. casei* was orally administered to mice using a standard gastric feeding tube. The suspension was given daily for two consecutive weeks before subcutaneous transplantation of adenocarcinoma breast tumor into mice. Thereafter, the treatment was continued for three weeks with a three-day interval between each week. Data analysis has shown that tumor growth rates were substantially decreased compared to controls, and there was a substantial increase in survival. The study has found that the promotion of Th1 cytokines in mouse spleen cell culture and NK cell cytotoxicity can be influenced when *L. casei* is administered orally. An immunoregulatory response was observed in *Lactobacillus BC*-breeding mice and suggested its use as an immune adjuvant therapy to protect against malignancy, by administering fermented milk with *helveticus R389*. There have been reports of an increased level of IFN-2 and IL-2 in mice treated with selenium nanoparticles enriched with *Lactobacillus flora*, but also the animals have shown enhanced NK cell activity. In addition, it appears that for highly metastatic breast tumor-bearing mice, oral administration of selenium enriched with nanoparticles containing *Lactobacillus brevis* may have a beneficial effect on disease prognosis. In the rat BC model, a Slovak group has demonstrated chemopreventive activity of an emerging probiotic bacterium, *Lactobacillus plantarum LS/07*. Mammary carcinogenesis was induced by 7, 12-dimethylbenz[a]anthracene (DMBA). *Lactobacillus plantarum LS/07* (at a dose of 8.4×10^8 CFU/rat) was administered daily before the first DMBA dose and given until the end of the experiment (16 weeks). The treatment with the probiotics significantly suppressed the tumor frequency, increased CD4+T-cells in tumor tissue, and reduced the serum tumor necrosis factor (TNF)-α concentration and the level of CD8+T-cells in the blood while increasing their number in tumor tissue. The unique cultured product that contains beneficial probiotics is a natural product, Kefir. There are different types of microbial flora in kefir grains, including the most commonly studied species with potential for bacterial growth, such as *L. acidophilus, L. casei,* and *Lactococcus lactis*. Kefir grains were fermented for a period of 24 h, at room temperature, during another animal experiment. In the meantime, grain was filtered from the fermentation medium. Mice with tumors were fed orally with kefir water (150 mg/kg body weight per day). All mice in the tumor group were inoculated with 1×10^5 4T1 triple-negative 4T1 murine BC cells subcutaneously, and treatment was administered orally on the same day as the tumor inoculation for 28 days. In mice receiving kefir water, the tumor size was not as large as in untreated mice. The tumor weight and volume were also reduced significantly by 29.53% and 32.92%, respectively, in kefir water-treated mice as compared to the untreated mice. By reducing 4Tangi1 cell mitosis, promoting apoptosis, inhibiting the immune system, inhibiting inflammation in the tumor microenvironment, and modulating proteins and genes regulating angiogenesis, kefir water has been shown to have antimetastatic and antiangiogenic effects in 4T1 cell-challenged mice in vivo.

3.9.2.2 Clinical Trials

The evidence of the activity of probiotics against BC has been shown in vitro and in vivo studies. But in these patients, there were just a few studies carried out. In a Japanese population-based case-control study, 306 cases with breast cancer (BC) and 662 controls aged 40–55 years were investigated about their diet, lifestyle, and other BC risk factors using a self-administered questionnaire and interview. In Japanese women, it was established that regular ingestion of *L. casei* Shirota and soy isoflavone since adolescence has been associated with a significant reduction in BC risk. Therefore, in order to achieve a chemopreventive effect on the development of cancer, long-term exposure was required. Only two ongoing studies looking at the benefit of probiotics for BC patients have been reported on this website, ClinicalTrials.gov. In the first trial, NCT03358511, a single experimental arm was conducted in patients receiving BC probiotics to evaluate the number of cytotoxic T lymphocytes, or CD8+ cells. Twenty post-menopausal BC subjects will be given 2–4 weeks of probiotics prior to surgery in operable stage I–III breast adenocarcinoma tumors ≥1.0 cm. The probiotics will be taken three times a day by the subjects. The probiotic under study is Primal Defense® ULTRA (Garden of Life LLC, USA), which is an over-the-counter probiotic that provides 15 billion colony-forming units of 13 species of beneficial bacteria. The second study, NCT03760653, is a randomized controlled pilot study to determine the effects of physical exercise together with supplementation of a probiotic on the GI bacteria balance, GI immune system, and quality of life in 30 BC survivors. BC survivors will be randomly assigned to each of the three groups: (i) probiotic supplementation + supervised combined physical exercise, (ii) probiotic supplementation and habitual sedentary lifestyle, and (iii) a control group following the usual lifestyle and receiving a placebo. Three capsules of *L. rhamnosus, Lactobacillus paracasei, L. acidophilus,* and *B. bifidum* will be taken daily at night before bedtime for 12 weeks in the two supplemented groups (Mendoza 2019).

3.9.3 Probiotic Using in Gastrointestinal Cancer

The intestinal microbiota has been linked to GI cancer development, also by the production of toxic and genotoxic bacterial metabolites that can lead to mutations by binding specific cell surface receptors and affecting intracellular signal transduction. *Streptococcus bovis, Bacterotics, Clostridia,* and *Helicobacter pylori* are some of the bacteria that form part of cancer's pathogenesis. By contrast, cancer growth in the colon is suppressed by certain types of bacteria, such as *L. acidophilus* and *B. longum.* Therefore, in determining the stages of cancer, it is important to balance between detrimental and "beneficial" bacteria. The change in the number of microbes has been reported as influencing carcinogenic bioactivation and thus cancer risk. This balance can be significantly altered by dietary components, and it is becoming increasingly apparent. Furthermore, the bacteria also have an effect on host biological characteristics as they improve intestinal barrier function, inhibit growth of pathogens, and reduce procarcinogenic substances' metabolism. In addition to the prevention and inhibition of carcinogenic agents, the benefits of probiotics may include therapeutic effects and the prevention of side effects of cancer treatment. Production of antimicrobial agents such as bacteriocins or antibiotics may be responsible for the therapeutic effects of probiotics (Ashique et al. 2024).

Bacteriocins produced by LAB are peptides or small proteins that are frequently inhibitory toward many undesirable bacteria, including food-borne pathogens. In addition, it is suggested that LAB and a soluble compound produced by the bacteria can interact directly with tumor cells in culture to inhibit their growth. Adhesion of epithelial cells is influenced by the competitive behavior of probiotics against pathogens. The ability to cling to epithelial cells is dependent on the strain, according to a series of studies that characterized LAB from different origins. Also, the suppressive effect of probiotics was associated with production of SCFAs, which can be explained by an increase in their related pathway. Chronic inflammation has been recognized as a risk factor of cancer. For example, IBD is a risk factor of colon cancer, and the risk of HCC can be increased by inflammatory conditions, such as hepatitis B and C virus infection. Inflammation not only plays a role in colitis-associated colon cancer but may also happen in sporadic colon cancer and affect the progression of cancer. *L. rhamnosus* GG was reported to prevent colon carcinogenesis, accompanied by the suppression of NFkB pathway, a pro-inflammatory pathway that links IBD and colon cancer. In the HCC model, Li et al. have demonstrated a reduction of pro-inflammatory cytokines IL-17 by probiotics in which potential relationship between immunomodulatory actions and anticancer effects has been established (Javanmard et al. 2018) (Table 3.1).

TABLE 3.1
Gene Annotation of Biological Materials That Are Involved in Human-Probiotic Interactions Collected from https://www.ncbi, https://nlm.nih.gov

Gene	Name	Location	Protein Length	Exon/Intron	mRNA Sequence	Protein Sequence
AHR	aryl hydrocarbon receptor	**Chromosome 7 - NC_000007.14**	848 aa	–	–	–
BAK1	BCL2 antagonist/killer 1	**Chromosome 6 - NC_000006.12**	211 aa	6 of 6 Exon	GCCAGGACGTCTGGG[G] TGTGGGGATTGGTG	–
BCL2	BCL2 apoptosis regulator alpha	**Chromosome 18 - NC_000018.10**	239 aa	–	–	–
	BCL2 apoptosis regulator X1		2239 aa	–	–	–
BAX	BCL2 associated X, apoptosis regulator 1	**Chromosome 19 - NC_000019.10**	221 aa	4 of 5 Intron	CAAACTGGTGCTCAAG[...] GCCCTGTGCACCAA	GRVVALFYFASKLVLK[...] ALCTKVPELIRTIM
	BCL2 associated X, apoptosis regulator alpha		192 aa	4 of 5 Intron	CAAACTGGTGCTCAAG[...] GCCCTGTGCACCAA	GRVVALFYFASKLVLK[...] ALCTKVPELIRTIM
	BCL2 associated X, apoptosis regulator beta		218 aa	1 of 4 Intron	CAGCCCAGAGGCGGGG[...] GGCCCACCAGCTCT	MDGSGEQPRGG[G...G] PTSSEQIMKTGALL
	BCL2 associated X, apoptosis regulator delta		143 aa	3 of 4 Intron	CAAACTGGTGCTCAAG[...] GCCCTGTGCACCAA	GRVVALFYFASKLVLK[...] ALCTKVPELIRTIM
	BCL2 associated X, apoptosis regulator gamma		181 aa	3 of 3 Intron	AAACTGGTGCTCAAG[...] GCCCTGTGCACCAAG	RVVALFYFASKLVLK[...] ALCTKVPELIRTIMG
	BCL2 associated X, apoptosis regulator sigma		179 aa	4 of 5 Intron	CAAACTGGTGCTCAAG[...] GCCCTGTGCACCAA	GRVVALFYFASKLVLK[...] ALCTKVPELIRTIM
	BCL2 associated X, apoptosis regulator zeta		114 aa	2 of 3 Intron	CAAACTGGTGCTCAAG[...] GCCCTGTGCACCAA	GRVVALFYFASKLVLK[...] ALCTKVPELIRTIM
	BCL2 associated X, apoptosis regulator lambda		140 aa	2 of 2 Intron	AAACTGGTGCTCAAG[...] GCCCTGTGCACCAAG	RVVALFYFASKLVLK[...] ALCTKVPELIRTIMG
	BCL2 associated X, apoptosis regulator X1		202 aa	4 of 5 Intron	AAACTGGTGCTCAAG[...] GCCCTGTGCACCAAG	RVVALFYFASKLVLK[...] ALCTKVPELIRTIMG

(Continued)

TABLE 3.1 (Continued)
Gene Annotation of Biological Materials That Are Involved in Human-Probiotic Interactions Collected from https://www.ncbi, https://nlm.nih.gov

Gene	Name	Location	Protein Length	Exon/Intron	mRNA Sequence	Protein Sequence
CASP3	caspase 3 a	**Chromosome 4 - NC_000004.12**	277 aa	–	–	–
	caspase 3 b		251 aa	–	–	–
	caspase 3 c		182 aa	–	–	–
	caspase 3 d		191 aa	–	–	–
	caspase 3 e		156 aa	–	–	–
	caspase 3 X1		277 aa	–	–	–
CASP9	caspase 9 alpha	**Chromosome 1 - NC_000001.11**	416 aa	–	–	–
	Caspase 9 beta		266 aa	–	–	–
	Caspase 9 X1		333 aa	–	–	–
	Caspase 9 X2		267 aa	–	–	–
	Caspase 9 X3		421 aa	–	–	–
CAT	Catalase	**Chromosome 11 - NC_000011.10**	527 aa	–	–	–
CYC1	Cytochrome c1	**Chromosome 8 - NC_000008.11**	325 aa	4 of 7 exon	CCCCCTGACCTCAGC[T] ACATCGTGCGAGCT	AARAANNGALPPDLS[Y] IVRARHGGEDYVFS
EGFR	epidermal growth factor receptor a	**Chromosome 7 - NC_000007.14**	1,210 aa	–	–	–
	epidermal growth factor receptor b		628 aa	–	–	–
	epidermal growth factor receptor c		405 aa	–	–	–
	epidermal growth factor receptor d		705 aa	–	–	–
	epidermal growth factor receptor e		1,091 aa	–	–	–
	epidermal growth factor receptor f		1,136 aa	–	–	–
	epidermal growth factor receptor g		1,165 aa	–	–	–
	epidermal growth factor receptor h		1,157 aa	–	–	–
	epidermal growth factor receptor i		943 aa	–	–	–
	epidermal growth factor receptor X1		1,157 aa	–	–	–
FOXP3	forkhead box P3 a	**Chromosome X - NC_000023.11**	431 aa	–	–	–
	forkhead box P3 b		396 aa	–	–	–

GPX1	glutathione peroxidase 1 isoform 1	**Chromosome 3 - NC_000003.12**	–	2 of 2 Exon	AAGAGATTCTGAATT[C] CCTCAAGTACGTCC	QFGHQENAKNEEILN[S] LKYVRPGGGFEPNF
	glutathione peroxidase 1 isoform 2		98 aa	1 of 1 Exon	CTTGTTTGTGGTTAG[A] ACGTTTCTCTCCTC	–
	glutathione peroxidase 1 isoform 3		104 aa	2 of 2 Exon	CGATGTTGCCTGGAA[C] TTTGAGAAGTTCCT	–
	glutathione peroxidase 1 isoform 4		170 aa	2 of 2 Exon	CCCTGCCAGCTCCCA[G] CGACGACGCCACCG	GAPSLRLPAGGPASS[Q] RRRHRAYDRPQAHH
	glutathione peroxidase 1 isoform 5		157 aa	2 of 2 Exon	CGACCCCAAGCTCAT[C] ACCTGGTCTCCGGT	APSDDATALMTDPKL[I] TWSPVCRNDVAWNF
GPX2	glutathione peroxidase 2	**Chromosome 14 - NC_000014.9**	–	1 of 1 Intron	CAATTTGGACATCAG[...] GAGAACTGTCAGAAT	RLVVLGFPCNQFGHQ[...] ENCQNEEILNSLKYV
GPX3	glutathione peroxidase 3 isoform 1	**Chromosome 5 - NC_000005.10**	–	1 of 4 Intron	CAAGAGAAGTCGAAG[...] ATGGACTGCCATGGT	LAGFVSQSRGQEKSK[...] MDCHGGISGTIYEYG
	glutathione peroxidase 3 isoform 2		235 aa	2 of 5 Intron	ATGGGCAATCCCCAG[...] ATGGACTGCCATGGT	GQEKSKAPRQMGNPQ[...] MDCHGGISGTIYEYG
GPX4	glutathione peroxidase 4 isoform A	**Chromosome 19 - NC_000019.10**	–	1 of 6 Intron	CTGGCCGGGACCATG[...] TGCGCGTCCCGGGAC	LLCGALAAPGLAGTM[...] CASRDDWRCARSMHE
	glutathione peroxidase 4 isoform B		227 aa	4 of 7 Exon	GCGGGCTACAACGTC[A] AATTCGATATGTTC	GSNEEIKEFAAGYNV[K] FDMFSKICVNGDDA
	glutathione peroxidase 4 isoform C		234 aa	2 of 6 Intron	ACCTGGACAAGTACCG[...] GGGCTTCGTGTGCA	SAKDIDGHMVNLDKY[R...R] GFVCIVTNVASQUG
	glutathione peroxidase 4 isoform D		170 aa	2 of 6 Intron	CCTGGACAAGTACCG[...] GGGCTTCGTGTGCAT	–
IFNG	interferon gamma	**Chromosome 12 - NC_000012.12**	166 aa	3 of 3 Intron	GCTGACTAATTATTCG[...] GTAACTGACTTGAA	SNKKKRDDFEKLTNYS[...] VTDLNVQRKAIHEL
IL-1B	interleukin 1 beta	**Chromosome 2 - NC_000002.12**	269 aa	4 of 6 Intron	TCATCTTTGAAGAAG[...] AACCTATCTTCTTCG	QENDLSTFFPFIFEE[E...E] PIFFDTWDNEAYVH
IL-4	interleukin 4 isoform 1	**Chromosome 5 - NC_000005.10**	153 aa	3 of 3 Intron	GGGCCTGGCGGGCTTG[...] AATTCCTGTCCTGT	RFLKRLDRNLWGLAGL[...] NSCPVKEANQSTLE
	interleukin 4 isoform 2		137 aa	2 of 3 Exon	GGGTGCGACTGCACA[G] CAGTTCCACAGGCA	YSHHEKDTRCLGATA[Q] QFHRHKQLIRFLKR
	interleukin 4 isoform 3		136 aa	3 of 4 Intron	TACAAAGCCCAGAGAG[...] AACACAACTGAGAA	SARVPSTSTRLQSPE[R...R] TQLRRKPSAGLRLC

(Continued)

TABLE 3.1 (Continued)
Gene Annotation of Biological Materials That Are Involved in Human-Probiotic Interactions Collected from https://www.ncbi, https://nlm.nih.gov

Gene	Name	Location	Protein Length	Exon/Intron	mRNA Sequence	Protein Sequence
IL-10	interleukin 10 isoform 1	**Chromosome 1 - NC_000001.11**	178 aa	3 of 4 Intron	GAGGCTACGGCGCTGT[...] CATCGATTTCTTCC	SLGENLKTLRLRLRRC[...] HRFLPCENKSKAVE
	interleukin 10 isoform 2		93 aa	2 of 3 Exon	CTGTCATCGATTTCT[T] CCCTGTGAAAACAA	NLKTLRLRLRRCHRF[L] PCENKSKAVEQVKN
IL-12A	interleukin-12 subunit alpha isoform 4	**Chromosome 3 - NC_000003.12**	219 aa	2 of 6 Intron	AACATGCTCCAGAAG[...] GCCAGACAAACTCTA	HSQNLLRAVSNMLQK[...] ARQTLEFYPCTSEEI
IL-12B	interleukin-12 subunit beta	**Chromosome 5 - NC_000005.10**	328 aa			
IL-17A	interleukin 17	**Chromosome 6 - NC_000006.12**	155 aa	2 of 2 Intron	CACCTTGGAATCTCCA[...] CCGCAATGAGGACC	RSSDYYNRSTSPWNL[H...H] RNEDPERYPSVIWE
IL-22	interleukin 22	**Chromosome 12 - NC_000012.12**	179 aa	5 of 6 Exon	TCCAGAGGAATGTGC[A] AAAGCTGAAGGACA	TCHIEGDDLHIQRNV[Q] KLKDTVKKLGESGE
MAPK1 (ERK)	mitogen-activated protein kinase 1 isoform 1	**Chromosome 22 - NC_000022.11**	360 aa	–	–	–
	mitogen-activated protein kinase 1 isoform 2		360 aa	–	–	–
MAP2K7 (MEK)	mitogen-activated protein kinase kinase 7 isoform 1	**Chromosome 19 - NC_000019.10**	435 aa	–	–	–
	mitogen-activated protein kinase kinase 7 isoform 2		426 aa	–	–	–
	mitogen-activated protein kinase kinase 7 isoform 3		419 aa	–	–	–
	mitogen-activated protein kinase kinase 7 isoform X1		442 aa	–	–	–
MUC2	mucin 2, oligomeric mucus/gel-forming	**Chromosome 11 - NC_000011.10**	5,130 aa	–	–	–

NFKB1	nuclear factor kappa B subunit 1	**Chromosome 4 - NC_000004.12**	969 aa	–	–	–
NFKB2	nuclear factor kappa B subunit 2	**Chromosome 10 - NC_000010.11**	900 aa	14 of 22 Intron	ACGAGAACGGAGACAC[...] ACCACTGCACCTAG	AGQRHLLTAQDENGD[T...T] PLHLAIIHGQTSVI
PTGS2 (COX2)	Prostaglandin-endoperoxide synthase 2	**Chromosome 1 - NC_000001.11**	604 aa	7 of 10 Exon	ACGGTTTGCTGTGGG[G] CAGGAGGTCTTTGG	MIYPPQVPEHLRFAV[G] QEVFGLVPGLMMYA
SOD1	superoxide dismutase 1	**Chromosome 21 - NC_000021.9**	154 aa	–	–	–
SOD2	superoxide dismutase 2 isoform A	**Chromosome 6 - NC_000006.12**	222 aa	–	–	–
	superoxide dismutase 2 isoform B		183 aa	–	–	–
	superoxide dismutase 2 isoform C		162 aa	–	–	–
	superoxide dismutase 2 isoform D		140 aa	–	–	–
	superoxide dismutase 2 isoform E		176 aa	–	–	–
SOD3	superoxide dismutase 3	**Chromosome 4 - NC_000004.12**	2240 aa	2 of 2 Exon	GCAGCCGTCGGCCAC[G] CTGGACGCCGCGCA	DDGALHAACQVQPSA[T] LDAAQPRVTGVVLF
STAT3	signal transducer and activator of transcription 3	**Chromosome 17 - NC_000017.11**	770 aa	–	–	–
TNF	tumor necrosis factor	**Chromosome 6 - NC_000006.12**	233 aa	3 of 3 Intron	TAGCCCATGTTGTAG[...] CAAACCCTCAAGCTG	SSSRTPSDKPVAHVV[A...A] NPQAEGQLQWLNRR

3.10 CONCLUSION

Probiotics can have many effects on the host's body, and in return, they can be affected by the host's body, and their effectiveness on the host's body can be personalized. Some of the effects of probiotics on the host body are nonspecific and can include regulation of gastrointestinal flora, stabilization of the IEC barrier, and inhibition of pathogens.

The specific effects of probiotics include their anti-inflammatory activity; for example, they can produce substances such as acetate, propionate, or butyrate, which generally prevent inflammation through different pathways.

Unlike medicinal products, the effectiveness of probiotics varies widely between individuals. The variability of effects is the result of different ages, physical conditions, gut microbial composition, colonization permission, and dietary habits in the host.

In this chapter, we have tried to conduct an investigation into the mechanisms of probiotics' actions in cancer prevention, cancer treatment, and using them to improve the toxic effects of cancer therapy. For example, by producing specific components, probiotics prevent the transformation of pro-carcinogens into carcinogens, or they neutralize antioxidants that play an essential role in cancer initiation, or in some cases, they stick to them with their cell walls and degrade them. Also, the balance of digestive metabolism, which favors the prevention of cancer, or the production of SCFAs, which plays a very important role in the anti-cancer function of probiotics, are other functional aspects of probiotics. Finally, after a detailed reference about the use of probiotics in various cancers, we have designed a table of gene annotations of biological materials that are involved in human-probiotic interactions.

REFERENCES

Al Mohaini M, Farid A, Alsalman AJ, Al Hawaj MA, Alhashem YN, Ghazanfar S, … Ismail S. Screening of anticancer and immunomodulatory properties of recombinant pQE-HAS113 clone derived from *Streptococcus equi*. Pakistan Journal of Medical & Health Sciences. 2022; 16(02):1100.

Ashique S, Houshyari M, Islam A, Pal R, Ghazanfar S, Taghizadeh-Hesary F. The role of microbiota in nasopharyngeal cancer: Where do we stand? Oral Oncology. 2024; 158:106982.

Azad H, Akbar MY, Sarfraz J, Haider W, Ghazanfar S. Simulation studies to identify high-affinity probiotic peptides for inhibiting PAK1 gastric cancer protein: A comparative approach. Computational Biology and Chemistry. 2025; 115:108345.

Cristofori F, Dargenio VN, Dargenio C, Miniello VL, Barone M, Francavilla R. Anti-inflammatory and immunomodulatory effects of probiotics in gut inflammation: A door to the body. Frontiers in Immunology. 2021; 12:578386.

Drago L. Probiotics and colon cancer. Microorganisms. 2019; 7(3):66.

Gou H-Z, Zhang Y-L, Ren L-F, Li Z-J, Zhang L. How do intestinal probiotics restore the intestinal barrier? Frontiers in Microbiology. 2022; 13:929346.

Javanmard A, Ashtari S, Sabet B, Davoodi SH, Rostami-Nejad M, Akbari ME, et al. Probiotics and their role in gastrointestinal cancers prevention and treatment: An overview. Gastroenterology and Hepatology from Bed to Bench. 2018; 11(4):284.

Lu K, Dong S, Wu X, Jin R, Chen H. Probiotics in cancer. Frontiers in Oncology. 2021; 11:638148.

Mendoza L. Potential effect of probiotics in the treatment of breast cancer. Oncology Reviews. 2019; 13(2):422.

Śliżewska K, Markowiak-Kopeć P, Śliżewska W. The role of probiotics in cancer prevention. Cancers. 2021; 13(1):20.

Thilakarathna WW, Langille MG, Rupasinghe HV. Polyphenol-based prebiotics and synbiotics: Potential for cancer chemoprevention. Current Opinion in Food Science. 2018; 20:51–7.

Wang X, Zhang P, Zhang X. Probiotics regulate gut microbiota: An effective method to improve immunity. Molecules. 2021; 26(19):6076.

Wieërs G, Belkhir L, Enaud R, Leclercq S, Philippart de Foy JM, Dequenne I, et al. How probiotics affect the microbiota. Frontiers in Cellular and Infection Microbiology. 2019; 9:454.

4 The Relevance of CRISPR-Cas9 Technology in Probiotic Research

Samuel Adeniyi Oyegbade, Samia S. Alkhalil, Samson Olusola Adeboye, Gbotemi Jerry Oni, Joy Iziegbe Azeta, Sherifat Ozavize Enejiyon, Oluwafemi Adebayo Oyewole, Charles Oluwaseun Adetunji, Muhammad Lawal Attanda, and S.A. Ado

4.1 INTRODUCTION

A vast, intricate, dynamic microbial ecology naturally exists in human beings. Early in development, bacteria start to colonize the gut by copying their mother's patterns. Adults in good health have more diverse and complex gut microbiomes. The community of the microbiota can be influenced by a host's diet, genetics, lifestyle, usage of antibiotics, and ageing process (Backhed et al., 2005). From roughly 10^2 microorganisms/mL in the stomach to roughly 10^{11}–10^{12} bacteria/mL in the colon, these microbes create a population density gradient (Guarner and Malagelada, 2003). According to Tsai et al. (2019) and Jandhyala et al. (2015), gut microbiota can perform both preventive (via immunomodulation) and metabolic (fermenting non-digestible food residues) functions. Dysbiosis, i.e., changes in microbiological diversity and composition, is associated with several chronic diseases, including immune-mediated diseases, obesity, type 2 diabetes, oxidative stress disorders, non-alcoholic fatty liver diseases, and cardiovascular diseases (Qin, 2012; Selvanantham et al., 2016). Antibiotic resistance in microorganisms means that using medications therapeutically will only have a limited impact. Probiotics and their impact on human health have garnered attention recently in this regard (Saadullah et al., 2022). Non-pathogenic living microorganisms that help the host's health when consumed are known as probiotics. Through the colonization and restoration of an upset gut microbial community, probiotics demonstrate this special capacity to improve human health (Ghazanfar). They should also be able to boost the immune system and express their activities against infections (Liévin-Le Moal and Servin, 2014). Clinical research has linked them to co-adjuvants with superior qualities like anti-diabetic, anti-obesity, anti-inflammatory, and anti-angiogenic effects (Le Barz et al., 2015; Palumbo et al., 2016). The majority of probiotics are sold commercially as foods or medications that include live microorganisms. Consequently, it is crucial to think about the probiotics' safety evaluation before using them. It could be regarded as a biotherapeutic agent given the extensive list of health advantages (Mishra and Acharya, 2021). This chapter examines the applicability of CRISPR-Cas9 technology to studies on probiotics.

4.2 HISTORICAL PERSPECTIVE OF CRISPR TECHNOLOGY

The needed new probiotics are obtained by modifying the current probiotics through gene editing. By modifying the engineering of these unique microbes, we can confirm directly whether the intended changes have been made to their genetic makeup, proteins, and functional roles. In

DOI: 10.1201/9781003399797-5

recent years, advances in gene editing technology have included homologous recombination, first-generation zinc finger nucleases (ZFNs), second-generation transcription activator-like effector nucleases (TALENs), and the increasingly well-liked third-generation clustered regularly spaced short palindromic repeats CRISPR associated (CRISPR-Cas) technology (Zhang et al., 2020). A few of these include CRISPR gene editing technology, which Science News named the greatest discovery of 2015, ZFNs and TALENs, which Nature Methods named the method of the year, and CRISPR-Cas9 technology, which will likely win the 2020 Nobel Prize in Chemistry and have a profound effect on the biological sciences. This section will mostly provide a synopsis of these three technological generations. According to Yadav and Shukla (2020), these gene-editing instruments cause double strand break (DSB) to occur in target genes. Homology-directed repair (HDR) or the error-prone non-homologous end joining (NHEJ) pathway promotes DNA repair. According to Yadav et al. (2018), ZFNs and TALENs are two artificial restriction endonucleases that modify or cut certain target DNA by using zinc finger or TAL-effect DNA domains. In order to provide gene editing with a completely new path, researchers have focused on CRISPR technology since the end of 2012. This technology is easier to use, more efficient at editing, and has greater editing capability (Bak, Gomez-Ospina, and Porteus, 2018). CRISPR and similar Cas genes are considered to be components of bacterial immune systems as self-protective bacterial sequences that may offer phage resistance (Barrangou et al., 2007). The first invasion virus genes are integrated into the bacterial CRISPR spacer at the molecular level. When the virus invades again, the transcription of CRISPR gives rise to a precursor CRISPR RNA (pre-crRNA), which is then processed to contain the sequence of viral genes that correspond to crRNA. By identifying similar viral genome sequences, CrRNA facilitates the attachment of viral genes to Cas proteins, therefore cleaving them (McGinn and Marraffini, 2019; Mekler, Kuznedelov, and Severinov, 2020).

Generally speaking, the Cas protein gene that is next to CRISPR helps to categorize the CRISPR-Cas system. There are primarily two classes, six kinds, and 33 subtypes of the CRISPR-Cas system that are currently known (Makarova et al., 2020). The two most popular nucleases in the Cas protein family, Cas9 and Cas12a (Cpf1), are easily editable and can be used to recognize particular DNA sequences (Liu et al., 2021). Right now, the type II CRISPR-Cas9 system is the one that is employed for genome editing the most frequently among them (Burgess, 2020). Trans-activating RNA (tracrRNA), crRNA, and the Cas9 protein make up the CRISPR-Cas9 system. Among *Streptococcus thermophilus*'s several Cas proteins, the Cas9 protein is the only one with DNA catalytic activity (Adli, 2018).

According to Zhang et al. (2021), it uses crRNA matching to cut double-stranded DNA at specific sequence targets. Through synthetic engineering, small guide RNA (sgRNA) can be synthesized from crRNA and tracrRNA, enabling Cas9 to carry out site-specific DNA cutting, gene knockdown, and insertion (Knott and Doudna, 2018). According to Zhang et al. (2021), CRISPR technology can be used for both functional gain screening studies, such as CRISPR activation (CRISPRa), and functional loss studies, such as CRISPR knockout (CRISPRKO) and CRISPR interference or inhibition (CRISPRi). New generations of probiotics have emerged and are strongly supported by advances in genetic editing technologies, particularly CRISPR. The development of CRISPR/Cas9 systems and the development of life sciences have enabled new heights in the fields of research on gene functions, gene therapy, and genetic improvement (Zhu, 2022).

4.3 THE MECHANISM OF THE CRISPR/Cas9 SYSTEM

The three main steps of bacterial resistance to foreign nucleic acid invasion are interference, expression, and adaptation (Shah et al., 2018; Wang et al., 2019). When viruses enter bacteria for the first time, bacteria can remember the virus that invades, because they recognize its unique PAM sequence and integrate it into the CRISPR space region. When bacteria are again infected with the same virus, they can recognize the sequence of the spacers and translate them into pre-crRNA. The partner of the TracrRNA is CnsI and RNaseIII, and CnsI and RNaseIII convert the pre-crRNA into mature crRNA (Deltcheva et al., 2011). Since crRNA uses complementary sequences to bind and recognize foreign DNA, it is also known as guide RNA (gRNA) (Jiang et al., 2013). Cas nuclease is unable to cleave foreign DNA on its own, but when it joins forces with mature tracrRNA and crRNA to form the ribonucleoprotein complex, Cas protein can be directed by crRNA to cleave the invasive DNA by identifying its PAM site, thereby achieving self-defense and destroying the foreign DNA (Cong et al., 2013; Hale et al., 2009).

The CRISPR/Cas system makes gene editing possible due to crRNA's specific target sequence recognition abilities, the Cas nucleus' DNA splitting activity, and the cell's DNA-repair mechanism. In order to prevent cell death and repair DNA damage, when DSB occurs, cells begin their own repair pathways. Homology-directed repair (HDR) and non-homological end joining (NHEJ) are two ways of repair (Tang, 2017; Wyman and Kanaar, 2006). When there are homogeneous sequences, organisms often use a homogeneous recombination mechanism (HRM) to integrate homogeneous fragments into DNA and repair double-strand breaks (DSBs). This repair mechanism allows us to combine donor DNA fragments with identical arms and create a genetic editing vector to form a repair template, allowing the donor DNA to be integrated into the cell genome DNA and produce a knock-in gene (Menchaca et al., 2020). Cells usually begin the NHEJ pathway and connect damaged DNA directly to the cell when there is no homologous DNA. Due to the vulnerability of this repair method to insertion or deletion of the basic pair (indicated), a mutation of the gene frame shift resulting in a gene knockout was reported (Mladenov and Iliakis, 2011; Wang et al., 2013).

Cas9 and gRNA can be combined into a single vector for genetic editing or separated into two different vectors with appropriate promoters, terminators, reproduction, and selection markers. Yang et al. (2013) Synthetic gRNA is a chimera RNA that combines all necessary RNAs and tracrRNA components. sgRNA, or the front section of gRNA, is responsible for identifying the target location. According to Redman et al. (2016), the following section functions as a scaffold for binding to Cas9 proteins. Every time sgRNA must identify a new target, it must be modified. Since sgRNA is the preferred length of 20nt, it is simple to create a CRISPR/Cas9 vector. CRISPR/Cas9 targets many potential targets because, on average, each 8-bp DNA sequence contains potential targets (Cong et al., 2013). These elements make CRISPR/Cas9 the most widely used genetic editing tool. Target cells injected with CRISPR/Cas9 vectors express Cas9 and gRNA gene editing tools (Shah et al., 2018). In 2013, the first CRISPR/Cas9 gene modification was reported in eukaryotic cells. Using CRISPR/Cas9, researchers successfully modified mouse *Th* gene and human CCR5, EMX1, PVALB, PPP1R12C, and CLTA genes (Cho et al., 2013; Mali et al., 2013). Subsequently, CRISPR/Cas9 was discovered to be able to change genes. CRISPR/Cas9 is one of the most important scientific discoveries of the past ten years, and its simplicity of use prompted many scientists to study it in depth.

4.4 THE RELEVANCE OF CRISPR-Cas9 TECHNOLOGY IN PROBIOTICS RESEARCH

4.4.1 Improving Safety and Efficacy of Probiotics

Use of engineered live microorganisms for in vivo treatment of disease has focused on safety and biocontainment, and one of the reasons is that plasmid systems-based engineered probiotics demand regular selection pressure and possess microbial horizontal transfer risks (Zou et al., 2023). Over the last ten years, there has been a growing focus on the study and application of probiotic microorganisms, particularly with special interest focused on the improvement of their properties, and this can be accomplished by employing lactic acid bacteria and/or *Bifidobacterial* cryptic plasmids derived food-grade vectors (plasmids whose antibiotic resistance have been substituted (with resistance to bacteriocins or heavy metals)) or the complementation of auxotrophs (Bravo and Landete, 2017). A live microbe used as a feed supplement to benefit the host in creating an intestinal microecological balance is called a probiotic (Cooke, Gibb, and Harnett, 2021; Hidalgo-Cantabrana, O'Flaherty, and Barrangou, 2017; Niu and Xiao, 2020; Zeng et al., 2022). Strains from *Bifidobacterium* (such as *B. breve*, *B. longum, and B. infantis*) and *Lactobacillus* (such as *L. casei*, *L. plantarum*, *L. rhamnosus, and L. helveticus*) are the most frequently used genera for probiotics (Cooke, Gibb, and Harnett, 2021; Hidalgo-Cantabrana, O'Flaherty, and Barrangou, 2017; Niu and Xiao, 2020; Zeng et al., 2022; Zuo and Marcotte, 2021). Research attention has been drawn to *S. thermophilus* and *Escherichia coli* strains as well, and in horses, feed is supplemented with probiotic yeast (*Saccharomyces cerevisiae* and *Saccharomyces boulardii*), yielding beneficial outcomes like fiber digestion and reduced hospital duration for horses with acute encephalitis (Cooke, Gibb, and Harnett, 2021). In a bid to reduce interference of the artificial genetic elements with the body, causing undesirable side effects, scientists have tried the insertion of functional elements into the DNA genome rather than the use of plasmids for the expression and execution of functions (Zou et al., 2023). Addition, deletion, activation, and modulation of probiotic *Lactobacilli* and *Bifidobacterial* require the use of genome editing to produce safe, robust, and effective probiotics (Imran et al., 2024) for food and pharmaceutical applications because a lot of starter cultures and probiotics have antibiotic resistance genes, and the prevailing *tet(W)* gene is high in *Lactobacilli* and *Bifidobacterial* (Zuo and Marcotte, 2021).

4.5 USING CRISPR-Cas9 TO REMOVE UNDESIRABLE GENES FROM PROBIOTIC STRAINS

Several techniques can be used for gene editing in molecular research, one of which is the clustered regularly interspaced short palindromic repeats (CRISPR)/Cas9, known to be less expensive, time-saving, more effective, and more precise than previous genetic engineering techniques, and it is one of the most powerful tools for developing next-generation probiotics (Hasan, Das, and Satter, 2024; Zuo and Marcotte, 2021). Another advantage of CRISPR/Cas9 technique is the recoverability and reusability of plasmids for further mutagenesis (Hasan, Das, and Satter, 2024). Some other methods, including gene targeting based on homologous recombination, Cre-lox, introns, TALEN, bacteriophage integrases, two-plasmid integration systems, etc., all come with various challenges in handling, such as time-consuming, laborious, and more expensive (Vaid et al., 2022; Van Pijkeren and Barrangou, 2017; Zuo and Marcotte, 2021). CRISPR-Cas9 is a technique that uses a gRNA to cleave the target DNA sequence by Cas9 machinery (Hasan, Das, and Satter, 2024). The grouping of the CRISPR system into different classes, types (type I and type II systems), and subtypes is determined by their Cas genes (Cas1, Cas2, Cas9, and Csn2/Cas4) signature and their mode of action (Hidalgo-Cantabrana, O'Flaherty, and Barrangou, 2017; Van Pijkeren and Barrangou, 2017).

Understanding the probiotics' molecular processes, physiological features, host-probiotic interaction, and host-microbiota interaction can be greatly enhanced through efficient genome editing techniques like CRISPR-Cas-based tools and other synthetic biological technologies, ultimately shaping the future use of probiotics in improving health and promoting the advancement of next-generation probiotics for research and industrial applications in the food and pharmaceutical industries (Zuo and Marcotte, 2021). CRISPR-Cas system is encoded in a larger percentage in *Bifidobacterium* (77%), *Lactobacillus* (62.9%), and *Thermophilus* strains (100%); however, barely in *Lactococcus lactis* (Hidalgo-Cantabrana, O'Flaherty, and Barrangou, 2017). Bacteria are shielded from plasmids and viruses using this technique (Hasan, Das, and Satter, 2024; Van Pijkeren and Barrangou, 2017).

Aside from the wide adoption of this technique in plants and animals, the microbiology community is not left out in harnessing the potentials of these CRISPR-based tools for engineering organisms that have relevance in agriculture, medicine, and biotechnology (Van Pijkeren and Barrangou, 2017). Recently, various works on *Lactobacillus species* and *Bifidobacteria species* have been done through the use of CRISPR-Cas system for genome editing (Zuo and Marcotte, 2021).

Instances where this technology was used include:

i. Using CRISPR-Cas system to assist ssDNA-mediated recombineering through counterselection in *Lactobacillus reuteri* (Zuo and Marcotte, 2021),
ii. Hasan, Das, and Satter (2024) edited the D-lactate dehydrogenase (D-LDH) gene in *Lactobacillus bulgaricus* using a combination of CRISPR/Cas9 with γ Red machinery to enhance the sensory quality of yogurt.
iii. CRISPR-Cas9D10A nickase was used for the alteration of chromosomal DNA by generating single-strand breaks that trigger the homology-dependent mechanism of repair, permitting rapid and accurate manipulation at the chromosomal level in *L. casei* through one-step transformation (Zuo and Marcotte, 2021).
iv. Replacement of *ATG* (endogenous initiating codon) of target proteins with second-generation base editor (*BE2*)-editable sequence of *ACT*-tag to silence target protein transition using a designed recording and activating translations (RAT) (Zou et al., 2023).
v. The genomes of *Lactobacilli* and *Bifidobacterial* contain diverse prophages that can be activated under certain conditions to form active phages that can contaminate industrial fermentation. Enhancement of genetic stability of strains, and prevention of cell lysis can be curtailed through the deletion of related genes prophage (prophage curing), and this was recently shown in *Lactobacillus crispatus* by deletion of prophage DNA packaging *Nu1* (Zuo and Marcotte, 2021).

4.6 ENHANCING THE SAFETY PROFILE OF PROBIOTICS BY ELIMINATING POTENTIAL VIRULENCE FACTORS

Certain traits of probiotic bacteria can be improved upon to eliminate unwanted traits and better enhance their probiotic role and/or include new features of interest in the probiotics (Bravo and Landete, 2017) through the incorporation of CRISPR-Cas system into targeted strains of bacteria or through lytic phages of specific design (Vaid et al., 2022). Engineering both beneficial and pathogenic bacteria to eliminate antibiotic resistance genes or virulence genes will be in the right direction because it will help to reduce the transfer of antibiotic resistance genes that had become a major public health bottleneck (Hidalgo-Cantabrana, O'Flaherty, and Barrangou, 2017; Vaid et al., 2022). Virulence genes can be deactivated through a programmable deactivated Cas9 of the CRISPR/Cas nuclease system, which can also make a selective elimination based on genetic contents of bacterial taxa; for example, it is the suppression of *E. coli's* shiga toxin expression (Khan Mirzaei and Deng, 2022). Metabolites of extracellular probiotics containing wall components have been used to treat immune deficiency illnesses and limit pathogenic growth (Vaid et al., 2022).

4.7 ENGINEERING PROBIOTICS TO DELIVER THERAPEUTIC MOLECULES (E.G., ANTIMICROBIAL PEPTIDES AND ENZYMES)

Enhancing probiotic potentials and optimization of metabolic processes remains the main goal of probiotic engineering, which seeks to manipulate bacterial strains well adapted for gastrointestinal (GI) tract colonization with the ability of in situ production of therapeutic molecules (Mugwanda et al., 2023). Although not much nucleotides modifications have been done by CRISPR system in lactic acid bacteria, however, with the introduction and development of this novel technique, much longer gene sequences editing are possible, which gives room for the incorporation of heterologous genes into lactic acid bacteria and *Bifidobacterial* for the enhancement of health-promoting features (Bravo and Landete, 2017). Engineering the genomes of these organisms depends on the type of application and the choice of organism, which is the first step in developing LAB therapeutics (Van Pijkeren and Barrangou, 2017). Due to the inherent health benefit and their safety (Muhammad et al., 2023), *Lactobacillus* and *Bifidobacteria* became the best delivery tool to be engineered for therapeutic proteins and peptides like antigens, antibodies, nanobodies, and cytokines (equally known as single-domain antibody of VHH) through extracellular or cell surface display. (Bravo and Landete, 2017). This approach could produce a better delivery system of drugs compared to therapeutics of abiotic sources (Mugwanda et al., 2023). Selective alterations to microbiota composition of the GI tract can be achieved by engaging *Lactobacilli* as the transporting and delivery vehicle of selected molecules in the GI, because of the potentials of *Lactobacilli* to survive the GI tract and also their ability for gene manipulation (Van Pijkeren and Barrangou, 2017). Alternative approaches that can address the challenges of antibiotic resistance (a major concern in human health) can be sought using the therapeutic potentials of probiotics, either as native or phage-derived antimicrobials, for example, the use of bacteriocins to target a wide spectrum of both Gram-negative and Gram-positive bacteria (Table 4.1) (Van Pijkeren and Barrangou, 2017).

TABLE 4.1
Some Antimicrobial Peptides Produced by Probiotics

Antimicrobial Peptides Produced	Activity/Purpose	Engineered Probiotics	References
Bacteriosin	Broad spectrum (Gram-negative and Gram-positive)	*L. salivarius* UCC118	Van Pijkeren and Barrangou (2017)
Nisin (Nis A)	Broad spectrum (Gram-negative and Gram-positive)	*L. lactis* strains and streptococci	
Reuterin (known as 3hydroxypropionaldehyde (3-HPA))	Broad spectrum	*L. reuteri* strains	
holins and lysins	Broad spectrum	Bacteriophages	
Cholera toxin B subunit	Anti-Listeria monocytogenes activity	*Lacticaseibacillus* species	Mugwanda et al. (2023)
Metastasis-inhibiting peptide KISS1	Cancer therapy	*L. lactis*	
Microbiocidal cyanovirin-N	Prevent the transmission of human immunodeficiency virus type 1 (HIV-1) Neutralizes the infectivity of HIV-1 in vitro	*L. lactis* and *Lactiplantibacillus plantarum* (formerly *Lactobacillus plantarum*)	

Despite the wide advantages engineered probiotics offer in therapeutic applications, there are still some bottlenecks, such as (i) identification of suitable chassis and (ii) diminished therapeutic effects of engineered probiotics by better-adapted microorganisms through flushing out of these probiotics within days (Mugwanda et al., 2023).

4.8 APPLICATIONS IN DISEASE PREVENTION TREATMENT (E.G., INFLAMMATORY BOWEL DISEASE AND INFECTIONS)

The prevalence of irritable bowel syndrome (IBS), a common functional bowel disorder, is of 5–20 percent within the general populace, and it is commonly found among females and young individuals (Niu and Xiao, 2020). The roles of intestinal microorganisms have been evident in IBS due to the IBD patients' change in intestinal flora composition or metabolic activities, and suggestions have been put forward for using probiotics as a promising IBD treatment in regulating gut microbiota (Li et al., 2020). Applying probiotics to restore or regulate intestinal flora balance is a relatively new treatment for IBD (Chen et al., 2021). Intestinal immune response, intestinal barrier function improvement, and intestinal flora activity and composition are some of the mechanisms of action used by probiotics (Chen et al., 2021). According to the moderate evidence gathered by Li et al. (2020), while carrying out a review and meta-analysis, using probiotics for the treatment of IBS patients seems safe, although not free from evidence of heterogeneity of trials. They equally noticed that a single probiotic and a shorter treatment duration (<8 weeks) may be more beneficial. Elsewhere, using multi-strain supplements for ≥8 weeks had greater beneficial effects on the symptoms of IBS (Niu and Xiao, 2020), implying that the duration of this therapy between the <4 weeks and >4 weeks groups showed no significant difference (Niu and Xiao, 2020). The combination of specific strains and species of probiotics had beneficial effects on IBS symptoms and abdominal pain generally, although the identification of the most effective was not possible (Niu and Xiao, 2020).

Zou et al. (2023) used a CRISPR-based genome editing to develop an engineered probiotics called iROBOT (intelligent responsive bacteria for diagnosis and therapy), an *E. coli* Nissle 1917 (tetrathionate ($S_4O_6^{2-}$) and thiosulfate ($S_2O_3^{2-}$) respondent) that noninvasively monitors and documents IBD occurrence and progression in patients and equally, through self-tunable mechanism, releases treatments in response to biomarkers. An oral vaccine using a bacTRL-Spike (a spike protein) expressed by a recombinant *Bifidobacteria longum* was produced for preventing coronavirus disease 2019 (COVID-19) (Mugwanda et al., 2023). In another study, Zeng et al. (2022) reported the use of probiotics and their metabolites or probiotic-fermented foods in the intervention of several inflammatory arthritis in a number of randomized controlled trials (RCTs) of several clinical studies. The treatment of infantile colic problem (a disease that defies traditional medicine) is possible using probiotics (Ma et al., 2023). Precigen ActoBio developed *L. lactis* AG019 for treating Type 1 diabetes. As proven by most clinical trials, the feasibility of using live biotherapuetics produced by recombinant LAB is efficacious, although certain concerns for optimization exist (Mugwanda et al., 2023).

4.9 ENGINEERED PROBIOTICS AND TUMORS

Decades have passed since the topic of using bacteria to treat cancer was raised (Moghimipour et al., 2021; Yaghoubi et al., 2020). Certain probiotics have the ability to cause anti-cancer benefits by promoting cancer cell death and inhibiting oxidative stress (Aggarwal et al., 2020; Badgeley et al., 2021; Sehrawat et al., 2021). With the progress of modern genetic editing technology, it has become possible to develop beneficial probiotics in cancer prevention. Researchers are constantly seeking and creating new probiotics to combat cancer (Adetunji et al., 2023).

Researchers at Columbia University, led by Candice R. Gurbatri, employed synthetic biology techniques to manipulate bacteria, resulting in the creation of an engineered strain subtype known as "SLIC" of *E. coli* (Gurbatri et al., 2020). After colonizing tumor cells to a certain point, SLIC can lyse spontaneously and generate nanoantibodies against CTLA-4 and PD-L1. In order for SLIC to effectively prevent tumor development, induce tumor regression, and prevent tumor metastasis, it must first activate T cells, boost memory T cells, and improve the anti-tumor immune response. Their work has promise for improving the efficacy of tumor therapy and preventing the growth of malignant cells. In a related investigation, Chinese scientists genetically modified *E. coli* Nissle 1917 to direct the tumor suppressor p53 and angiogenic inhibitor TUM-5 toward the anaerobic tumor area (Ma et al., 2022). The growth of the transplanted tumor in mice was considerably suppressed by this treatment, and the concept of targeting tumors with certain bacteria's tumor tropism offers a new avenue for the development of anti-tumor medications. In addition, a modified *E. coli* strain Nissle 1917, which could be used to invade tumors, was released. It is able to effectively enhance the therapeutic effects of programmed cell death protein 1 and its ligand inhibitors on tumors in mice and transform metabolic waste from tumors into L-arginine, thereby increasing the immune response to cancers (Canale et al., 2021). In addition, a modified strain of 1917 *E. coli* Nissle was successfully created to treat cancer effectively by manipulating light color variations (9). According to Leventhal et al. (2020), the strain has the ability to generate a cyclic di-AMP, which activates the interferon gene pathway and promotes innate immunity by exposing tumor antigens to cells that present antigens. They are currently conducting phase 1 clinical trials, and we expect this modified probiotic drug to be used in medicine in the near future.

4.9.1 Future Prospect

The development of synthetically created probiotics through the application of developing gene-editing techniques has enormous promise for the diagnosis and treatment of various human diseases. In particular, engineered probiotics may prove to be a novel approach to the treatment of infections, inflammation, cancer, and other ailments. More patients and families will benefit from the researchers' tailored probiotics since they are more cost-efficient, less prone to adverse effects, and more effective than conventional therapies or wild-type strains. Probiotics have been engineered to have stability, low cost, specificity, preference, and relative safety. These attributes may open up new therapy options for many of the disorders listed above and more. Nonetheless, there are still a number of obstacles in the way of using modified probiotics. For instance, only a small portion of the large family of microbes is now employed for study design, and many of them have few genetic tools that may be changed. As a result, probiotic modification is limited to a few strains of bacteria. Some researchers' attempts to create tailored probiotics are hampered by the fact that only a small number of these products make it from the lab to the clinic, and even if they do, there is no guarantee that they will have the desired impact. There are many who contend that the use of genetically modified organisms (GMOs) poses risks and hinders the successful application of designed probiotic medicines.

4.10 CONCLUSION

In modern research, the CRISPR–Cas9 system's rigorous development is yielding incredibly positive outcomes. Simple and highly effective transcriptional regulation, genomic alterations, and epigenetic editing are all made possible by it. We cannot exclude the potential benefits of modified probiotics for enhancing human health, even in the face of research and application restrictions. The study and creation of genetically modified probiotics will inevitably bring both success and failure. In order to increase our chances of success, we need to further study various techniques for the modification of probiotics and to better understand the pathophysiology of disease. This will help us to create the probiotics we want more easily. Although the effectiveness of genetically modified probiotics is a concern, safety concerns must be addressed before these probiotics are clinically developed. Ensure that compliance with the requirements of the various health regulators is crucial to the development and production of safe and effective probiotics. We think that as our knowledge of the human microbiome and disease pathways grows and gene editing technology advances quickly, safe, stable, and efficient tailored probiotics will undoubtedly excel in a variety of disciplines.

REFERENCES

Adetunji, C.O., Akram, M., Adetuyi, B.O., Laila, U., Saeed, M.M., Olaniyan, O.T., … Ghazanfar, S. (2023) Role of genetical conservation for the production of important biological molecules derived from beneficial algae. *Next-Generation algae*, *1*, pp. 263–279.

Adli, M. (2018) The CRISPR tool kit for genome editing and beyond. *Nature communications*, *9*(1), p. 1911.

Aggarwal, N., Breedon, A.M.E., Davis, C.M., Hwang, I.Y. and Chang, M.W. (2020) Engineering probiotics for therapeutic applications: Recent examples and translational outlook. *Current opinion in biotechnology*, *65*, pp. 171–179.

Backhed, F., Ley, R.E., Sonnenburg, J.L., Peterson, D.A. and Gordon, J.I. (2005) Host-bacterial mutualism in the human intestine. *Science*, *307*(5717), pp. 1915–1920.

Badgeley, A., Anwar, H., Modi, K., Murphy, P. and Lakshmikuttyamma, A. (2021) Effect of probiotics and gut microbiota on anti-cancer drugs: Mechanistic perspectives. *Biochimica et biophysica acta (BBA)-reviews on cancer*, *1875*(1), p. 188494.

Bak, R.O., Gomez-Ospina, N. and Porteus, M.H. (2018) Gene editing on center stage. *Trends in genetics*, *34*(8), pp. 600–611.

Barrangou, R., Fremaux, C., Deveau, H., Richards, M., Boyaval, P., Moineau, S., Romero, D.A. and Horvath, P. (2007) CRISPR provides acquired resistance against viruses in prokaryotes. *Science*, *315*(5819), pp. 1709–1712.

Bravo, D. and Landete, J.M. (2017) Genetic engineering as a powerful tool to improve probiotic strains. *Biotechnology and genetic engineering reviews*, *33*(2), pp. 173–189. Available at: https://doi.org/10.1080/02648725.2017.1408257.

Burgess, D.J. (2020) CRISPR screens beyond Cas9. *Nature reviews genetics*, *21*(5), pp. 273–274.

Canale, F.P., Basso, C., Antonini, G., Perotti, M., Li, N., Sokolovska, A., Neumann, J., James, M.J., Geiger, S., Jin, W. and Theurillat, J.P. (2021) Metabolic modulation of tumours with engineered bacteria for immunotherapy. *Nature*, *598*(7882), pp. 662–666.

Cho, S.W., Kim, S., Kim, J.M. and Kim, J.S. (2013) Targeted genome engineering in human cells with the Cas9 RNA-guided endonuclease. *Nature biotechnology*, *31*(3), pp. 230–232.

Chen, M., Feng, Y., & Liu, W. (2021). Efficacy and safety of probiotics in the induction and maintenance of inflammatory bowel disease remission: a systematic review and meta-analysis. *Annals of palliative medicine*, *10*(11), p. 118211829.

Cong, L., Ran, F.A., Cox, D., Lin, S., Barretto, R., Habib, N., Hsu, P.D., Wu, X., Jiang, W., Marraffini, L.A. and Zhang, F. (2013) Multiplex genome engineering using CRISPR/Cas systems. *Science*, *339*(6121), pp. 819–823.

Cooke, C.G., Gibb, Z. and Harnett, J.E. (2021) The safety, tolerability and efficacy of probiotic bacteria for equine Use. *Journal of equine veterinary science*, *99*, p. 103407. Available at: https://doi.org/10.1016/j.jevs.2021.103407.

Deltcheva, E., Chylinski, K., Sharma, C.M., Gonzales, K., Chao, Y., Pirzada, Z.A., Eckert, M.R., Vogel, J. and Charpentier, E. (2011) CRISPR RNA maturation by trans-encoded small RNA and host factor RNase III. *Nature*, *471*(7340), pp. 602–607.

Ghazanfar, S. Effect of Lactobacillus plantarum and Pediococcus pentosaceus on the Growth Performance and Morphometry of the Genetically Improved Farmed Tilapia (Oreochromis niloticus).

Guarner, F. and Malagelada, J.R. (2003) Gut flora in health and disease. *The lancet*, *361*(9356), pp. 512–519.

Gurbatri, C.R., Lia, I., Vincent, R., Coker, C., Castro, S., Treuting, P.M., Hinchliffe, T.E., Arpaia, N. and Danino, T. (2020) Engineered probiotics for local tumor delivery of checkpoint blockade nanobodies. *Science translational medicine*, *12*(530), p.eaax0876.

Hale, C.R., Zhao, P., Olson, S., Duff, M.O., Graveley, B.R., Wells, L., Terns, R.M. and Terns, M.P. (2009) RNA-guided RNA cleavage by a CRISPR RNA-Cas protein complex. *Cell*, *139*(5), pp. 945–956.

Hasan, G.M.M.A., Das, A.K. and Satter, M.A. (2024) CRISPR/Cas9-mediated Gene editing of *L. bulgaricus*-lactate dehydrogenase (LDH) Gene to monitor yogurt characteristics. *Food chemistry advances*, *5*, p. 100772. Available at: https://doi.org/10.1016/j.focha.2024.100772.

Hidalgo-Cantabrana, C., O'Flaherty, S. and Barrangou, R. (2017) CRISPR-based engineering of next-generation lactic acid bacteria. *Current opinion in microbiology*, *37*, pp. 79–87. Available at: https://doi.org/10.1016/j.mib.2017.05.015.

Imran, K., Majgaonkar, A., Abid, R., Bhandari, S., Khan, B., Gillani, S.Q., … Duary, P. (2024) *Genetics and proteomics of polyextremophiles existing in high temperature*. Trends in biotechnology of polyextremophiles, pp. 33–62.

Jandhyala, S.M., Talukdar, R., Subramanyam, C., Vuyyuru, H., Sasikala, M. and Reddy, D.N. (2015) Role of the normal gut microbiota. *World journal of gastroenterology: WJG*, *21*(29), p. 8787.

Jiang, W., Bikard, D., Cox, D., Zhang, F. and Marraffini, L.A. (2013) RNA-guided editing of bacterial genomes using CRISPR-Cas systems. *Nature biotechnology*, *31*(3), pp. 233–239.

Khan Mirzaei, M. and Deng, L. (2022) New technologies for developing phage-based tools to manipulate the human microbiome. *Trends in microbiology*, *30*(2), pp. 131–142. Available at: https://doi.org/10.1016/j.tim.2021.04.007.

Knott, G.J. and Doudna, J.A. (2018) CRISPR-Cas guides the future of genetic engineering. *Science*, *361*(6405), pp. 866–869.

Le Barz, M., Anhê, F.F., Varin, T.V., Desjardins, Y., Levy, E., Roy, D., Urdaci, M.C. and Marette, A. (2015) Probiotics as complementary treatment for metabolic disorders. *Diabetes & metabolism journal*, *39*(4), pp. 291–303.

Leventhal, D.S., Sokolovska, A., Li, N., Plescia, C., Kolodziej, S.A., Gallant, C.W., Christmas, R., Gao, J.R., James, M.J., Abin-Fuentes, A. and Momin, M. (2020) Immunotherapy with engineered bacteria by targeting the STING pathway for anti-tumor immunity. *Nature communications*, *11*(1), p. 2739.

Li, B. et al (2020) Efficacy and safety of probiotics in irritable bowel syndrome: A systematic review and meta-analysis. *Frontiers in pharmacology*, *11*, p. 332. Available at: https://doi.org/10.3389/fphar.2020.00332.

Liévin-Le Moal, V. and Servin, A.L. (2014) Anti-infective activities of lactobacillus strains in the human intestinal microbiota: From probiotics to gastrointestinal anti-infectious biotherapeutic agents. *Clinical microbiology reviews*, *27*(2), pp. 167–199.

Ma, J., Lyu, Y., Liu, X., Jia, X., Cui, F., Wu, X., Deng, S. and Yue, C. (2022) Engineered probiotics. *Microbial cell factories*, *21*(1), p. 72.

Ma, T. et al (2023) Targeting gut microbiota and metabolism as the major probiotic mechanism – an evidence-based review. *Trends in food science & technology*, *138*, pp. 178–198. Available at: https://doi.org/10.1016/j.tifs.2023.06.013.

Makarova, K.S., Wolf, Y.I., Iranzo, J., Shmakov, S.A., Alkhnbashi, O.S., Brouns, S.J., Charpentier, E., Cheng, D., Haft, D.H., Horvath, P. and Moineau, S. (2020) Evolutionary classification of CRISPR-Cas systems: A burst of class 2 and derived variants. *Nature reviews microbiology*, *18*(2), pp. 67–83.

Mali, P., Yang, L., Esvelt, K.M., Aach, J., Guell, M., DiCarlo, J.E., Norville, J.E. and Church, G.M. (2013) RNA-guided human genome engineering via Cas9. *Science*, *339*(6121), pp. 823–826.

McGinn, J. and Marraffini, L.A. (2019) Molecular mechanisms of CRISPR-Cas spacer acquisition. *Nature reviews microbiology*, *17*(1), pp. 7–12.

Mekler, V., Kuznedelov, K. and Severinov, K. (2020) Quantification of the affinities of CRISPR-Cas9 nucleases for cognate protospacer adjacent motif (PAM) sequences. *Journal of biological chemistry*, *295*(19), pp. 6509–6517.

Menchaca, A., Dos Santos-Neto, P.C., Mulet, A.P. and Crispo, M. (2020) CRISPR in livestock: From editing to printing. *Theriogenology*, *150*, pp. 247–254.

Mishra, S. and Acharya, S. (2021) A brief overview on probiotics: The health-Friendly microbes. *Biomedical and pharmacology journal*, *14*(4), pp. 1869–1880.

Mladenov, E. and Iliakis, G. (2011) Induction and repair of DNA double strand breaks: The increasing spectrum of non-homologous end joining pathways. *Mutation research/fundamental and molecular mechanisms of mutagenesis*, *711*(1–2), pp. 61–72.

Moghimipour, E., Abedishirehjin, S., Baghbadorani, M.A. and Handali, S. (2021) Bacteria and archaea: A new era of cancer therapy. *Journal of controlled release*, *338*, pp. 1–7.

Mugwanda, K. et al (2023) Recent advances in genetic tools for engineering probiotic lactic acid bacteria. *Bioscience reports*, *43*(1), p.BSR20211299. Available at: https://doi.org/10.1042/BSR20211299.

Muhammad, Z., Anjum, M.Z., Akhter, S., Irfan, M., Amin, S., Jamal, Y., … Ghazanfar, S. (2023) Effect of *Lactobacillus plantarum* and *Pediococcus pentosaceus* on the growth performance and morphometry of the genetically improved farmed tilapia (*Oreochromis niloticus*). *Pakistan journal of zoology*, *56*(1).

Niu, H.-L. and Xiao, J.-Y. (2020) The efficacy and safety of probiotics in patients with irritable bowel syndrome: Evidence based on 35 randomized controlled trials. *International journal of surgery*, *75*, pp. 116–127. Available at: https://doi.org/10.1016/j.ijsu.2020.01.142.

Palumbo, V.D., Romeo, M., Marino Gammazza, A., Carini, F., Damiani, P., Damiano, G., Buscemi, S., Ignazio, A., Monte, L. and Gerges-Geagea, A. (2016) The long-term effects of probiotics in the therapy of ulcerative colitis: A clinical study. *Biomed pap med fac univ palacky olomouc Czech repub*, *160*(3), pp. 372–377.

Qin, J., Li, Y., Cai, Z., Li, S., Zhu, J., Zhang, F., Liang, S., Zhang, W., Guan, Y., Shen, D. and Peng, Y. (2012) A metagenome-wide association study of gut microbiota in type 2 diabetes. *Nature*, *490*(7418), pp. 55–60.

Redman, M., King, A., Watson, C. and King, D. (2016) What is CRISPR/Cas9? *Archives of disease in childhood-education and practice*, *101*(4), pp. 213–215.

Saadullah, M., Asif, M., Farid, A., Naseem, F., Rashid, S.A., Ghazanfa, S., … Arif, S. (2022) A novel distachionate from *Breynia distachia* treats inflammations by modulating COX-2 and inflammatory cytokines in rat liver tissue. *Molecules*, *27*(8), p. 2596.

Sehrawat, N., Yadav, M., Singh, M., Kumar, V., Sharma, V.R. and Sharma, A.K. (2021) Probiotics in microbiome ecological balance providing a therapeutic window against cancer. In *Seminars in cancer biology* (Vol. 70, pp. 24–36). Academic Press.

Selvanantham, T., Lin, Q., Guo, C.X., Surendra, A., Fieve, S., Escalante, N.K., Guttman, D.S., Streutker, C.J., Robertson, S.J., Philpott, D.J. and Mallevaey, T. (2016) NKT cell–deficient mice harbor an altered microbiota that fuels intestinal inflammation during chemically induced colitis. *The journal of immunology*, *197*(11), pp. 4464–4472.

Shah, S.Z., Rehman, A., Nasir, H., Asif, A., Tufail, B., Usama, M. and Jabbar, B. (2018) Advances in research on genome editing CRISPR-Cas9 technology. *Journal of ayub medical college Abbottabad*, *31*(1), pp. 108–122.

Tang, X., Lowder, L.G., Zhang, T., Malzahn, A.A., Zheng, X., Voytas, D.F., Zhong, Z., Chen, Y., Ren, Q., Li, Q. and Kirkland, E.R. (2017) A CRISPR-Cpf1 system for efficient genome editing and transcriptional repression in plants. *Nature plants*, *3*(3), pp. 1–5.

Tsai, Y.L., Lin, T.L., Chang, C.J., Wu, T.R., Lai, W.F., Lu, C.C. and Lai, H.C. (2019) Probiotics, prebiotics and amelioration of diseases. *Journal of biomedical science*, *26*(1), p. 8.

Vaid, S. et al (2022) A concise review on development of probiotics from *Lactobacillus* using CRISPR-Cas technology of Gene editing. *Food chemistry advances*, *1*, p. 100099. Available at: https://doi.org/10.1016/j.focha.2022.100099.

Van Pijkeren, J.-P. and Barrangou, R. (2017) Genome editing of food-grade *Lactobacilli* to develop therapeutic Probiotics. *Microbiology spectrum*. Edited by R.A. Britton and P.D. Cani, *5*(5), pp. 5–5.09. Available at: https://doi.org/10.1128/microbiolspec.BAD-0013-2016.

Wang, F., Wang, L., Zou, X., Duan, S., Li, Z., Deng, Z., Luo, J., Lee, S.Y. and Chen, S. (2019) Advances in CRISPR-Cas systems for RNA targeting, tracking and editing. *Biotechnology advances*, *37*(5), pp. 708–729.

Wang, H., Yang, H., Shivalila, C.S., Dawlaty, M.M., Cheng, A.W., Zhang, F. and Jaenisch, R. (2013) One-step generation of mice carrying mutations in multiple genes by CRISPR/Cas-mediated genome engineering. *Cell*, *153*(4), pp. 910–918.

Wyman, C. and Kanaar, R. (2006) DNA double-strand break repair: All's well that ends well. *Annual review of genetics*, *40*(1), pp. 363–383.

Yadav, M. and Shukla, P. (2020) Efficient engineered probiotics using synthetic biology approaches: A review. *Biotechnology and applied biochemistry*, *67*(1), pp. 22–29.

Yadav, R., Kumar, V., Baweja, M. and Shukla, P. (2018) Gene editing and genetic engineering approaches for advanced probiotics: A review. *Critical reviews in food science and nutrition*, *58*(10), pp. 1735–1746.

Yaghoubi, A., Khazaei, M., Jalili, S., Hasanian, S.M., Avan, A., Soleimanpour, S. and Cho, W.C. (2020) Bacteria as a double-action sword in cancer. *Biochimica et biophysica acta (BBA)-reviews on cancer*, *1874*(1), p. 188388.

Yang, H., Wang, H., Shivalila, C.S., Cheng, A.W., Shi, L. and Jaenisch, R. (2013) One-step generation of mice carrying reporter and conditional alleles by CRISPR/Cas-mediated genome engineering. *Cell*, *154*(6), pp. 1370–1379.

Zeng, L., Deng, Y., He, Q., Yang, K., Li, J., Xiang, W., Liu, H., Zhu, X. and Chen, H. (2022) Safety and efficacy of probiotic supplementation in 8 types of inflammatory arthritis: A systematic review and meta-analysis of 34 randomized controlled trials. *Frontiers in immunology*, *13*, p. 961325.

Zhang, M., Eshraghian, E.A., Al Jammal, O., Zhang, Z. and Zhu, X. (2021) CRISPR technology: The engine that drives cancer therapy. *Biomedicine & pharmacotherapy*, *133*, p. 111007.

Zhu, Y. (2022) Advances in CRISPR/Cas9. *BioMed research international*, *2022*(1), p. 9978571.

Zou, Z.P., Du, Y., Fang, T.T., Zhou, Y. and Ye, B.C. (2023) Biomarker-responsive engineered probiotic diagnoses, records, and ameliorates inflammatory bowel disease in mice. *Cell host & microbe*, *31*(2), pp. 199–212.

Zuo, F. and Marcotte, H. (2021) Advancing mechanistic understanding and bioengineering of probiotic *Lactobacilli* and *Bifidobacteria* by genome editing. *Current opinion in biotechnology*, *70*, pp. 75–82. Available at: https://doi.org/10.1016/j.copbio.2020.12.015.

5 Relationship Between Genus-, Species- and Strain-Specific and Disease-Condition Specific Effects, Issues, and Applications of Probiotics

Haala Marikar, Maliha Rashid, Hadia Afzal, Maryam Gohar, Saiqa Noureen, Shayista Jan, Shakira Ghazanfar, Maria Fareed Siddiqui, and Khudija Tul Kubra

5.1 INTRODUCTION

5.1.1 Background

Deciphering the relationship between genus, species, and particular strains becomes like cracking a microbial code in the complex world of probiotics. This code is the key to gaining access to personalized therapies for a variety of medical concerns. The small distinctions between probiotic strains are critical for designing targeted treatments for a variety of medical ailments, ranging from immunological fortification to digestive difficulties (Sanders et al., 2013). In this chapter, we untangle the complexities of probiotic diversity, revealing how the specific characteristics of genus, species, and individual strains shape their specific effects and uses in the complicated game of preserving health and curing certain disorders.

5.1.2 Introducing Our Most Used Probiotic Companions

Lactobacillus: Consider *Lactobacillus* as your digestive system's superhero. *Lactobacillus* makes sure our stomachs feel good and aids in digestion, especially in cases where they are not fond of milk (Anon) (Mazhar et al., 2024).

Bifidobacterium: This one functions in our guts as a kind of balance keeper. It helps digestion and may alleviate stomach discomfort in persons suffering from irritable bowel syndrome (IBS) (Anon).

Saccharomyces: It works as an immune system defense. Saccharomyces helps to strengthen our immune system and avoid diarrhea (Abid et al., 2022).

Enterococcus: Think of it as the protector of intestinal resilience. By supporting a balanced microbial environment and improving digestive health, *Enterococcus* promotes general well-being (Dahiya & Nigam, 2022).

5.1.3 Introducing the Relationship between Genus and Species of Probiotics

The identification of probiotics' genus and species is essential to comprehending their effects on health. Consider the species as individual members of a family and the genus as

DOI: 10.1201/9781003399797-6

a wide group. For example, there are differences between *Lactobacillus acidophilus* and *Lactobacillus casei* within the *Lactobacillus* genus. Certain characteristics may influence the ailments that each species treats. For instance, *L. casei* may be beneficial for intestinal problems, while *L. acidophilus* is linked to digestive health. In keeping with the notion that distinct microbial strains within the same family might have a range of therapeutic effects, the relationship between the genus and species aids in the identification of probiotics designed to treat disorders.

5.2 EFFECTS OF PROBIOTICS ON HOST HEALTH

5.2.1 Probiotics as Prospective Therapeutic Interventions

Successful probiotics exhibit antagonism against pathogens, activation and optimization of the immune system, and other therapeutic benefits in the host organism (Plaza-Díaz et al., 2017). Prior to administration, the mechanism of action, viability, and efficiency of the probiotic must be studied thoroughly (Plaza-Díaz et al., 2018).

Demonstration of improvement of host health by ingestion of *Lactobacillus* species (LAB) conducted by Ilya Ilyich Mechnikov was one of the preliminary experiments on the effects of probiotics (Mechnikov & Mitchell, 1910). Subsequently, several studies have shown the positive effects of probiotic engraftment or ingestion on host health (Didari et al., 2014; Szajewska, 2015) in gastrointestinal (Martin et al., 2008), inflammatory (Sáez-Lara et al., 2015), neurological (Bear et al., 2021), and dermatological disorders (Rather et al., 2016). Probiotics have also been shown in pilot studies to be efficient in improving plant stress response, antioxidant production, and overall development (Pandey et al., 2022).

Despite several probiotic strains being regarded as "Generally Recognized as Safe" (GRAS), the European Food Safety Authority (EFSA) has not permitted the circulation of probiotics being pitched as a therapeutic or clinical tool with more than 60 benefits (EFSA Panel on Dietetic Products, Nutrition and Allergies (NDA), 2016). Likewise, the FDA has not approved any probiotics that claim to provide any health benefit source. The reasons behind this refusal range from inadequate characterization of the strain to lack of related studies in human models (Ko, 2020).

However, Health Canada permits the circulation of a consortium of *L. acidophilus* (SD5212), *Streptococcus salivarius* subsp. *thermophilus* (SD5207), *Lactobacillus plantarum* (SD5209), *Bifidobacterium breve* (SD5206), *Lacticaseibacillus paracasei* (SD5218), *Bifidobacterium animalis* subsp. *lactis* (SD5220, SD5219), and *Lactobacillus helveticus* (SD5210) and a single strain of *B. animalis* spp. *lactis* LAFTI B94 as a natural therapeutic intervention for symptomatic relief of IBS (Ko, 2020).

5.2.2 Mechanisms of Action across Probiotics of Different Genii and Strains

Changes in the quantitative and qualitative composition of the gut microbiome disturb host-microbial interactions, which promote inflammatory disorders leading to non-communicable diseases (Byndloss & Bäumler, 2018). Maintenance of homeostasis in the gut is mediated by limitation of oxygen (hypoxia) at the epithelial level. This balances the gut microbiota and prompts the production of metabolites that regulate host immunity, nutrition, and niche security (Byndloss & Bäumler, 2018; Cani, 2017).

Probiotics restore microbial diversity to mitigate gut dysbiosis (Vieira et al., 2016). The following mechanisms (Ko, 2020) have been proposed over the years: refurbishment of perturbed intestinal microbiota; competitive exclusion of pathogens and bacteriocin production; enzyme action and incidence of volatile fatty acids; cell adhesion and mucin production; modulation of the immune system; and interaction with the brain-gut axis.

5.2.2.1 Refurbishment of Perturbed Intestinal Microbiota

Refurbishment includes the recolonization of the gut and return to functional normalcy (Yang et al., 2016). The assembly of the gut microbiome begins before birth and develops throughout childhood. It is decided by factors such as genetic makeup, birthing method, type of feeding, use of antibiotics, and whether the mother experiences stress or some inflammatory condition (Hesla et al., 2014). The bacteria isolated from the umbilical cord, placenta, meconium (Aagaard et al., 2014; Jiménez et al., 2005) and more importantly, the vagina and human milk are crucial for the establishment of the gut microbiome (Jiménez-Truque et al., 2012). These include *Enterococcus faecium, Propionibacterium acnes, Staphylococcus epidermidis*, and *Escherichia coli* (Aagaard et al., 2014; Jiménez et al., 2005).

Dendritic cells or macrophages engulf these bacteria and carry them to extra-digestive areas, as well as take them along to circulate through the blood (MartíN et al., 2004). Some probiotics facilitate lactose digestion in lactose-intolerant hosts by increasing the overall hydrolytic potential in the small intestine and increasing fermentation in the colon (Dhama et al., 2017). Such probiotics can decrease the lactose content of the fermented matter in the small intestine and increase the concentration of lactase enzyme that moves with the fermented matter (He et al., 2007). It is important that the bacteria adhere to host surfaces (Ribet & Cossart, 2015). These adhesive substances, termed Adhesins, include Pili, a hair-like protrusion from the bacterial surface that binds bacteria to host cells; molecules that recognize transmembrane proteins like integrins or cadherins; molecules that recognize extracellular matrix proteins like fibronectin, laminin, etc.

In adults, probiotic use has been shown to increase the concentration of short-chain fatty acids (SCFAs), stool moisture content, frequency of defecation, and volume of stools (Olivares et al., 2006).

5.2.2.2 Competitive Exclusion of Pathogens and Bacteriocin Production

Competitive exclusion refers to the race amongst two or more bacterial species or strains to attach to the intestinal receptors vigorously (Bermúdez-Brito et al., 2012). The bacteria that successfully attach to the receptor colonize the gut against the bacteria that could not attach as vigorously. Proposed mechanisms include lowering of the luminal pH, competition for nutrients, and production of bacteriocins or bacteriocin-like substances (Collado et al., 2009). Probiotic metabolome (amines, organic acids, bacteriocins, etc.) has been shown to target several substances in the metabolic pathways that regulate cellular proliferation, differentiation, apoptosis, inflammation, angiogenesis, and metastasis (Kumar et al., 2012).

Bacteriocins are small (approximately 30–60 amino acids long), cationic, antimicrobial peptides that inhibit the proliferation of selected pathogens (Umu et al., 2017). These act at the energized membrane vesicles to disrupt the proton motive pumps, thus destroying the bacterial membrane. Bacteriocins derived from *L. plantarum* (Idrees et al., 2023) and *L. acidophilus* have been reported to inhibit the growth of *Helicobacter, Clostridioides difficile*, rotaviruses, and multidrug-resistant *Shigella* spp. and *E. coli* in gut-related issues (Kumar et al., 2016) and show inhibitory capacity against some uropathogens as well (Mokoena, 2017).

5.2.2.3 Enzyme Action and Incidence of Volatile Fatty Acids

5.2.2.3.1 Enzyme Action

The enzymatic interactions of probiotic bacteria with the lumen can interfere with their biological activity. *Lactobacillus* and *Bifidobacterium* strains show more than 20 reported enzymatic activities, with β-galactosidase being the most common. Some enzymes can be used as presymptomatic markers of colon carcinogenesis (Kim & Jin, 2001).

Probiotics that produce Bile Salt Hydrolases (BSHs) promote a hypocholesterolemic effect attributed to such bacteria. These enzymes catalyze the first reaction of the deconjugation of bile salts (Pavlović et al., 2012).

5.2.2.3.2 Volatile Fatty Acids

Probiotic usage has been shown to increase the content of SCFAs in the wet fecal matter of hosts (Schneider et al., 2005). They have also been linked to the increase in incidence of branched-chain fatty acids (BCFAs) in wet fecal content (Hemalatha et al., 2017).

SCFAs are key signaling molecules in gut health and are the food source of enterocytes. They regulate energy maintenance and metabolism (Canfora et al., 2015). SCFAs stimulate satiety-enhancing proteins when they couple with G-protein-coupled receptors (Hur & Lee, 2015). They interact with receptors in the adipose tissue to decrease fat accumulation, which diminishes lipolysis and inflammation. SCFAs play a major role in increasing insulin sensitivity by activating AMP kinase (Canfora et al., 2015).

5.2.2.4 Cell Adhesion and Mucin Production

Gut bacteria require strict adhesion to the lumen for colonization (Bermúdez-Brito et al., 2012). However, the intestine naturally releases a mucous layer to prevent this adhesion (Collado et al., 2009). Probiotic strains, for example, *Lactobacillus*, express surface adhesins (Van Tassell & Miller, 2011) to attach to the lumen and overcome the resistance that the mucus layer poses (Buck et al., 2005).

5.2.2.5 Modulation of Immune System

Probiotic strains induce the development of regulatory T-cells and modulate the immune system to maintain gut homeostasis (D'Amelio & Sassi, 2017). These bacteria interact with the gut epithelium, dendritic cells, macrophages, lymphocytes, immunoglobulins, etc., to induce the production of cytokines or other immunomodulatory or anti-inflammatory molecules (Gómez-Llorente et al., 2010).

5.2.2.6 Interaction with the Brain-Gut Axis

Gut microbiome, the brain-gut axis, and the interaction of the microbiota with genetic receptors have been shown to be associated with short- and long-term behavior (Tillisch et al., 2013). Stress activates the hypothalamic-pituitary-adrenal axis along with the sympathetic nervous system, thus increasing the permeability of the lumen that allows bacteria or its antigenic determinants to cross the intestinal barrier. This activates an immune response, which, in turn, alters the composition of the microbiome (Bonaz & Bernstein, 2013). The mechanisms that relay the effects of probiotic strains in the central nervous system may be neuronal, hormonal, or immunological. The bottom line is that bacterial metabolites mediate all such interactions.

5.3 PROBIOTICS AND THEIR TAILORED TRIUMPHS: EXPLORING THE EFFICACY AND APPLICATIONS OF GENUS-, SPECIES-, AND DISEASE-CONDITION SPECIFIC PROBIOTICS

5.3.1 Mental Well-being

Probiotics recognized for their diverse health benefits, have emerged as potential contributors to mental well-being. The beneficial bacteria play a crucial role in maintaining a healthy gut microbiome, which, in turn, is intricately linked to optimal brain function. The gut microbiome's influence on neurotransmitter and hormone production, regulating mood and behavior, and underscores its significance in mental health. In exploring this connection, research has uncovered strong evidence associating gut health with psychological wellness and the onset of various mental diseases. Neurodegenerative illnesses like Alzheimer's, Parkinson's, and Huntington's diseases exhibit characteristics such as neuroinflammation, specifically microgliosis and astrogliosis. Despite the apparent differences in the etiologies of modern-day disorders—ranging from obesity to anxiety, depression, Parkinson's disease, diabetes, and Alzheimer's disease (AD)—there are commonalities. They shared the status of being non-communicable, non-transmissible ailments lacking a comprehensive cure. The literature highlights the role of the microbiota in these conditions, and various probiotics have demonstrated the ability to enhance microbiome stability, positively impacting brain health and potentially preventing neurodegenerative diseases such as AD. The study not only provides a wealth of research supported by evidence regarding the contribution of probiotics to slowing AD progression but also highlights positive outcomes from clinical trials and in vivo research. Importantly, the consumption of probiotics for AD treatment has shown no reported negative effects. Moving beyond AD, studies propose a broader positive impact of probiotics on mental health. This encompasses the reduction of inflammation, regulation of the gut microbiota, and enhancement of stress response, as outlined in Table 5.1. Probiotics show promise in alleviating symptoms of depression and anxiety, curbing stress-related eating behaviors, and improving cognitive outcomes. Moreover, these beneficial bacteria may enhance sleep quality, reduce fatigue, and uplift mood.

The intricate relationship between the microbiota residing in the gastrointestinal tract (GIT) and the central nervous system (CNS) is encapsulated by the term gut-brain axis. This communication, facilitated through the endocrine system, inflammatory responses, and the neural system, is bidirectional. Clinical trials have elucidated the role of the vagus nerve in maintaining the microbiota-gut-brain axis and its significant function in neurotransmitter signaling. As a result, the microbiota-gut-brain axis becomes a two-way mechanism, influencing cognitive functions such as memory, creativity, and emotional control. This exploration underscores the multifaceted impact of probiotics on mental well-being, positioning them as promising contributors to holistic health.

5.3.1.1 Dermatology Studies

Probiotics are recognized for their therapeutic potential, emerge as a secure and alternative technique for managing various cutaneous ailments. Extensive literature suggests their efficacy in treating skin conditions, including psoriasis, atopic dermatitis, and acne vulgaris. Probiotics exhibit promising results by modulating the immune response through alterations in the gut microbiome. Specific dermatological conditions in which probiotics play a pivotal role include seborrheic dermatitis, acne vulgaris, psoriasis vulgaris, rosacea, atopic dermatitis, bacterial vaginosis (BV), vulvovaginal candidiasis (VVC), aging, oral mucositis, and wound healing.

The skin, undergoing continuous self-renewal, sheds resident microbial cells. While most skin microbes are commensal and harmless, some are considered mutualistic, conferring health benefits. Recent research underscores the intricate connection between the skin's microbial population and its barrier functions. The resident microbiota plays a crucial role in competitive exclusion of pathogens, processing skin proteins, free fatty acids, and sebum. However, disturbances in skin integrity can transform once-beneficial microorganisms into potential threats, contributing to conditions like atopic dermatitis, eczema, rosacea, psoriasis, and acne.

Topical application of probiotic strains capable of producing antimicrobial substances has gained attention as a preventive measure. This includes bacteriocins, bacteriocin-like substances, organic acids, and hydrogen peroxide. In 2005, scientists laid the foundation for topical compositions containing probiotic bacteria, spores, and extracellular products, inhibiting the growth of various pathogens. These probiotic microorganisms, whether bacteria, yeast, or mold, are selected based on specific properties such as competitive exclusion, adherence to human tissue, sensitivity to antibiotics, antimicrobial activity, acid tolerance, and high oxygen tolerance. The method involves diverse application modalities, such as lotions, spraying, and wipe paper, to prevent contamination, colonization, and growth of pathogenic bacteria on surfaces like human skin and hospital equipment. Probiotics, therefore, hold promise in the realm of dermatological studies, offering innovative and effective approaches to address cutaneous conditions.

5.3.2 Vaginal Health

Understanding the crucial link between probiotics and female vaginal health involves unraveling the specific role of *Lactobacillus* strains in establishing and maintaining a harmonious balance within the vaginal microbiota, thereby reinforcing defenses against infections. The composition of the vaginal microbiota and mycobiota significantly influences the overall health of the vaginal ecosystem. Vaginal microbiome profiling, especially over the past decade, has revealed that healthy vaginal communities are characterized by the dominance of LAB, particularly *Lactobacillus*, low diversity of anaerobic bacteria, and a balanced immune system. These lactobacilli play a pivotal role in preventing the invasion of pathogens by regulating their population. Disruptions to the vaginal ecosystem can lead to the overgrowth of pathogens, resulting in complex infections such as BV, sexually transmitted infections (STIs), and VVC.

The colonization and dominance of *Lactobacilli*, particularly species like *Lactobacillus* crispatus, *Lactobacillus gasseri, Lactobacillus iners*, and *Lactobacillus jensenii*, are essential features of a healthy vaginal microbiota. In the core vaginal microbiota of a majority of Asian and white women, *Lactobacilli* dominate at percentages of 80.2% and 89.7%, respectively. The prevailing genera of LAB used as probiotics include *Lactobacillus, Bifidobacterium, Streptococcus, Enterococcus*, and *Pediococcus* (Ghazanfar et al., 2019).

Numerous studies highlight that pathogens causing aerobic vaginitis (AV), such as *E. coli, E. faecalis, S. aureus, S. epidermidis,* and *S. agalactiae*, commonly inhabit the vagina, leading to inflammatory vaginitis. Prolonged use of antimicrobial drugs for vaginitis treatment can result in the development of drug resistance. Hence, there is growing interest in the probiotic lactobacilli-based approach as an alternative to conventional antimicrobial treatments. The role of *Lactobacillus* spp. in maintaining vaginal health is attributed to their antimicrobial activity, pH regulation, hydrogen peroxide production, and bacteriocin synthesis. Their capacity to adhere to and compete for adhesion sites in the vaginal epithelium, along with the production of antimicrobial compounds, is crucial for preventing the colonization of pathogens. This holistic understanding underscores the potential of probiotics as a valuable strategy in promoting and sustaining female vaginal health.

5.3.3 Immune System and Food Allergies

Understanding the intricate relationship between probiotics and the immune system reveals promising avenues for preventing food allergies. The global surge in allergic diseases has become a significant medical and socio-economic challenge, with imbalanced diets, irregular lifestyles, and environmental factors emerging as common causes. Processed foods, coupled with stressful routines and limited exposure to natural environments, contribute to reduced contact with beneficial bacteria. These factors play a role in the development of allergies, as the absence of a well-established gut microbiome weakens immunity. The diversity of lactic acid bacteria in the gut is influenced by various factors, rendering the immune system more sensitive to food allergens.

Probiotic products, particularly those based on nondairy substrates, hold promise as nutraceuticals for lactose-intolerant consumers allergic to dairy milk products. Allergens, recognized by the consumer's immune system, can trigger varying reactions, from mild and temporary to severe anaphylactic responses, such as those caused by nuts—one of the most familiar food allergens. Exploring the application of characterized individual strains or consortia of probiotic bacteria in renovating nutritive food matrices through fermentation into synbiotic products is a viable approach. Functional foods or beverages fortified with bioactive compounds extracted from edible plant sources offer safer alternatives for flavor enhancement, coloring, and antimicrobial preservation, reducing the risk of allergic reactions associated with chemical additives.

The utilization of specifically selected probiotic strains in commercial products designed as supplements provides consumer-friendly therapeutic options. These options aim to minimize gut

TABLE 5.1
***In Vivo* Antimicrobial Activity of Probiotics against Human Pathogens**

Pathogens	Probiotics	Observation
Staphylococcus aureus	*B. subtilis* MB40	Reduces *S. aureus* colonization in human gut
Candida albicans	*L. casei*	In vulvovaginal candidiasis (VVC) murine model, *L. casei* showed fungicidal effect
Staphylococcus aureus, *Staphylococcus pyogenes* M1, *Cutibacterium acnes* AS12, *P. aeruginosa*	*L. reuteri* DSM 17938	Reduces pro-inflammatory IL-8 & IL-6 in native skin models and reconstructs human epidermis, shows antimicrobial activity against skin bacteria
Pseudomonas aeruginosa	*Pediococcus acidilactici* HW01	Decreases virulence factors, such as pyocyanin, rhamnolipid and protease, and prevent biofilm formation
L. monocytogenes	*Ped. acidilactici* HW01	Prevents biofilm formation, invasion, and adhesion of HT-29 cells by bacterial lysate

disorders and discomfort caused by certain foods. Examples of such products include probiotic beverages like Kefir made from cow's milk, natural bio-yogurt from free-range milk, and formulations of germinated brown rice or fermented products functionalized with probiotics. These products, enriched with bioactive compounds and characterized probiotic strains, represent a step toward mitigating the impact of food allergies and enhancing overall digestive health. Further details are indicated in Table 5.1.

5.3.4 Gastrointestinal Diseases Like Irritable Bowel Syndrome

Exploring the potential of probiotics, specifically *Bifidobacterium infantis* (*B. infantis*), in alleviating symptoms of irritable bowel syndrome (IBS) provides a promising avenue for addressing abdominal discomfort. IBS, characterized by gut-brain interaction dysfunction, manifests as abdominal pain and alterations in stool form or frequency. Evidence suggests the efficacy of certain probiotics, including *L. plantarum* DSM 9843 and *Bifidobacterium bifidum* MIMBb75, in managing IBS symptoms.

This chronic condition lacks apparent structural or anatomical causes but involves recurring episodes of stomach pain, bloating, and irregular bowel habits. Traditional treatments for IBS, such as laxatives, antispasmodics, and antidiarrheal medications, can lead to serious adverse effects with prolonged usage. In contrast, synbiotics, which combine probiotics and prebiotics, offer a holistic approach to enhance host health. Symptoms of IBS encompass constipation, abnormal pain, bloating, gases, nausea/vomiting, and diarrhea. Conventional pharmacological treatments, including mesalazine, corticosteroids, immunosuppressive drugs, and monoclonal antibodies targeting tumor necrosis factor (TNF), may present potential side effects. Consequently, non-pharmacological therapies, such as cognitive-behavioral therapy, hypnotherapy, psychotherapy, meditation, yoga, acupuncture, and exercise, have gained popularity among individuals with IBS.

Probiotic strains exhibit several advantages, including improved stool consistency, alleviation of nonspecific symptoms in healthy individuals, and positive effects in inflammatory gastrointestinal diseases. Research by Dimidi et al. highlighted lower levels of Lactobacilli and Bifidobacterium, along with higher levels of breath methane, in people with painful constipation. Commercially available probiotics, administered in a combination pill containing bacterial strains of *Bifidobacterium*, *Lactobacillus*, and *Streptococcus*, demonstrated potential efficiency in individuals with IBS. Overall, these findings emphasize the valuable role of probiotics in managing IBS symptoms and promoting gut health when appropriately administered.

5.3.5 Alternative to Antibiotics against Human and Animal Pathogens

The inappropriate and excessive use of antibiotics poses a significant threat to human and animal health, leading to the rise of pathogen resistance and dysbiosis phenomena. In response to this challenge, probiotics emerge as promising alternatives due to their multifunctional properties, offering a reliable option compared to traditional antibacterial substances. The potential of probiotics as alternatives to antibiotics has been substantiated through various in vivo and ex vivo experiments, as detailed in Table 5.1. Notably, lactic acid- and soil-based bacteria exhibit bacteriostatic and bactericidal activities against pathogens such as *S. aureus*, *L. monocytogenes*, *P. aeruginosa*, and *Candida albicans*, effectively reducing their colonization in the human body. Clinical trials have further demonstrated the efficacy of probiotics in the treatment of various diseases, providing a compelling case for their use as antibiotic substitutes, as highlighted in Table 5.1. This shift toward probiotics represents a crucial step in addressing the challenges posed by antibiotic resistance and fostering a healthier balance in microbial ecosystems.

5.4 ISSUES ASSOCIATED WITH USING PROBIOTICS ACROSS DIFFERENT GENERA, SPECIES, AND DISEASE-CONDITIONS

The commonly administered probiotic species are unlikely to cause harm in healthy people. Side effects are usually minor and consist of self-limited gastrointestinal symptoms, such as gas. However, gaps in available safety evidence do exist, especially in cases of commercial probiotics that have undergone inadequately designed clinical trials. These oversights have resulted in great harm to certain vulnerable populations, such as preterm infants and immunocompromised individuals.

Therefore, just as probiotics have a plethora of benefits associated with their use as a therapeutic modality, so too do they have their drawbacks and issues. Here, we highlight some of the issues associated with using probiotics, across several common genera, and disease conditions.

5.4.1 Lactobacillus

L. rhamnosus, L. acidophilus, and *L. casei,* have all been involved in cases of bacteremia and/or sepsis, pneumonia, abscesses, and endocarditis in populations including immunocompromised patients (due to HIV/AIDS infection, organ transplantations, etc.), preterm infants, patients with short-gut syndrome, and patients under cardio-surgery. (Mantegazza et al., 2018).

Another concern surrounding the use of *Lactobacillus* as a probiotic is their prevalent antibiotic resistance, which may represent a food safety concern. The in vivo transfer of resistance determinants to (or from) potentially harmful microbes is also a concern, although it should be noted that transfer to lactobacilli from genera such as enterococci is quite rare (Gueimonde et al., 2013).

Lactobacillus also has some diagnostic difficulties associated with it, arising from misidentificiation events of the probiotic strain. Therefore, characterizing clinical isolates and probiotic strains often requires the use of culture-independent molecular-based phenotypic and genomic characterization (Castro-González et al., 2019).

5.4.2 Bifidobacterium

Commensal *Bifidobacteria* have only rarely been implicated in cases of dental and other infections, with only one strain, *Bifidobacterium dentium,* being recognized as pathogenic. According to several safety assessment studies on *Bifidobacterium*, it is generally recognized as not having any health effects associated with its use as a probiotic.

Vigilance must be maintained with regard to possible rare cases of infection, and it must be noted that antibiotic resistance is a great concern that must continue to be monitored, because

Bifidobacterium, like other inhabitants of the GIT, does have the potential to act as reservoirs for antibiotic resistance genes (Gueimonde et al., 2013).

5.4.3 Enterococcus

Enterococci is not a probiotic that is GRAS, and it has not been included in the Qualified Presumption of Safety (QPS) list—and for good reason, as we will elaborate here. *Enterococci* express many virulence traits, including cytolysin, proteases, aggregation substance, capsular polysaccharide, enterococcal surface protein, biofilm formation, extracellular superoxide, intestinal translocation, and resistance to innate immunity that can lead to serious hospital-acquired infections. As such, isolates of this genus have emerged as opportunistic pathogens for humans (Fiore et al., 2019).

5.4.4 Saccharomyces

Many cases of fungemia due to *S. cerevisiae* or *S. boulardii* have been reported (Thygesen et al., 2012). The question of whether these probiotics should be administered to critically ill or immunocompromised patients is one that remains to be answered (Ghazanfar, 2021).

Just as evidence has been found linking together the efficacy between genus-, species/strain-, and disease-condition specificity of probiotics, so too can we observe a link between issues surrounding genus-, strain-, and disease-condition specific probiotics. Care must be taken, therefore, to use only those strains which are chosen from the commensal flora of humans, and which do not carry intrinsic resistance to antibiotics that would prevent treatment of a rare probiotic infection. Health-care providers need to take genus-, species/strain- and disease-condition specificity into consideration when recommending the appropriate probiotic for their patients.

REFERENCES

Aagaard, K. M., Ma, J., Antony, K. M., Ganu, R., Petrosino, J. F., & Versalovic, J. (2014). The placenta harbors a unique microbiome. *Science Translational Medicine*, *6*(237). https://doi.org/10.1126/scitranslmed.3008599

Abid, R., Waseem, H., Ali, J., Ghazanfar, S., Muhammad Ali, G., Elasbali, A. M., & Alharethi, S. H. (2022). Probiotic yeast saccharomyces: Back to nature to improve human health. *Journal of Fungi*, *8*, 444. https://doi.org/10.3390/jof8050444

Bear, T., Dalziel, J. E., Coad, J., Roy, N. C., Butts, C. A., & Gopal, P. K. (2021). The microbiome-gut-brain axis and resilience to developing anxiety or depression under stress. *Microorganisms*, *9*(4), 723. https://doi.org/10.3390/microorganisms9040723

Bermúdez-Brito, M., Plaza-Díaz, J., Muñoz-Quezada, S., Gómez-Llorente, C., & Gil, Á (2012). Probiotic mechanisms of action. *Annals of Nutrition and Metabolism*, *61*(2), 160–174. https://doi.org/10.1159/000342079

Bonaz, B., & Bernstein, ÇN. (2013). Brain-gut interactions in inflammatory bowel disease. *Gastroenterology*, *144*(1), 36–49. https://doi.org/10.1053/j.gastro.2012.10.003

Buck, B. L., Altermann, E., Svingerud, T., & Klaenhammer, T. R. (2005). Functional analysis of putative adhesion factors in lactobacillus acidophilus NCFM. *Applied and Environmental Microbiology*, *71*(12), 8344–8351. https://doi.org/10.1128/aem.71.12.8344-8351.2005

Byndloss, M. X., & Bäumler, A. J. (2018). The germ-organ theory of non-communicable diseases. *Nature Reviews Microbiology*, *16*(2), 103–110. https://doi.org/10.1038/nrmicro.2017.158

Canfora, E. E., Jocken, J. W. E., & Blaak, E. E. (2015). Short-chain fatty acids in control of body weight and insulin sensitivity. *Nature Reviews Endocrinology*, *11*(10), 577–591. https://doi.org/10.1038/nrendo.2015.128

Cani, P. D. (2017). Gut microbiota—At the intersection of everything? *Nature Reviews Gastroenterology & Hepatology*, *14*(6), 321–322. https://doi.org/10.1038/nrgastro.2017.54

Castro-González, J. M., De Castro, P., Sandoval, H., & Castro-Sandoval, D. (2019). Probiotic lactobacilli precautions. *Frontiers in Microbiology*, *10*. https://doi.org/10.3389/fmicb.2019.00375

Collado, M. C., Isolauri, E., Salminen, S., & Sanz, Y. (2009). The impact of probiotic on gut health. *Current Drug Metabolism*, *10*(1), 68–78. https://doi.org/10.2174/138920009787048437

D'Amelio, P., & Sassi, F. (2017). Gut microbiota, immune system, and bone. *Calcified Tissue International, 102*(4), 415–425. https://doi.org/10.1007/s00223-017-0331-y

Dahiya, D., & Nigam, P. S. N. (2022). The gut microbiota influenced by the intake of probiotics and functional foods with prebiotics can sustain wellness and alleviate certain ailments like gut-inflammation and colon-cancer. *Microorganisms, 10*: 665. https://doi.org/10.3390/microorganisms10030665

Dhama, K., Latheef, S. K., Munjal, A., Khandia, R., Samad, H. A., Iqbal, H. M., & Joshi, S. K. (2017). Probiotics in curing allergic and inflammatory conditions—Research progress and futuristic vision. *Recent Patents on Inflammation & Allergy Drug Discovery, 10*(2), 105–118. https://doi.org/10.2174/1872213x10666161226162229

Didari, T., Solki, S., Mozaffari, S., Nikfar, S., & Abdollahi, M. (2014). A systematic review of the safety of probiotics. *Expert Opinion on Drug Safety, 13*(2), 227–239. https://doi.org/10.1517/14740338.2014.872627

Fiore, E., Van Tyne, D., & Gilmore, M. S. (2019). Pathogenicity of enterococci. *Microbiology Spectrum, 7*(4). https://doi.org/10.1128/microbiolspec.gpp3-0053-2018

Ghazanfar, S. (2021). Understanding the mechanism of action of indigenous target probiotic yeast: Linking the manipulation of gut microbiota and performance in animals. In *Saccharomyces*. IntechOpen.

Ghazanfar, S., Riaz, A., Ali, G. M., Naveed, S., Arif, I., Irshad, S., … Manzoor, K. N. (2019). Common methods to understand and develop indigenous probiotics yeast for ruminant. In *Yeasts in biotechnology*. IntechOpen.

Gómez-Llorente, C., Muñoz, S. G., & Gil, Á (2010). Role of toll-like receptors in the development of immunotolerance mediated by probiotics. *Proceedings of the Nutrition Society, 69*(3), 381–389. https://doi.org/10.1017/s0029665110001527

Gueimonde, M., Sánchez, B., De Los Reyes-Gavilán, C. G., & Margollés, A. (2013). Antibiotic resistance in probiotic bacteria. *Frontiers in Microbiology, 4*. https://doi.org/10.3389/fmicb.2013.00202

He, T., Priebe, M., Zhong, Y., Huang, C., Harmsen, H. J. M., Raangs, G. C., Antoine, J., Welling, G. W., & Vonk, R. J. (2007). Effects of yogurt and bifidobacteria supplementation on the colonic microbiota in lactose-intolerant subjects. *Journal of Applied Microbiology, 0*(0). https://doi.org/10.1111/j.1365-2672.2007.03579.x

Hemalatha, R., Ouwehand, A. C., Saarinen, M., Prasad, U., Swetha, K., & Bhaskar, V. V. (2017). Effect of probiotic supplementation on total lactobacilli, bifidobacteria and short chain fatty acids in 2–5-year-old children. *Microbial Ecology in Health and Disease, 28*(1), 1298340. https://doi.org/10.1080/16512235.2017.1298340

Hesla, H. M., Stenius, F., Jäderlund, L., Nelson, R. M., Engstrand, L., Alm, J., & Dicksved, J. (2014). Impact of lifestyle on the gut microbiota of healthy infants and their mothers—The ALADDIN birth cohort. *FEMS Microbiology Ecology, 90*(3), 791–801. https://doi.org/10.1111/1574-6941.12434

Hur, K. Y., & Lee, M. (2015). Gut microbiota and metabolic disorders. *Diabetes & Metabolism Journal, 39*(3), 198. https://doi.org/10.4093/dmj.2015.39.3.198

Idrees, M., Atiq, N., Zahra, R., Imran, M., & Ghazanfar, S. (2023). Draft genome sequence of *Lactiplantibacillus plantarum* subsp. Plantarum strain HF43, a human gut-associated potential probiotic. *Microbiology Resource Announcements, 12*(9), e00945-22.

Jiménez, E., Fernández, L., Marín, M., MartíN, R., Odriozola, J. M., Nueno-Palop, C., Narbad, A., Olivares, M., Xaus, J., & Rodríguez, J. M. (2005). Isolation of commensal bacteria from umbilical cord blood of healthy neonates born by cesarean section. *Current Microbiology, 51*(4), 270–274. https://doi.org/10.1007/s00284-005-0020-3

Jiménez-Truque, N., Tedeschi, S. K., Saye, E. J., McKenna, B. D., Langdon, W., Wright, J. P., Alsentzer, A., Arnold, S. R., Saville, B. R., Wang, W., Thomsen, I., & Creech, C. B. (2012). Relationship between maternal and neonatal *Staphylococcus aureus* colonization. *Pediatrics, 129*(5), e1252–e1259. https://doi.org/10.1542/peds.2011-2308

Kim, D., & Jin, Y. (2001). Intestinal bacterial β-glucuronidase activity of patients with colon cancer. *Archives of Pharmacal Research, 24*(6), 564–567. https://doi.org/10.1007/bf02975166

Ko, J. S. (2020). Mechanisms of action of probiotics. *Advances in Nutrition, 11*(4), 1054. https://doi.org/10.1093/advances/nmaa042

Kumar, M., Dhaka, P., Vijay, D., Vergis, J., Mohan, V., Kumar, A., Kurkure, N. V., Barbuddhe, S. B., Malik, S., & Rawool, D. B. (2016). Antimicrobial effects of *Lactobacillus plantarum* and *Lactobacillus acidophilus* against multidrug-resistant enteroaggregative *Escherichia coli*. *International Journal of Antimicrobial Agents, 48*(3), 265–270. https://doi.org/10.1016/j.ijantimicag.2016.05.014

Kumar, M., Nagpal, R., Verma, V., Kumar, A., Kaur, N., Hemalatha, R., Gautam, S. K., & Singh, B. (2012). Probiotic metabolites as epigenetic targets in the prevention of colon cancer. *Nutrition Reviews, 71*(1), 23–34. https://doi.org/10.1111/j.1753-4887.2012.00542.x

Mantegazza, C., Molinari, P., D'Auria, E., Sonnino, M., Morelli, L., & Zuccotti, G. V. (2018). Probiotics and antibiotic-associated diarrhea in children: A review and new evidence on *Lactobacillus rhamnosus* GG during and after antibiotic treatment. *Pharmacological Research, 128*, 63–72. https://doi.org/10.1016/j.phrs.2017.08.001

Martin, P., Ford, A. C., Talley, N. J., Cremonini, F., Foxx–Orenstein, A. E., Brandt, L. J., & Quigley, E. M. (2008). The efficacy of probiotics in the treatment of irritable bowel syndrome: A systematic review. *Gut, 59*(3), 325–332. https://doi.org/10.1136/gut.2008.167270

MartíN, R., Langa, S., Reviriego, C., Jiménez, E., Marín, M., Olivares, M., Bozá, J., Jiménez, J., Fernández, L., Xaus, J., & Rodríguez, J. M. D. (2004). The commensal microflora of human milk: New perspectives for food bacteriotherapy and probiotics. *Trends in Food Science and Technology, 15*(3–4), 121–127. https://doi.org/10.1016/j.tifs.2003.09.010

Mazhar, M. U., Naz, S., Khan, J. Z., Khalid, S., Ghazanfar, S., Selim, S., … Ansari, M. Y.. (2024). Safety evaluation and antioxidant potential of new probiotic strain bacillus subtilis (NMCC-path-14) in Balb/c mice by sub-acute repeated dose toxicity. *Heliyon, 10*(19), e38581.

Mechnikov, I. I., & Mitchell, P. C. (1910). *The Prolongation of Life : Optimistic Studies.* http://ci.nii.ac.jp/ncid/BA13679693

Mokoena, M. P. (2017). Lactic acid bacteria and their bacteriocins: Classification, biosynthesis and applications against uropathogens: A mini-review. *Molecules, 22*(8), 1255. https://doi.org/10.3390/molecules22081255

Olivares, M., Díaz-Ropero, M., Gómez, N., Lara-Villoslada, F., Sierra, S., Maldonado, J. A., MartíN, R., López-Huertas, E., Rodríguez, J. M., & Xaus, J. (2006). Oral Administration of two probiotic strains, *Lactobacillus gasseri* CECT5714 and *Lactobacillus coryniformis* CECT5711, enhances the intestinal function of healthy adults. *International Journal of Food Microbiology, 107*(2), 104–111. https://doi.org/10.1016/j.ijfoodmicro.2005.08.019

Pandey, S. S., Jain, R., Bhardwaj, P., Thakur, A. S., Kumari, M., Bhushan, S., & Kumar, S. (2022). Plant probiotics—Endophytes pivotal to plant health. *Microbiological Research, 263*, 127148. https://doi.org/10.1016/j.micres.2022.127148

Pavlović, N., Stankov, K., & Mikov, M. (2012). Probiotics—Interactions with bile acids and impact on cholesterol metabolism. *Applied Biochemistry and Biotechnology, 168*(7), 1880–1895. https://doi.org/10.1007/s12010-012-9904-4

Plaza-Díaz, J., Robles-Sanchez, C., Abadía-Molina, F., Sáez-Lara, M. J., Vilchez-Padial, L. M., Gil, Á, Gómez-Llorente, C., & Fontana, L.. (2017). Gene expression profiling in the intestinal mucosa of obese rats administered probiotic bacteria. *Scientific Data, 4*(1). https://doi.org/10.1038/sdata.2017.186

Plaza-Díaz, J., Ruiz-Ojeda, F. J., Gil-Campos, M., & Gil, Á (2018). Immune-mediated mechanisms of action of probiotics and synbiotics in treating pediatric intestinal diseases. *Nutrients, 10*(1), 42. https://doi.org/10.3390/nu10010042

EFSA Panel on Dietetic Products, Nutrition and Allergies (NDA) (2016). Guidance on the scientific requirements for health claims related to the immune system, the gastrointestinal tract and defence against pathogenic microorganisms. *EFSA Journal, 14*(1). https://doi.org/10.2903/j.efsa.2016.4369

Rabetafika, H., Razafindralambo, A., Ebenso, B., & Razafindralambo, H. (2023). Probiotics as antibiotic alternatives for human and animal applications. *Encyclopedia, 3*(2), 561–581. https://doi.org/10.3390/encyclopedia3020040

Rather, I. A., Bajpai, V. K., Kumar, S., Lim, J., Paek, W. K., & Park, Y. (2016). Probiotics and atopic dermatitis: An overview. *Frontiers in Microbiology, 7.* https://doi.org/10.3389/fmicb.2016.00507

Ribet, D., & Cossart, P. (2015). How bacterial pathogens colonize their hosts and invade deeper tissues. *Microbes and Infection, 17*(3), 173–183. https://doi.org/10.1016/j.micinf.2015.01.004

Sáez-Lara, M. J., Gómez-Llorente, C., Plaza-Díaz, J., & Gil, Á (2015). The role of probiotic lactic acid bacteria and bifidobacteria in the prevention and treatment of inflammatory bowel disease and other related diseases: A systematic review of randomized human clinical trials. *BioMed Research International, 2015*, 1–15. https://doi.org/10.1155/2015/505878

Sanders, M. E., Guarner, F., Guerrant, R. L., Holt, P. R., Quigley, E. M. M., Balfour Sartor, R., Sherman, P. M., & Mayer, E. A. (2013). An update on the use and investigation of probiotics in health and disease. *Gut, 62*, 787–796. https://doi.org/10.1136/gutjnl-2012-302504

Schneider, S., Girard-Pipau, F., Filippi, J., Hébuterne, X., Moyse, D., Hinojosa, G. C., Pompei, A., & Rampal, P. (2005). Effects of *Saccharomyces boulardiion* fecal short-chain fatty acids and microflora in patients on long-term total enteral nutrition. *World Journal of Gastroenterology, 11*(39), 6165.

Szajewska, H. (2015). What are the indications for using probiotics in children? *Archives of Disease in Childhood, 101*(4), 398–403. https://doi.org/10.1136/archdischild-2015-308656

Thygesen, J. B., Glerup, H., & Tarp, B. (2012). *Saccharomyces boulardii* fungemia caused by treatment with a probioticum. *Case Reports*, *2012*, bcr0620114412. https://doi.org/10.1136/bcr.06.2011.4412

Tillisch, K., Labus, J. S., Kilpatrick, L. A., Jiang, Z., Stains, J., Ebrat, B., Guyonnet, D., Legrain–Raspaud, S., Trotin, B., Naliboff, B. D., & Mayer, E. A. (2013). Consumption of fermented milk product with probiotic modulates brain activity. *Gastroenterology*, *144*(7), 1394–1401.e4. https://doi.org/10.1053/j.gastro.2013.02.043

Umu, ÖC. O., Rudi, K., & Diep, D. B. (2017). Modulation of the gut microbiota by prebiotic fibres and bacteriocins. *Microbial Ecology in Health and Disease*, *28*(1), 1348886. https://doi.org/10.1080/16512235.2017.1348886

Van Tassell, M. L., & Miller, M. J. (2011). Lactobacillus adhesion to mucus. *Nutrients*, *3*(5), 613–636. https://doi.org/10.3390/nu3050613

Vieira, A. T., Fukumori, C., & Ferreira, C. M. (2016). New insights into therapeutic strategies for gut microbiota modulation in inflammatory diseases. *Clinical & Translational Immunology*, *5*(6). https://doi.org/10.1038/cti.2016.38

Yang, I., Corwin, E. J., Brennan, P. A., Jordan, S., Murphy, J., & Dunlop, A. L. (2016). The infant microbiome. *Nursing Research*, *65*(1), 76–88. https://doi.org/10.1097/nnr.0000000000000133

6 Recent Advances on New Classes of Probiotic Applications viz. Postbiotics, Parabiotic, Pharmacobiotics, and Immunobiotics

Noshaba Rashid, Maliha Rashid, Abdul Nafeh, Saiqa Noureen, Sharjeel Khalid, Shakira Ghazanfar, Mehraj Gul, and Hamza Ali

6.1 INTRODUCTION

Probiotics are defined as "live **microorganisms which when administered in adequate amounts confer health benefits to the host**" (Swanson et al. 2020). Probiotics have various clinical uses. They are non-pathogenic microbes that are sometimes part of our gastrointestinal tract flora. They are used for both gastrointestinal and non-gastrointestinal diseases. For instance, recent studies established the role of probiotics in the treatment of acute diarrhea (Ghazanfar et al. 2019).

Probiotics have been found to have a role in the management of dermatitis, genitourinary tract infections, and food allergies (Islam 2016). The use of probiotics for the promotion of health and wellness has increased worldwide (Suez et al. 2019). Their use in acute gastroenteritis and diarrhea due to *Clostridium difficile*, inflammatory bowel disease, neonatal sepsis, and respiratory tract infections has been established (Suez et al. 2019).

Probiotics are also used in the cosmetic industry and personal care products (Puebla-Barragan and Reid 2021). Inflammatory bowel disease, which is a functional disorder of the gut due to alterations in gut microbiota composition, generates symptoms. Probiotics improve the symptoms of inflammatory bowel disease by changing gut microbiota as compared to placebo (Dale et al. 2019). Probiotics have an important effect on the immune mechanisms, although the exact mechanism is not known (Ashique et al. 2024). The use of probiotics improves gut function, increases mucus secretion, and prevents the destruction of tight junctions by decreasing the number of lipopolysaccharides (Cristofori et al. 2021).

The International Scientific Association of Probiotics and Prebiotics (ISAPP) defined postbiotics as the "preparation of inanimate microorganisms and/or their components that confers a health benefit on the host." Postbiotics are significant primarily in pediatric settings because they don't contribute to local inflammation. Both postbiotics and parabiotics promote health and guard against gastrointestinal illnesses (Swanson et al. 2020). A new class of probiotics called immunobiotics works to improve health by altering the immune system (Zaman et al. 2019). Pharmacobiotics are the targeted administration of live probiotic cultures (Vitetta et al. 2015). This chapter focuses on the recent advances in the new categories of probiotics.

DOI: 10.1201/9781003399797-7

6.2 ROLE OF PROBIOTICS IN GASTROINTESTINAL DISORDERS AND RECENT PROGRESS IN THE NEW AVENUES OF PROBIOTICS

Probiotics have efficacy in the treatment of many gastrointestinal diseases (Zaman et al. 2019). They are effective in *Clostridium difficile* diarrhea and other acute diarrheal illnesses in children. Probiotics also have a role in the remission of ulcerative colitis and pouchitis. Moreover, probiotics have a role in the prevention of Crohn's disease after surgical treatment (Cristofori et al. 2021). Strains of probiotics, *Firmicutes* and *Bacteroidetes*, improve the health of gastrointestinal tract. Due to their robust role in defense mechanisms, they promote oral as well as nutritional health. All these properties depend on the type of antimicrobial strain. Lactic acid bacteria and Bifidobacteria are the most common microbes used as probiotics (Ranjha et al. 2021). Their role as probiotics is established, as lactic acid bacteria have antibacterial potential and have no adverse effects (Li et al. 2020).

Moreover, probiotics also help in protein digestion. For instance, probiotics have a role in regulating intestinal microbiota and hence influence intestinal microbes related to proteolysis. Probiotics can also enhance the digestive activity and peptidase activity in the host. Probiotics reduce the dangerous fermentation of proteins and decrease their toxicity (Wang and Ji 2019).

The probiotic industry has surged in the last few years due to a rapid increase in knowledge about gut microflora. Use of commensal bacteria as probiotics ensures a healthy gut, opening up a new kind of probiotic called *next-generation probiotics* (Martín and Langella 2019).

6.3 NEXT-GENERATION PROBIOTICS

Next-generation probiotic is defined as "live microorganism determined by the comparative evaluations when administered in sufficient amounts provide health benefits to the host" (Martín and Langella 2019). *Faecalibacterium prausnitzii*, *Akkermansia muciniphila*, and *Eubacterium hallii* have important roles in next-generation probiotics. They have a promising role in the prevention and management of imbalances in microbiota-related diseases (Almeida et al. 2020).

Next-generation probiotics are found to have a role in the management of obesity. Obesity is multifactorial, but recently it has been found that gut microflora is the key contributing factor responsible for its pathogenesis. Modulation of intestinal flora is the key factor in the multifarious management of obesity (Breton et al. 2022). Gut microflora and its relation to cancer have been studied in recent years. Next-generation probiotics, *Bacteroides fragilis,* are non-toxigenic strains having a beneficial role in the development of cancer (Kaźmierczak et al. 2022).

Many gut microflora have been found to have positive contributions toward health that are considered to be next-generation probiotics. Four of these, *Ruminococcus bromii, Akkermansia muciniphila, Roseburia intestinalis,* and *Faecalibacterium prausnitzii*, next-generation probiotics that are derived from the gut, have a role in the treatment of metabolic diseases (Kumari et al. 2021).

New avenues for probiotic research are being made possible by recent and continuing advancements in the field of microbiome science. The scientific understanding of these interventions, as well as their applications in nutrition and healthcare, could be altered by novel types, processes, and uses that are presently being investigated. The burgeoning domains of microbiome-focused therapies and the dynamic landscape of implementation in the regulatory, policy, prescriber, and consumer domains herald a momentous period of transformation (Cunningham et al. 2021).

6.4 NEW CATEGORIES OF PROBIOTICS

Probiotics definition is evolved over time. Nowadays, the most accepted definition is that of the International Scientific Association of Probiotics and Prebiotics (ISAPP) in 2014. It states "live microorganisms that, when administered in adequate amounts, confer a health benefit on the host." The concept has evolved. So, various categories of probiotics have been created according to their potential health benefits and the systems involved (Martín and Langella 2019).

Postbiotics are defined as "non-viable bacterial products or metabolic products from microorganisms that have biologic activity in the host." Parabiotic states: "Non-viable microbial cells (intact or broken) or crude cell extracts which when administered (orally or topically) in adequate amounts, confer a benefit to the human or animal consumer." Pharmabiotics are "bacterial cells of human origin, or their products, with a proven pharmacological role in health or disease." Immunobiotics improve health by targeting the mucosal immune level (Martín and Langella 2019).

6.5 PROBIOTICS AND POSTBIOTICS

The use of probiotics and its role in health maintenance cannot be negated. Probiotics provide health benefits and treat various diseases but have adverse effects in people with immunosuppression because they are biologically active substances. To overcome this difficulty, postbiotics were used. Postbiotics, metabolites of nonliving probiotics, boost the immune system and diminish the adverse effects of probiotics. Postbiotics are used in the food and pharmaceutical industries and also play an immune-boosting role (Yeşilyurt et al. 2021). Postbiotics have antioxidant, immunomodulation, and anti-inflammatory as well as anticancer properties (Żółkiewicz et al. 2020). Probiotics improve oral health. They decrease tooth decay, fungal infections in the oral cavity, and inflammation of gums. Studies suggest that the use of postbiotics instead of probiotics is equally effective against periodontal infections (Homayouni Rad et al. 2023).

Postbiotics are also used in the treatment of iron deficiency anemia. They increase the bioavailability of iron absorption, decreasing the damage to intestinal microflora. Moreover, the use of postbiotics in colorectal cancer is promising. Colorectal carcinoma is the leading cause of death worldwide. Colorectal cancer is associated with microbiota imbalance (dysbiosis). Administration of probiotics and postbiotics is a better anticancer strategy that functions by regeneration of intestinal microflora and enhances the intestinal barrier, boosting immune functions (Kvakova et al. 2022).

Postbiotics improve dermal health by maintaining the barrier function of the skin, reducing inflammation, and improving the appearance of acne-prone skin (De Almeida et al. 2023). Postbiotics improve sleep and decrease sleep disorders, although more studies are required in this field. Despite the role of probiotics in the betterment of human health, they had certain limitations at therapeutic levels. So, in recent years, the focus has shifted from live "*probiotics*" to nonliving, biologically activated "*Parabiotic and Postbiotics*."

6.6 PARABIOTIC, IMMUNOBIOTICS, AND PHARMABIOTICS

The terms "*Postbiotics*" and "*Paraprobiotics*," which refer to nonviable microorganisms or bacterial-free extracts that may help the host by providing extra bioactivities beyond probiotics, have gained popularity in recent years. The components of bacterial cell walls that make up "*paraprobiotics*" include peptidoglycans, teichoic acids, polysaccharides, and cell surface proteins. A wide range of processes, including heat treatments, high pressure, UV radiation, sonication, ionizing radiation, and pH manipulation, are used to create paraprobiotics. Parabiotics have anti-inflammatory, antioxidant, immunomodulatory, anticancer, and antimicrobial activity (Cuevas-González et al. 2020). Probiotics have certain limitations in viability control. So, focus is shifting to "*Parabiotics*" in the food and drug industry.

Immunobiotics are specific antimicrobial strains that drive mucosal immune mechanisms. Viral respiratory diseases have high mortality and morbidity. Best example is the SARS-CoV-2 virus, which has infected more than 4 million people. Research has reported the role of intestinal microbiota in the immunity of the respiratory system. Immunobiotics have a favorable response against the SARS-CoV-2 virus. They worked by increasing interferon, stimulating type 1 helper T-lymphocyte response, producing antibody, and regulating coagulation cascade (Villena and Kitazawa 2020). Immunobiotics are effective against the influenza virus, respiratory syncytial virus, and secondary pneumococcal pneumonia (Tonetti et al. 2023). Immunobiotic *Lactobacillus rhamnosus* CRL1505 prevents inflammatory response in the respiratory tract. In short, immunobiotics modulate the immune response in infectious and inflammatory diseases of bacterial, viral, and fungal etiology.

Pharmabiotics are probiotic products with proven pharmacological action on health or disease (Chaudhari and Dwivedi 2022). Pharmacobiotics play an important role in the maintenance of health. Imbalance in gut microbiota is related to various diseases, including Crohn's and ulcerative colitis and other functional disorders of the gastrointestinal tract. Pharmacobiotic, *Lactobacillus acidophilus*, is effective and safe in the treatment of chronic diarrheal illness, acute diarrhea, and antibiotic-associated diarrhea even in patients with suppressed immune systems (Remes Remes-Troche et al. 2020). Pharmacobiotics also play a role in managing coronavirus and Parkinson's by regulating micobiome through various mechanisms (Danilenko et al. 2021). These days, the goal of probiotics for human health has shifted from gut stability to a broader definition of Pharmabiotics, such as managing disease (such as obesity, immune system, and mental stabilization, among others).

6.7 CONCLUSION

In recent years, research about probiotics has been increased tremendously. The probiotic is now shifting toward Parabiotic, Immunobiotics, and Pharmabiotics, and researchers worldwide are actively looking at new technologies in next-generation probiotics and prebiotics. There is a global surge in the use of probiotics as a component of food, pharmaceutical products, and animal feed. Probiotics have a potential role in the therapeutics of acute infections and many diseases. New subdivisions are discovered. Last, probiotics can be used to treat a lot of health conditions, and the discovery of new classes of probiotics is even more promising.

REFERENCES

Almeida, Diana, Daniela Machado, José Carlos Andrade, Sónia Mendo, Ana Maria Gomes, and Ana Cristina Freitas. "Evolving trends in next-generation probiotics: A 5W1H perspective." Critical reviews in food science and nutrition 60, no. 11 (2020): 1783–96.

Ashique, S., M. Faiyazuddin, M. Bhowmick, S. Ghazanfar, N. Kumar, P. Kumar, and R Sah11. (2024). Unlocking the Secrets of Probiotics – A Therapeutic Breakthrough for Major Depressive Disorders.

Breton, Jonathan, Marie Galmiche, and Pierre Déchelotte. "Dysbiotic gut bacteria in obesity: An overview of the metabolic mechanisms and therapeutic perspectives of next-generation probiotics." Microorganisms 10, no. 2 (2022): 452.

Chaudhari, Archana, and Mitesh Kumar Dwivedi. "The concept of probiotics, prebiotics, postbiotics, synbiotics, nutribiotics, and pharmabiotics." In Probiotics in the prevention and management of human diseases, 1–11. Academic Press, 2022.

Cristofori, Fernanda, Vanessa Nadia Dargenio, Costantino Dargenio, Vito Leonardo Miniello, Michele Barone, and Ruggiero Francavilla. "Anti-inflammatory and immunomodulatory effects of probiotics in gut inflammation: A door to the body." Frontiers in immunology 12 (2021): 578386.

Cuevas-González, P. F., A. M. Liceaga, and J. E. Aguilar-Toalá. "Postbiotics and paraprobiotics: From concepts to applications." Food research international 136 (2020): 109502.

Cunningham, Marla, M. Andrea Azcarate-Peril, Alan Barnard, Valerie Benoit, Roberta Grimaldi, Denis Guyonnet, and Hannah D. Holscher et al. "Shaping the future of probiotics and prebiotics." Trends in microbiology 29, no. 8 (2021): 667–85.

Dale, Hanna Fjeldheim, Stella Hellgren Rasmussen, Özgün Ömer Asiller, and Gülen Arslan Lied. "Probiotics in irritable bowel syndrome: An up-to-date systematic review." Nutrients 11, no. 9 (2019): 2048.

Danilenko, Valery, Andrey Devyatkin, Mariya Marsova, Madina Shibilova, Rustem Ilyasov, and Vladimir Shmyrev. "Common inflammatory mechanisms in COVID-19 and Parkinson's diseases: The role of microbiome, pharmabiotics and postbiotics in their prevention." Journal of inflammation research 14 (2021): 6349.

De Almeida, Carolina Vieira, Emiliano Antiga, and Matteo Lulli. "Oral and topical probiotics and postbiotics in skincare and dermatological therapy: A concise review." Microorganisms 11, no. 6 (2023): 1420.

Ghazanfar, S., A. Riaz, M. N. Tahir, S. Maqbool, G. M. Ali, F. Tariq, and I Arif. Probiotic supplement improves the health status and lactation performance in dairy animals. In Lactation in farm animals-biology, physiological basis, nutritional requirements, and modelization. IntechOpen, 2019.

Homayouni Rad, Aziz, Hadi Pourjafar, and Esmaeel Mirzakhani. "A comprehensive review of the application of probiotics and postbiotics in oral health." Frontiers in cellular and infection microbiology 13 (2023): 1120995.

Islam Saif, Ul. "Clinical uses of probiotics." Medicine 95, no. 5 (2016): e2658.

Kaźmierczak-Siedlecka, Karolina, Karolina Skonieczna-Żydecka, Theodore Hupp, Renata Duchnowska, Natalia Marek-Trzonkowska, and Karol Połom. "Next-generation probiotics – Do they open new therapeutic strategies for cancer patients?" Gut microbes 14, no. 1 (2022): 2035659.

Kumari, Manorama, Parul Singh, Basavaprabhu H. Nataraj, Anusha Kokkiligadda, Harshita Naithani, Syed Azmal Ali, Pradip V. Behare, and Ravinder Nagpal. "Fostering next-generation probiotics in human gut by targeted dietary modulation: An emerging perspective." Food research international 150 (2021): 110716.

Kvakova, Monika, Anna Kamlarova, Jana Stofilova, Veronika Benetinova, and Izabela Bertkova. "Probiotics and postbiotics in colorectal cancer: Prevention and complementary therapy." World journal of gastroenterology 28, no. 27 (2022): 3370.

Li, Miao, Yi Wang, Hongyu Cui, Yongfeng Li, Yuan Sun, and Hua-Ji Qiu. "Characterization of lactic acid bacteria isolated from the gastrointestinal tract of a wild boar as potential probiotics." Frontiers in veterinary science 7 (2020): 49.

María Remes-Troche, José, Enrique Coss-Adame, Miguel Ángel Valdovinos-Díaz, Octavio Gómez-Escudero, José María Eugenia Icaza-Chávez, Flora Antonio Chávez-Barrera, and Zárate-Mondragón et al. "*Lactobacillus acidophilus* LB: A useful pharmabiotic for the treatment of digestive disorders." Therapeutic advances in gastroenterology 13 (2020): 1756284820971201.

Martín, Rebeca, and Philippe Langella. "Emerging health concepts in the probiotics field: Streamlining the definitions." Frontiers in microbiology 10 (2019): 1047.

Puebla-Barragan, Scarlett, and Gregor Reid. "Probiotics in cosmetic and personal care products: Trends and challenges." Molecules 26, no. 5 (2021): 1249.

Ranjha, Muhammad Modassar Ali Nawaz, Bakhtawar Shafique, Maria Batool, Przemysław Łukasz Kowalczewski, Qayyum Shehzad, Muhammad Faisal Muhammad Usman, Syeda Mahvish Manzoor, Shazia Zahra, Rana Muhammad Yaqub, and Aadil. "Nutritional and health potential of probiotics: A review." Applied sciences 11, no. 23 (2021): 11204.

Suez, Jotham, Niv Zmora, Eran Segal, and Eran Elinav. "The pros, cons, and many unknowns of probiotics." Nature medicine 25, no. 5 (2019): 716–29.

Swanson, Kelly S., Glenn R. Gibson, Robert Hutkins, Raylene A. Reimer, Gregor Reid, Kristin Verbeke, and Karen P. Scott et al. "The international scientific association for probiotics and prebiotics (ISAPP) consensus statement on the definition and scope of synbiotics." Nature reviews gastroenterology & hepatology 17, no. 11 (2020): 687–701.

Tonetti, F. Raya, M. Tomokiyo, K. Fukuyama, M. Elean, R. Ortiz Moyano, H. Yamamuro, and R. Shibata et al. "Post-immunobiotics increase resistance to primary respiratory syncytial virus infection and secondary pneumococcal pneumonia." Beneficial microbes (2023): 1–14.

Villena, Julio, and Haruki Kitazawa. "The modulation of mucosal antiviral immunity by immunobiotics: Could they offer any benefit in the SARS-CoV-2 pandemic." Frontiers in physiology 11 (2020): 699.

Vitetta, Luis, Talia Palacios, Sean Hall, and Samantha Coulson. "Gastrointestinal tract commensal bacteria and probiotics: Influence on end-organ physiology." Novel natural products: Therapeutic effects in pain, arthritis and gastro-intestinal diseases (2015): 1–33.

Wang, Jing, and Haifeng Ji. "Influence of probiotics on dietary protein digestion and utilization in the gastrointestinal tract." Current protein and peptide science 20, no. 2 (2019): 125–31.

Yeşilyurt, Neslihan, Birsen Yılmaz, Duygu Ağagündüz, and Raffaele Capasso. "Involvement of probiotics and postbiotics in the immune system modulation." Biologics 1, no. 2 (2021): 89–110.

Zaman, S., M. Gohar, M. Tubassam, M. Qubtia, S. Ghazanfar, and M Imran. (2019). Isolation and Characterization of Potential Probiotic Bacteria from Sahiwal Cow Gut.

Żółkiewicz, Jakub, Aleksandra Marzec, Marek Ruszczyński, and Wojciech Feleszko. "Postbiotics – A step beyond pre-and probiotics." Nutrients 12, no. 8 (2020): 2189.

7 Challenges Faced by Researchers, Clinicians, and People in General Regarding Decision-Making in Choosing or Prescribing a Probiotic for Special Health Reasons

Ayesha Khan, Shakira Ghazanfar, and Mohammad Moeen Badar

7.1 INTRODUCTION

Human gut microbiota has been known to mediate the internal environment of the gut and intestine and plays a major role in maintaining human health by increasing metabolism and immunity. Such benefactory bacteria have often been termed as "beneficial bacteria" due to their health-promoting effects (Ghazanfar et al. 2024). There are about 100–1000 bacterial species comprising the gut microbiota (Kerry et al. 2018). To enhance the abilities of gut microbiota in an efficient yet cost-effective manner, recent research has focused on the formulation of probiotics.

The word "Probiotics" is a Greek word that means "for life" (Soccol et al. 2010). The term "Probiotic" was first introduced in 1953 by German scientist Werner Kollath to describe the active compounds that are necessary for a healthy development of life (Gasbarrini, Bonvicini, and Gramenzi 2016). Then, it was used by Lilly and Stillwell in 1965 (Soccol et al. 2010). They used it to describe the secretions of one microorganism that would favor the growth of another microorganism (Soccol et al. 2010). The present-day definition of probiotics is very much close to what Fuller adopted in 1992, that is, a live microbial feed that benefits the host by improving its intestinal microbial balance (Gasbarrini, Bonvicini, and Gramenzi 2016). The modern-day recognition of the potential health benefits of fermented foods is often attributed to Elie Metchnikoff, a Russian immunologist, in the early 20th century. He tried to find out the effect of probiotics on human health (Gasbarrini, Bonvicini, and Gramenzi 2016). Metchnikoff observed that Bulgarian peasants, who consumed large amounts of fermented milk, tended to live longer. He hypothesized that the lactic acid bacteria in fermented milk contributed to their health and longevity (Gasbarrini, Bonvicini, and Gramenzi 2016).

The goal of probiotic use is to reestablish microbial balance. Within the adult gastrointestinal (GI) tract, there are over 500 distinct types of bacteria. While many microorganisms are thought to be harmful to humans, others are thought to be advantageous. Although surgery, immunosuppressive drugs, antibiotics, and radiation can increase the number of harmful bacteria and upset the balance of the gut flora, overall, the equilibrium is kept in check. Probiotics have the potential to correct the microbial balance in the GI system since they include yeast and helpful bacteria. For probiotics to be effective, they need to have specific qualities. Probiotic organisms have to be resilient enough to endure digestion (i.e., be able to withstand the breakdown of bile and acid), colonize and multiply in the gut, cling to the intestinal epithelium, and maintain homeostasis of

DOI: 10.1201/9781003399797-8

TABLE 7.1
Historical Fermented Products

Food Origin	Region	Year
Mushrooms	China	4000 BC
Soy sauce	China, Korea, Japan	3000 BC
Wine	North Africa, Europe, Middle East	3000 BC
Fermented milk	Middle East	10,000 BC
Fermented rice	Asia	2000 BC
Fermented honey	North Africa, Middle East	2000 BC
Fermented milk products	Egypt, Italy, Greece	7000–5000 BC
Bread	Europe, Egypt	1500 BC
Fermented meat	Middle East	1500 BC
Cheese	Middle East, China	2000 BC
Pickled vegetables	Europe, China	1000 BC
Tea	China	200 BC

the gut microbiota (Ghazanfar and Arif 2019). Probiotic strains must also not possess pathogenic qualities, be effective and secure in humans, and continue to exist during the product's shelf life. Goods with many organisms are especially desirable for two reasons: probiotic mixes may work in concert to reduce infections, and certain patients may only colonize with one strain of the organism (Williams 2010).

Bifidobacterium and *Lactobacillus* are among the popular strains that are used as probiotics (Soccol et al. 2010). Some strains of yeast, *Pediococcus* sp., and *Bacillus* sp. can also be used. The history of probiotics dates back to ancient civilizations of China, Rome, Greece, Middle East, Egypt, and the Indus Valley where there have been accounts of manufacture of cheese, yogurt, pickled foods, beer, etc. (Gasbarrini, Bonvicini, and Gramenzi 2016). Undoubtedly, the creation of several conventional sour milk or fermented dairy products, such as leben, kefir, koumiss, and dahi, was aided by favorable climate circumstances. These products, the majority of which continue to be widely consumed, were frequently used medicinally prior to the discovery of germs (Soccol et al. 2010). A list of some traditionally fermented foods is given in Table 7.1.

Probiotics can be derived from both dairy and nondairy sources and are usually consumed after antibiotic consumption in order to aggravate their effects on the human immune system by replenishing the population of gut microbiota, as the beneficial bacteria are indiscriminately destroyed by antibiotics while targeting the pathogenic bacteria, and for speedy recovery of bodily functions (Soccol et al. 2010).

However, even with the encouraging findings, further research is essential to explore the role of probiotics in human health and ensure the safety of their use. The specific characteristics required for their optimal functionality in the gut is not fully understood yet (Kechagia et al. 2013).

The swift advancement of high-throughput sequencing technology and bioinformatics has enhanced our comprehension of the connection between the human microbiome, intestinal microbiome, and human diseases. Consequently, numerous gut microbes, challenging to cultivate in vitro, have been successfully isolated and recognized. Several strains showing potential health benefits are steadily coming to light. These promising strains can serve as contenders for next-generation probiotics (NGPs) (Shahid et al. 2024) and may act as active biological agents in clinical applications for the precise treatment of specific diseases. Nevertheless, the precise mechanism underlying the interaction of NGPs with the host remains uncertain (Zhang et al. 2022) (Figure 7.1).

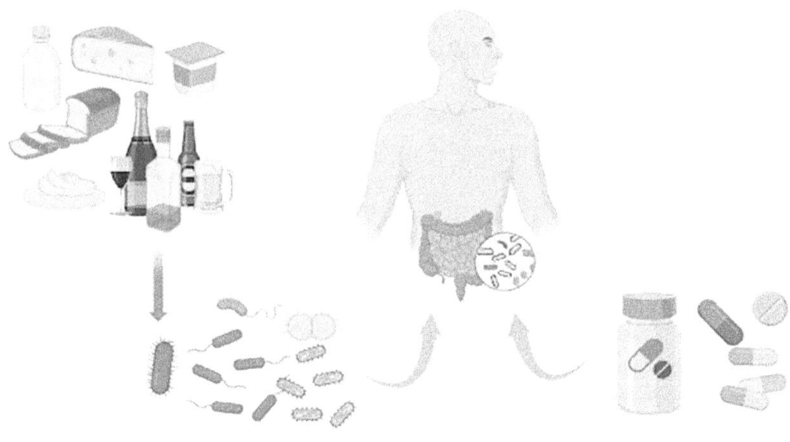

FIGURE 7.1 Sources of probiotics; Naturally occurring probiotics and next-generation probiotics (synthetic).

7.2 SIGNIFICANCE OF NGPs

In the past 15 years, advancements in microbiome research have transformed our understanding of human microbiology. We can now precisely characterize the composition and roles of microflora in various body regions, linking them to potential diseases, risks, and evident disease. Notably, probiotics and fecal microbiota transplantation (FMT) are recognized as promising solutions (Zhang et al. 2022).

Many of the probiotics available in the market belong to conventional strains such as *Lactobacillus*, *Streptococcus*, *Bifidobacterium*, yeast, and *Bacillus*. For example, *Streptococcus thermophilus*, found in yogurt, secretes β-galactosidase, which aids in breaking down lactose into easily digestible glucose and galactose—beneficial for individuals with lactose intolerance. However, it's important to note that in many countries, probiotics are not classified as drugs and are largely unregulated. Consequently, a variety of probiotic products are often directly marketed to consumers without robust evidence of their efficacy.

Despite the promise of probiotics, the existing research supporting their claims is variable, and a few clinical verifications have been undertaken. Additionally, since each individual possesses a unique intestinal microflora, the effects of different bacteria on diverse individuals are likely to vary. Consequently, further investigations are necessary to fully understand the potential effects of probiotics in the future.

The impact of traditional probiotics on the human intestinal microbiota is limited, and evaluating the efficacy of flora intervention should be conducted at the strain level. Approximately 80% of intestinal bacteria remain unidentified, and over 25 types of immune metabolic diseases are associated with the abundance of flora in the body. Certain species demonstrate greater resilience to inflammation, tumors, and metabolic diseases. However, minimal research has been conducted on them due to their extreme sensitivity to oxygen, making it challenging to obtain pure cultures (Zhang et al. 2022) (Table 7.2).

Through the advancements in culturomics, it has become possible to obtain pure bacterial cultures, including *Akkermansia muciniphila, Faecalibacterium prausnitzii, Bacteroides fragilis, Eubacterium hallii*, and *Roseburia* spp., in vitro. Termed next-generation probiotics (NGPs), these cultures contribute to an expanded understanding of the underlying mechanisms of probiotic-host health interactions. The formal introduction of the NGP concept occurred in 2017, with Nature Microbiology being the first to suggest it (Singh and Natraj 2021). NGPs are distinguished from

TABLE 7.2
Microorganisms Used as Probiotics

Lactobacillus Species	*Bifidobacterium* Species	Other Lactic Acid Bacteria	Non-Lactic Acid Bacteria
L. acidophilus	*B. adolescentis*	*Enterococcus faecalis*	*Saccharomyces cerevisiae*
L. casei	*B. animalis*	*E. faecim*	*S. boulardii*
L. crispatus	*B. bifidum*	*Lactococcus lactis*	
L. gasseri	*B. breve*	*Leuconostoc mesenteroides*	
L. johnsonii	*B. infantis*	*Pediococcus acidilactici*	
L. paracasei	*B. lactis*	*Sporolactobacillus inulinus*	
L. plantarum	*B. longum*	*Streptococcus thermophilus*	

traditional probiotics as they align with the attributes of "active biological agents," following the guidelines of the US Food and Drug Administration (FDA):

1. containing live organisms, such as bacteria
2. being applicable for the prevention, treatment, or cure of a disease or condition in human beings
3. not being a vaccine.

NGPs are required to undergo clinical trials (phases I–IV) and gain approval from relevant regulatory authorities before entering the market (Singh and Natraj 2021).

Presently, candidate strains of NGPs are extensively researched and documented. Each of these mechanisms may contribute to the onset and progression of various inflammatory diseases, though many are still in the early stages of mechanistic investigation. NGPs are reported to address the limitations of existing probiotic preparations and hold significant potential for intervening or treating human diseases in the future (Zhang et al. 2022).

7.3 ISOLATION OF NGPs

To identify NGPs, the first step is to determine the correlation between microbiota and host, such as the diseased, healthy, and experimental groups. This can be achieved through clinical and animal studies. After this, potential probiotics are characterized, and related metabolites are isolated for validation of studies. Till now, a large number of microbiota bacteria still cannot be cultured efficiently in vitro. Nevertheless, culturomics technology is advancing rapidly, making the specific isolation and culturing of target bacteria possible. Specific pathogen free (SPF) or germ-free (GF) mice models will then be used to study the in vivo efficacy of probiotics. This is done for proof of concept. After the studies are compiled, the functions, safety, and molecular mechanisms of the probiotics in animals are addressed, paving the way for clinical human studies.

Currently, the US FDA has initiated a program for live biotherapeutic products (LBPs) specifically to regulate the application, clinical trial, and commercialization of emerging probiotics. It is anticipated that in the near future, NGPs will become a reality in the health bio-industry (Chang et al. 2019).

7.4 UNDERSTANDING NGPs AS PERSONALIZED MEDICINE

In recent years, gut microbiota has significantly performed the function of biomarkers and indicators to describe GI health. The linkage between dysbiosis in gut microbiota and both GI and non-GI diseases is apparent in studies that examine associations across the entire microbiome (Singh and Natraj 2021). As the therapeutic and diagnostic approaches adopted for every patient are different, the response to particular therapy is not the same in each patient (Singh and Natraj 2021). This is where the concept of personalized medicine shows its perks.

NGPs are the tools for tailored medicines and precision therapeutics that can alleviate the disease by mediating gut microbiota as per the patient's needs. Two general approaches are used to develop NGPs for a patient.

The first approach involves identifying the strain linked to a specific health phenotype and confirming its ability to replicate the health phenotype in suitable experimental models. The second strategy entails identifying a potent molecule capable of alleviating the disease phenotype and incorporating it into a well-characterized probiotic strain, which can act as a delivery vehicle. Various bacterial strains have been studied as candidates for NGPs in diseases like cancer, obesity, GI diseases, etc.

7.5 POTENTIAL NGP CANDIDATES

7.5.1 *Akkermansia muciniphila*

A. muciniphila belongs to the phylum *Verrucomicrobia* and, due to its unique properties, is a key probiotic candidate. This bacterium is responsible for occupying up to 5% of the total microbiota, and it uses mucin as a source of nutrition for its proliferation (Cani and de Vos 2017). Various study reports show that we have observed that the composition of these microbiota can differ between the gut of obese/T2D individuals and those of lean/nondiabetic individuals. Currently, it is widely accepted that the usage of prebiotics such as the inulin-type fructans increases the abundance of *A. muciniphila*, which leads to the improvement of metabolic disorders and obesity (Everard et al. 2013). *A. muciniphila's* fundamental action is revealed to operate via an immunomodulatory protein called "Amuc_1100," situated in the bacterial outer membrane (Plovier et al. 2017). Other study reports show that *A. muciniphila* also regulates the endocannabinoid (eCB) system (Plovier et al. 2017). This is a crucial regulatory system that targets mainly obesity, type 2 diabetes and inflammation. The eCB system is also involved in the control of glucose levels and energy metabolism (Plovier et al. 2017). *A. muciniphila* may also involve anticancer immunotherapy, such as anti-PD-1 treatment (Imhann et al. 2018). In addition to clinical associations, transferring fecal microbiota from responders or non-responders to GF mice demonstrated that *A. muciniphila* enhanced the effectiveness of anti-PD-1 blockade, particularly when compared to mice receiving microbiota from non-responders (Wang et al. 2018). The mechanisms and reasons behind *A. muciniphila's* impactful effects are still not fully understood or characterized. Despite having many beneficial effects, some studies have shown a reverse situation where there is an increase in the abundance of *A. muciniphila* in HFD mice (Schneeberger et al. 2015) and the in diseases multiple sclerosis and Parkinson's disease (Heintz-Buschart et al. 2018).

7.5.2 *Faecalibacterium prausnitzii*

This is a Gram-positive bacterium of the family *Ruminococcaceae* (Heintz-Buschart et al. 2018). This species is the sole recognized member of the *Faecalibacterium* genus, constituting approximately 5% of the gut microbiota in healthy adults (Heintz-Buschart et al. 2018). It has the ability to ferment glucose and produce short chain fatty acids (SCFAs) like formic acid, D-lactate, butyrate, etc. Butyrate is essential for maintaining gut health and integrity, as it acts as a cellular mediator and promotes cell and tissue development (Bedford and Gong 2018). Moreover, there is a direct relation between the abundance of this *F. prausnitzii* and longer durations of progression-free survival (PFS) (Singh and Natraj 2021).

In contrast to *A. muciniphila, F. prausnitzii* lacks the ability to utilize mucin from the intestinal epithelium, but mucin can improve its abundance (Sadaghian Sadabad et al. 2015). High abundance of this bacterium can promote the resynthesis of mucin and tight junction proteins, repairing damaged intestinal mucosa (Martín et al. 2015). *F. prausnitzii* in association with *Bacteroides thetaiotaomicron* is involved in the regulation of the differentiation process of intestinal goblet cells and maintaining the mucous barrier (Wrzosek et al. 2013). This is why it is used as a biomarker for diagnosis of adolescents with ulcerative colitis and obesity (Chierico et al. 2018). *F. prausnitzii* A2-165 enhances the proliferation of ovalbumin-specific T cells, decreases the number of interferon-γ+ T cells, and causes the secretion of IL-10 by both human and mouse dendritic cells (Rossi et al. 2016). These findings generate enthusiasm for considering *F. prausnitzii* as an NGP. In individuals experiencing obesity, adipocytes undergo enlargement, resulting in local tissue hypoxia, cellular demise, infiltration of leukocytes, and the onset of systemic inflammation (Kanda et al. 2006). Mice treated with *F. prausnitzii* ATCC 27766 exhibited a decrease in the inflammatory marker CD45+, elevated adiponectin expression in visceral adipose tissues, and a reduction in inflammation within adipose tissues (Munukka et al. 2017). Nevertheless, due to the high sensitivity of this bacterium to oxygen and the stringent requirements for its cultivation, the exploration and application of *F. prausnitzii* face limitations. Consequently, researchers have sought to substitute live bacteria with the supernatant cultured by *F. prausnitzii* in numerous studies.

7.5.3 *Eubacterium hallii*

E. hallii is another candidate with great potential for probiotic formulations. It belongs to the phylum Firmicutes and is a Gram-positive anaerobic bacterium that can be found in murine and human feces (Louis et al. 2010). *E. hallii* is one of the few Clostridium cluster XIVa butyrate-producing bacterial strains capable of butyrate production from both acetate and lactate in a low pH environment (Duncan, Louis, and Flint 2004). Due to this very unique property, this bacterium plays a crucial role in the intestinal metabolic balance, considering that the accumulation of lactate can be linked to various intestinal disorders and might also endow competitive dominance over other microorganisms, as observed in batch fermentations of fecal and colon microbiota (Duncan, Louis, and Flint 2004; Fekry et al. 2016). Administering *E. hallii* has the potential to modify the function of the intestinal microbiome, with microbial metabolites likely playing a role in enhancing the metabolic phenotype. Despite considerable progress in understanding the relevance of this species, there's still much work to be done to grasp its true beneficial or potentially deleterious effects (Almeida et al. 2020).

7.5.4 *Bifidobacterium*

Bifidobacterium belongs to the phylum Actinobacteria and has been exploited as an NGP candidate (Turroni, Van Sinderen, and Ventura 2011). The oral delivery of a combination of *Bifidobacterium breve* and *Bifidobacterium longum* markedly enhances the effectiveness of anti-PD-L1 immunotherapy in a mouse model of anti-tumor treatment (Sivan et al. 2015). This increased efficacy is due to the activation of dendritic cells, leading to increased binding of CD8+ T-cell and their accumulation in the tumor microenvironment. *Bifidobacterium* species have also gained significant attention as candidates for cancer treatment, colitis, and colorectal cancer (CRC) therapy (Kahouli, Tomaro-Duchesneau, and Prakash 2013). In an alternative investigation, the use of symbionts (resistant starch and *Bifidobacterium lactis*) displayed mitigating effects on CRC development in a rat model induced with azoxymethane (Leu et al. 2010). Likewise, an intervention involving oligofructose-enriched inulin, *Bifidobacterium lactis*, and Lactobacillus *rhamnosus* significantly enhanced immune function in individuals with polypectomies and colon cancer (Roller et al. 2007).

7.5.5 *Prevotella copri*

As a prospective NGPs, *P. copri*, a member of the Bacteroidetes phylum, has been identified for its ability to ameliorate abnormal glucose tolerance syndromes and augment hepatic glycogen storage in animals. This effect is attributed to the production of succinate, an intermediate in the tricarboxylic acid (TCA) cycle, derived from the fermentation of dietary fibers. Succinate plays a crucial role in glucose homeostasis by influencing intestinal gluconeogenesis (Vadder et al. 2016). While *P. copri* has demonstrated positive effects in improving prediabetic syndromes, it has recently been indicated that *P. copri* and *Bacteroides vulgatus* can aggravate glucose tolerance and increase insulin resistance. These effects occur prior to the onset of ischemic cardiovascular disease and type 2 diabetes. The observed exacerbation is attributed to the synthesis of branched-chain fatty acids (BCAA) (Pedersen et al. 2016).

7.5.6 *Bacteroides fragilis*

B. fragilis is emerging as a noteworthy probiotic (Erturk-Hasdemir and Kasper 2018). This Gram-negative, strictly anaerobic commensal bacterium belongs to the genus Bacteroides in the phylum Bacteroidetes, constituting around 1% of the human intestinal microbiota (Huang, Lee, and Mazmanian 2011). *B. fragilis* strains containing enterotoxin (non-enterotoxin-producing *B. fragilis*, ETBF) significantly contribute to abdominal infections, leading to clinical abscesses from fecal spillage or bacteremia (Sears et al. 2008). Interestingly, ETBF strains have demonstrated various beneficial effects (Sears et al. 2008). Decades of research have highlighted *B. fragilis* polysaccharide A (PSA) as a prototypical bacterial capsular polysaccharide influencing interactions between microbiota and hosts (Sears et al. 2008). PSA stands out for its zwitterionic motifs, leading to the enrichment of a specific subset of anti-inflammatory memory CD4+FoxP3+ T cells through direct interactions with antigen-presenting cells like plasmacytoid dendritic cells (Sears et al. 2008). This mechanism contributes to the systemic alleviation of inflammation-related conditions such as abscesses, neuro-inflammations, and cancers (Lukiw 2016).

7.5.7 *Christensenella minuta*

C. minuta demonstrates potential probiotic effects in combating obesity and related metabolic disorders (Goodrich et al. 2014). This Gram-positive, nonspore-forming, anaerobic probiotic candidate has shown efficacy in reducing obesity and associated syndromes (Goodrich et al. 2014). The abundance of *Christensenellaceae* was observed to be higher in individuals with a low body mass index (BMI) (Goodrich et al. 2016). Further investigation revealed that *C. minuta* has the ability to mitigate an obesity-associated microbiome, leading to reduced body weight gain and alterations in the microbiome pattern of recipient mice (Singh and Natraj 2021).

7.5.8 *Parabacteroides goldsteinii*

The escalating obesity epidemic necessitates novel, safe, and effective anti-obesity measures that are widely accessible. *P. goldsteinii*, a bacterium, is emerging as a promising probiotic for combating obesity (Chang et al. 2015; Wu et al. 2019). Recent findings indicate that a water extract of the medicinal mushroom *Hirsutella sinensis* has notable anti-obesity effects, reducing obesity, inflammation, and insulin resistance in high-fat diet (HFD)-fed mice (Singh and Natraj 2021). A high molecular weight polysaccharide fraction (>300 kDa) from the water extract not only significantly decreases body weight by 50% but also mitigates intestinal permeability, metabolic endotoxemia, inflammation, and insulin resistance (Singh and Natraj 2021). Further investigation, involving horizontal fecal transfer and antibiotic-induced depletion of specific gut bacteria, reveals that the effects of *H. sinensis* polysaccharides are reliant on neomycin-sensitive bacteria. Gut microbiota analysis highlights a reduction in the Gram-negative bacterium *P. goldsteinii* in the microbiota of HFD-fed mice, whereas this bacterium is enriched in polysaccharide-treated mice (Singh and Natraj 2021). Remarkably, oral administration of live (but not heat-killed) *P. goldsteinii* bacteria to HFD-fed mice significantly curtails weight gain and mitigates obesity-associated metabolic disorders (Singh and

Natraj 2021). These findings suggest that the anti-obesity, anti-inflammatory, and insulin-sensitizing effects of mushroom polysaccharides are mediated by the gut microbiota and involve the newly identified probiotic *P. goldsteinii* (Singh and Natraj 2021). Consequently, mushroom polysaccharides and *P. goldsteinii* may be employed as prebiotics and probiotics, respectively, for addressing obesity and related metabolic complications (Singh and Natraj 2021).

7.5.9 *Roseburia* spp.

Roseburia spp. is an obligate anaerobe that produces butyrate, presenting a Gram-positive, slightly curved rod-like structure. It falls within the Clostridium XIVa subgroup (Duncan et al. 2006), constituting approximately 3–15% of the intestinal tract's healthy population (Leth et al. 2018; Parthasarathy et al. 2016). This bacterium is linked to resistance against inflammation (Parthasarathy et al. 2016) and is associated with conditions such as Parkinson's disease (Nishiwaki et al. 2020) and inflammatory bowel disease (IBD) (Imhann et al. 2018), making it a potential candidate for NGPs. Individuals with cardiovascular disease tend to exhibit lower levels of Roseburia and other butyrate-producing bacteria. Intriguingly, studies on mice demonstrated reduced atherosclerosis when they were fed a combination of *Roseburia* and a high-fiber diet. In this context, the high-fiber diet facilitated the production of butyrate from *Roseburia*, contributing to the reduction of atherosclerosis (Kasahara et al. 2018).

7.6 DESIGNER PROBIOTICS AS NGPs

As mentioned earlier, the term NGPs encompasses genetically modified microorganisms (GMMs) created through gene addition, deletion, or overexpression. These modifications aim to establish a resilient probiotic strain with benefits like improved metabolism, technological stress tolerance, GI survivability, and targeted delivery of therapeutic molecules. This concept introduces the idea of a "bio-drug" for preventing or treating various diseases (Jesus et al. 2019). By employing molecular editing tools, established probiotics can be further modified and investigated for the development of personalized medicine. The recent emergence of the CRISPR/Cas9 genome editing tool has provided a better platform for the editing of the genome more precisely to binate gene expression or provides novel features so as to enhance host colonization and promote human health. But still the diting outcomes can be different across different strains and methods (Jesus et al. 2019; Leenay et al. 2019).

7.7 IMMUNE CHECKPOINTS BLOCKADE (ICB) THERAPY AND OTHER POTENTIAL CANDIDATES

Several other bacterial species have demonstrated promising effects in boosting the effectiveness of anticancer immunotherapies. These bacteria include *Eubacterium hirae, Enterococcus faecium, Eubacterium limosum, Enterococcus hirae, Collinsella aerofaciens,* and *Burkholderia cepacia. E. limosum,* classified in the class Clostridia, order Clostridiales, and phylum Firmicutes, is an anaerobic, non-spore-forming, Gram-positive rod, notably abundant in the elderly human population (Kanauchi et al. 2006). This bacterium shows effects on ameliorating DSS-induced colitis in mice, probably due to the production of SCFAs to present anti-inflammation effects (Kanauchi et al. 2006). Additionally, a higher abundance of *E. limosum* in the intestines has been linked to a decreased risk of disease relapse and progression following allogeneic hematopoietic stem cell transplantation (HSCT) in individuals with hematological malignancies (Mulanovich, Desai, and Popat 2016). The underlying mechanisms of how *E. limosum* contributes to favorable outcomes after allogeneic hematopoietic stem cell transplantation (allo-HSCT) are still awaiting further characterization. *E. hirae* belongs to the *Enterococcaceae* and is categorized under the class of bacilli and the phylum Firmicutes. Studies showed that it was more abundant in feces from the responder patients with various carcinomas to PD-1/PD-L1 blockade than in samples from the non-responders (Routy et al. 2018). The potential ameliorative mechanism might involve the enhancement of memory responses of T cells through the administration of *E. hirae* 13144. (Singh and Natraj 2021).

7.8 CHALLENGES IN DEVELOPING NEXT-GENERATION PROBIOTICS WITH ADVANCED FUNCTIONALITIES

To bring probiotic strains into commercial use, it is crucial to thoroughly assess their techno-functional robustness, including their ability to withstand various processing treatments, enrichment processes, freeze-drying, product incorporation, and viability control during product shelf life (Chang et al. 2019). While many traditional probiotics from bacterial and yeast domains are already produced at an industrial scale, handling these strains remains relatively straightforward, as they are microaerophilic to aerobic. However, the extreme sensitivity to oxygen exhibited by many potential NGP candidates poses a significant challenge for large-scale production. Existing industrial plant designs and process flow lines often lack the capacity to create the necessary anaerobic microenvironment to achieve maximum viable bacterial cells per unit time (Chang et al. 2019).

Moreover, incorporating these bacteria into food systems without microencapsulation becomes challenging, as the technique involves mechanical agitation, leading to the incorporation of air and the development of an unfavorable toxic niche. Producing bacterial biomass as a freeze-dried concentrate requires freeze-drying under obligate anaerobic conditions, including the incorporation of potential cryoprotective agents, necessitating microbial quality control steps to monitor contaminants and viable cell counts. Additionally, for effective microencapsulation under anaerobic conditions, the freeze-dried bacterial mass needs low moisture content and must be milled to meet powder characteristics such as wettability, dispersibility, and sinkability (Chang et al. 2019).

While specific nutritional requirements for all NGPs are yet to be fully understood, well-studied species demand complex nutritional conditions, including mucin, rumen fluid, haemin broth, etc., at a laboratory scale. Providing such specific growth requirements at an industrial scale presents a considerable challenge. Therefore, further research is essential to develop economically feasible and suitable cultivation media for NGPs on a large scale. These technological factors collectively present potential hurdles for industrialists seeking to develop NGP therapeutics from commensal gut bacteria. (Chang et al. 2019).

7.9 REGULATORY GUIDELINES

The commercial utilization of probiotics necessitates regulatory approval from various apex bodies worldwide, each with its distinct categorization based on intended use. Probiotics are predominantly marketed as dietary food supplements, and different countries have established categories such as Dietary Supplements, Drugs, Live Biotherapeutic Agents and Medical Food in the USA, Functional Foods in Japan, India, China, Malaysia, Biotherapeutic Agent in European countries, Belgium, Germany, Dietary Food in Italy, Natural Health Products in Canada, and Food Supplements in Denmark, Sweden, and Finland. These countries have devised regulatory systems to ensure the approval of candidate probiotics while addressing safety concerns.

Regulatory authorities, such as the Food for Specified Health Use (FOSHU) in Japan; Joint Health Claims Initiative (JHCI) in the UK; Food and Drug Administration (FDA) and Dietary Supplement Health and Education Act (DSHEA) in the USA; European Food Safety Authority (EFSA) in Europe; State Food and Drug Administration (SFDA) in China; Food Safety and Standards Authority of India (FSSAI) in India; Canadian Food and Drugs Act under Natural Health Products Regulations in Canada; Food Safety and Quality Division (FSQD); Drug Control Authority, the National Pharmaceutical Control Bureau (NPCB) in Malaysia; and National Health Surveillance Agency in Brazil, are globally recognized.

The joint expert consultants of FAO/WHO collaborate to ensure the health and nutritional properties of probiotics according to scientific recommendations (Arora and Baldi 2015). In the European Union, the EFSA Scientific Committee provides a generic concept of qualified presumption of safety (QPS) status for microorganisms intended for food and feed applications. However, this list currently does not include gut-derived NGPs or engineered probiotics, except for generic *Bifidobacteria*.

In the USA, live microbes are classified as foods or dietary supplements when used as such and as live biotherapeutic products (LBPs) when intended for use as a drug. NGPs, especially those derived from the human host or engineered probiotics, fall under the category of LBPs, with engineered constructs or designer probiotics specifically categorized as recombinant LBPs. The Centre for Biologics Evaluation and Research (CBER) of the FDA regulates the LBP section, demanding stringent in vitro characterization of probiotics at both phenotypic and genotypic levels. This includes identity, antimicrobial resistance, mutations, virulence factors, and the mode of action in disease mitigation. LBPs must be produced following good manufacturing practice (GMP) standards, similar to vaccines. For successful marketing approval, LBPs must undergo well-designed preclinical and clinical trials to evaluate safety, efficacy, and effectiveness (Ross et al. 2008).

7.10 CONCLUSION

Conventional probiotics exhibit generic actions and are limited to a narrow spectrum of organisms. The evolving understanding of gut microbiota and its intricate relationship with host health has transformed this paradigm, broadening the exploration of a diverse range of organisms across the phylogenetic spectrum. In-depth scientific investigations into the health benefits associated with specific commensals may unveil novel microbes that could be harnessed for designing personalized medicine tailored to specific diseases. These studies might uncover unidentified microorganisms with unique properties, including some belonging to well-known pathogenic groups, posing challenges for research, industrial applications, and regulatory oversight. Such health-promoting non-traditional microorganisms are commonly referred to as NGPs, subject to regulation under the LBPs guidelines of the FDA when used for pharmaceutical applications.

Identifying NGPs is a formidable task, relying on an assumption-based approach and primarily grounded in correlation studies comparing the relative abundance levels of microorganisms between healthy and diseased subjects. Once identified, these organisms must undergo validation for various functionalities, safety considerations, and technological parameters using appropriate in silico, in vitro, and in vivo models. Uncommon growth requirements must be addressed to enable large-scale production of LBPs, and formulations must ensure the viability and bioactivity of the culture until consumption. The developed product undergoes scrutiny for safety, efficacy, and efficiency and dose confirmation through phase 1–3 clinical trials, navigating the regulatory approval process. In conclusion, the concept of NGPs is in its infancy but holds significant promise, requiring extensive research for its exploitation as a preventive medicine.

REFERENCES

Almeida, Diana, Daniela Machado, José Carlos Andrade, Sónia Mendo, Ana Maria Gomes, and Ana Cristina Freitas. 2020. "Evolving Trends in Next-Generation Probiotics: A 5W1H Perspective." *Critical Reviews in Food Science and Nutrition* 60 (11). Taylor & Francis: 1783–96.

Arora, M, and A Baldi. 2015. "Regulatory Categories of Probiotics across the Globe: A Review Representing Existing and Recommended Categorization." *Indian Journal of Medical Microbiology* 33. Elsevier: S2–10.

Bedford, Andrea, and Joshua Gong. 2018. "Implications of Butyrate and Its Derivatives for Gut Health and Animal Production." *Animal Nutrition* 4 (2). Elsevier: 151–9.

Cani, Patrice D, and Willem M de Vos. 2017. "Next-Generation Beneficial Microbes: The Case of *Akkermansia Muciniphila*." *Frontiers in Microbiology* 8. Frontiers Media SA: 1765.

Chang, Chih-Jung, Chuan-Sheng Lin, Chia-Chen Lu, Jan Martel, Yun-Fei Ko, David M Ojcius, Shun-Fu Tseng, Tsung-Ru Wu, Yi-Yuan Margaret Chen, and John D Young. 2015. "*Ganoderma Lucidum* Reduces Obesity in Mice by Modulating the Composition of the Gut Microbiota." *Nature Communications* 6 (1). Nature Publishing Group: 1–19.

Chang, Chih-Jung, Tzu-Lung Lin, Yu-Ling Tsai, Tsung-Ru Wu, Wei-Fan Lai, Chia-Chen Lu, and Hsin-Chih Lai. 2019. "Next Generation Probiotics in Disease Amelioration." *Journal of Food and Drug Analysis* 27 (3). Elsevier: 615–22.

Chierico, Federica Del, Francesca Abbatini, Alessandra Russo, Andrea Quagliariello, Sofia Reddel, Danila Capoccia, Romina Caccamo, Stefano Ginanni Corradini, Valerio Nobili, and Francesco De Peppo. 2018. "Gut Microbiota Markers in Obese Adolescent and Adult Patients: Age-Dependent Differential Patterns." *Frontiers in Microbiology* 9. Frontiers Media SA: 1210.

Duncan, Sylvia H, Petra Louis, and Harry J Flint. 2004. "Lactate-Utilizing Bacteria, Isolated from Human Feces, That Produce Butyrate as a Major Fermentation Product." *Applied and Environmental Microbiology* 70 (10). Am Soc Microbiol: 5810–17.

Duncan, Sylvia H, Rustam I Aminov, Karen P Scott, Petra Louis, Thaddeus B Stanton, and Harry J Flint. 2006. "Proposal of *Roseburia Faecis* Sp. *Nov., Roseburia Hominis* Sp. *Nov. and Roseburia Inulinivorans* Sp. Nov., Based on Isolates from Human Faeces." *International Journal of Systematic and Evolutionary Microbiology* 56 (10). Society for General Microbiology: 2437–41.

Erturk-Hasdemir, Deniz, and Dennis L Kasper. 2018. "Finding a Needle in a Haystack: Bacteroides Fragilis Polysaccharide A as the Archetypical Symbiosis Factor." *Annals of the New York Academy of Sciences* 1417 (1). Wiley Online Library: 116–29.

Everard, Amandine, Clara Belzer, Lucie Geurts, Janneke P Ouwerkerk, Céline Druart, Laure B Bindels, Yves Guiot, Muriel Derrien, Giulio G Muccioli, and Nathalie M Delzenne. 2013. "Cross-Talk between *Akkermansia Muciniphila* and Intestinal Epithelium Controls Diet-Induced Obesity." *Proceedings of the National Academy of Sciences* 110 (22). National Acad Sciences: 9066–71.

Fekry, Mostafa I, Christina Engels, Jianbo Zhang, Clarissa Schwab, Christophe Lacroix, Shana J Sturla, and Christophe Chassard. 2016. "The Strict Anaerobic Gut Microbe *Eubacterium Hallii* Transforms the Carcinogenic Dietary Heterocyclic Amine 2-amino-1-methyl-6-phenylimidazo [4, 5-b] Pyridine (PhIP)." *Environmental Microbiology Reports* 8 (2). Wiley Online Library: 201–9.

Gasbarrini, Giovanni, Fiorenza Bonvicini, and Annagiulia Gramenzi. 2016. "Probiotics History." *Journal of Clinical Gastroenterology* 50. LWW: S116–9.

Ghazanfar, S., & Arif, I. (2019). Isolation and Identification of Probiotic to Improve Nutritional Value of Diet Supplements. *Health*, 3, 139–43.

Ghazanfar, S., Abid, R., Ajmal, W., Haider, Z., Khan, B., Rani, S., … Jyoti, S. U. (2024). An Introduction to Food Safety and Its Impact on Human Health. *Food Safety: Contaminants and Risk Assessment*, 1.

Goodrich, Julia K, Emily R Davenport, Michelle Beaumont, Matthew A Jackson, Rob Knight, Carole Ober, Tim D Spector, Jordana T Bell, Andrew G Clark, and Ruth E Ley. 2016. "Genetic Determinants of the Gut Microbiome in UK Twins." *Cell Host & Microbe* 19 (5). Elsevier: 731–43.

Goodrich, Julia K, Jillian L Waters, Angela C Poole, Jessica L Sutter, Omry Koren, Ran Blekhman, Michelle Beaumont, William Van Treuren, Rob Knight, and Jordana T Bell. 2014. "Human Genetics Shape the Gut Microbiome." *Cell* 159 (4). Elsevier: 789–99.

Heintz-Buschart, Anna, Urvashi Pandey, Tamara Wicke, Friederike Sixel-Döring, Annette Janzen, Elisabeth Sittig-Wiegand, Claudia Trenkwalder, Wolfgang H Oertel, Brit Mollenhauer, and Paul Wilmes. 2018. "The Nasal and Gut Microbiome in Parkinson's Disease and Idiopathic Rapid Eye Movement Sleep Behavior Disorder." *Movement Disorders* 33 (1). Wiley Online Library: 88–98.

Huang, Julie Y, S Melanie Lee, and Sarkis K Mazmanian. 2011. "The Human Commensal Bacteroides Fragilis Binds Intestinal Mucin." *Anaerobe* 17 (4). Elsevier: 137–41.

Imhann, Floris, Arnau Vich Vila, Marc Jan Bonder, Jingyuan Fu, Dirk Gevers, Marijn C Visschedijk, Lieke M Spekhorst, Rudi Alberts, Lude Franke, and Hendrik M Van Dullemen. 2018. "Interplay of Host Genetics and Gut Microbiota Underlying the Onset and Clinical Presentation of Inflammatory Bowel Disease." *Gut* 67 (1). BMJ Publishing Group: 108–19.

Jesus, Luís Cláudio Lima de, Fernanda Alvarenga Lima, Nina Dias Coelho-Rocha, Tales Fernando da Silva, Júlia Paz, Vasco Azevedo, Pamela Mancha-Agresti, and Mariana Martins Drumond. 2019. "Recombinant Probiotics and Microbiota Modulation as a Good Therapy for Diseases Related to the GIT." In *The Health Benefits of Foods-Current Knowledge and Further Development*. IntechOpen.

Kahouli, Imen, Catherine Tomaro-Duchesneau, and Satya Prakash. 2013. "Probiotics in Colorectal Cancer (CRC) with Emphasis on Mechanisms of Action and Current Perspectives." *Journal of Medical Microbiology* 62 (8). Microbiology Society: 1107–23.

Kanauchi, Osamu, Masanobu Fukuda, Yoshiaki Matsumoto, Shino Ishii, Toyokazu Ozawa, Makiko Shimizu, Keiichi Mitsuyama, and Akira Andoh. 2006. "*Eubacterium Limosum* Ameliorates Experimental Colitis and Metabolite of Microbe Attenuates Colonic Inflammatory Action with Increase of Mucosal Integrity." *World Journal of Gastroenterology: WJG* 12 (7). Baishideng Publishing Group Inc: 1071.

Kanda, Hajime, Sanshiro Tateya, Yoshikazu Tamori, Ko Kotani, Ken-ichi Hiasa, Riko Kitazawa, Sohei Kitazawa, Hitoshi Miyachi, Sakan Maeda, and Kensuke Egashira. 2006. "MCP-1 Contributes to Macrophage Infiltration into Adipose Tissue, Insulin Resistance, and Hepatic Steatosis in Obesity." *The Journal of Clinical Investigation* 116 (6). Am Soc Clin Investig: 1494–505.

Kasahara, Kazuyuki, Kimberly A Krautkramer, Elin Org, Kymberleigh A Romano, Robert L Kerby, Eugenio I Vivas, Margarete Mehrabian, John M Denu, Fredrik Bäckhed, and Aldons J Lusis. 2018. "Interactions between Roseburia Intestinalis and Diet Modulate Atherogenesis in a Murine Model." *Nature Microbiology* 3 (12). Nature Publishing Group UK London: 1461–71.

Kechagia, Maria, Dimitrios Basoulis, Stavroula Konstantopoulou, Dimitra Dimitriadi, Konstantina Gyftopoulou, Nikoletta Skarmoutsou, and Eleni Maria Fakiri. 2013. "Health Benefits of Probiotics: A Review." *International Scholarly Research Notices* 2013. Hindawi: 481651.

Kerry, Rout George, Jayanta Kumar Patra, Sushanto Gouda, Yooheon Park, Han-Seung Shin, and Gitishree Das. 2018. "Benefaction of Probiotics for Human Health: A Review." *Journal of Food and Drug Analysis* 26 (3). Elsevier: 927–39.

Leenay, Ryan T, Justin M Vento, Malay Shah, Maria Elena Martino, François Leulier, and Chase L Beisel. 2019. "Genome Editing with CRISPR-Cas9 in *Lactobacillus Plantarum* Revealed That Editing Outcomes Can Vary across Strains and between Methods." *Biotechnology Journal* 14 (3). Wiley Online Library: 1700583.

Leth, Maria Louise, Morten Ejby, Christopher Workman, David Adrian Ewald, Signe Schultz Pedersen, Claus Sternberg, Martin Iain Bahl, Tine Rask Licht, Finn Lillelund Aachmann, and Bjørge Westereng. 2018. "Differential Bacterial Capture and Transport Preferences Facilitate Co-Growth on Dietary Xylan in the Human Gut." *Nature Microbiology* 3 (5). Nature Publishing Group UK London: 570–80.

Leu, Richard K Le, Ying Hu, Ian L Brown, Richard J Woodman, and Graeme P Young. 2010. "Synbiotic Intervention of *Bifidobacterium Lactis* and Resistant Starch Protects against Colorectal Cancer Development in Rats." *Carcinogenesis* 31 (2). Oxford University Press: 246–51.

Louis, Petra, Pauline Young, Grietje Holtrop, and Harry J Flint. 2010. "Diversity of Human Colonic Butyrate-producing Bacteria Revealed by Analysis of the Butyryl-CoA: Acetate CoA-transferase Gene." *Environmental Microbiology* 12 (2). Wiley Online Library: 304–14.

Lukiw, Walter J. 2016. "Bacteroides Fragilis Lipopolysaccharide and Inflammatory Signaling in Alzheimer's Disease." *Frontiers in Microbiology* 7. Frontiers Media SA: 1544.

Martín, Rebeca, Sylvie Miquel, Florian Chain, Jane M Natividad, Jennifer Jury, Jun Lu, Harry Sokol, Vassilia Theodorou, Premysl Bercik, and Elena F Verdu. 2015. "*Faecalibacterium Prausnitzii* Prevents Physiological Damages in a Chronic Low-Grade Inflammation Murine Model." *BMC Microbiology* 15 (1). BioMed Central: 1–12.

Mulanovich, Victor E, Parth A Desai, and Uday R Popat. 2016. "Allogeneic Stem Cell Transplantation for HIV-Positive Patients with Hematologic Malignancies." *AIDS (London, England)* 30 (17). NIH Public Access: 2653.

Munukka, Eveliina, Anniina Rintala, Raine Toivonen, Matts Nylund, Baoru Yang, Anna Takanen, Arno Hänninen, Jaana Vuopio, Pentti Huovinen, and Sirpa Jalkanen. 2017. "*Faecalibacterium Prausnitzii* Treatment Improves Hepatic Health and Reduces Adipose Tissue Inflammation in High-Fat Fed Mice." *The ISME Journal* 11 (7). Nature Publishing Group: 1667–79.

Nishiwaki, Hiroshi, Mikako Ito, Tomohiro Ishida, Tomonari Hamaguchi, Tetsuya Maeda, Kenichi Kashihara, Yoshio Tsuboi, Jun Ueyama, Teppei Shimamura, and Hiroshi Mori. 2020. "Meta-analysis of Gut Dysbiosis in Parkinson's Disease." *Movement Disorders* 35 (9). Wiley Online Library: 1626–35.

Parthasarathy, Gopanandan, Jun Chen, Xianfeng Chen, Nicholas Chia, Helen M O'Connor, Patricia G Wolf, H Rex Gaskins, and Adil E Bharucha. 2016. "Relationship between Microbiota of the Colonic Mucosa vs Feces and Symptoms, Colonic Transit, and Methane Production in Female Patients with Chronic Constipation." *Gastroenterology* 150 (2). Elsevier: 367–79.

Pedersen, Helle Krogh, Valborg Gudmundsdottir, Henrik Bjørn Nielsen, Tuulia Hyotylainen, Trine Nielsen, Benjamin A H Jensen, Kristoffer Forslund, Falk Hildebrand, Edi Prifti, and Gwen Falony. 2016. "Human Gut Microbes Impact Host Serum Metabolome and Insulin Sensitivity." *Nature* 535 (7612). Nature Publishing Group UK London: 376–81.

Plovier, Hubert, Amandine Everard, Céline Druart, Clara Depommier, Matthias Van Hul, Lucie Geurts, Julien Chilloux, Noora Ottman, Thibaut Duparc, and Laeticia Lichtenstein. 2017. "A Purified Membrane Protein from *Akkermansia muciniphila* or the Pasteurized Bacterium Improves Metabolism in Obese and Diabetic Mice." *Nature Medicine* 23 (1). Nature Publishing Group US New York: 107–13.

Roller, Monika, Yvonne Clune, Kevin Collins, Gerhard Rechkemmer, and Bernhard Watzl. 2007. "Consumption of Prebiotic Inulin Enriched with Oligofructose in Combination with the Probiotics *Lactobacillus rhamnosus* and *Bifidobacterium lactis* Has Minor Effects on Selected Immune Parameters in Polypectomised and Colon Cancer Patients." *British Journal of Nutrition* 97 (4): 676–84.

Ross, Jennifer J, Philip E Boucher, Siba P Bhattacharyya, Dennis J Kopecko, Elizabeth M Sutkowski, Patricia J Rohan, Donna K F Chandler, and Julienne Vaillancourt. 2008. "Considerations in the Development of Live Biotherapeutic Products for Clinical Use." In Current Issues in Molecular Biology 10 (1–2). Multidisciplinary Digital Publishing Institute: 13–16.

Rossi, Oriana, Lisette A Van Berkel, Florian Chain, M Tanweer Khan, Nico Taverne, Harry Sokol, Sylvia H Duncan, Harry J Flint, Hermie J M Harmsen, and Philippe Langella. 2016. "*Faecalibacterium Prausnitzii* A2-165 Has a High Capacity to Induce IL-10 in Human and Murine Dendritic Cells and Modulates T Cell Responses." *Scientific Reports* 6 (1). Nature Publishing Group UK London: 18507.

Routy, Bertrand, Emmanuelle Le Chatelier, Lisa Derosa, Connie P M Duong, Maryam Tidjani Alou, Romain Daillère, Aurélie Fluckiger, Meriem Messaoudene, Conrad Rauber, and Maria P Roberti. 2018. "Gut Microbiome Influences Efficacy of PD-1-Based Immunotherapy against Epithelial Tumors." *Science* 359 (6371). American Association for the Advancement of Science: 91–7.

Sadaghian Sadabad, Mehdi, Julius Z H von Martels, Muhammed Tanweer Khan, Tjasso Blokzijl, Giuseppe Paglia, Gerard Dijkstra, Hermie J M Harmsen, and Klaas Nico Faber. 2015. "A Simple Coculture System Shows Mutualism between Anaerobic Faecalibacteria and Epithelial Caco-2 Cells." *Scientific Reports* 5 (1). Nature Publishing Group UK London: 17906.

Schneeberger, Marc, Amandine Everard, Alicia G Gómez-Valadés, Sébastien Matamoros, Sara Ramírez, Nathalie M Delzenne, Ramon Gomis, Marc Claret, and Patrice D Cani. 2015. "*Akkermansia Muciniphila* Inversely Correlates with the Onset of Inflammation, Altered Adipose Tissue Metabolism and Metabolic Disorders during Obesity in Mice." *Scientific Reports* 5 (1). Nature Publishing Group UK London: 16643.

Sears, Cynthia L, Salequl Islam, Amit Saha, Maleka Arjumand, Nur Haque Alam, A S G Faruque, M A Salam, Jai Shin, David Hecht, and Andrej Weintraub. 2008. "Association of *Enterotoxigenic Bacteroides Fragilis* Infection with Inflammatory Diarrhea." *Clinical Infectious Diseases* 47 (6). The University of Chicago Press: 797–803.

Shahid, S., Abid, R., Ajmal, W., Almuqbil, M., Almadani, M. E., Khan, Y., … Ghazanfar, S. (2024). Antibiotic Resistance Genes Prediction via Whole Genome Sequence Analysis of *Stenotrophomonas maltophilia. Journal of Infection and Public Health*, 17(2), 236–44.

Singh, Tejinder Pal, and Basavaprabhu Haranahalli Natraj. 2021. "Next-Generation Probiotics: A Promising Approach towards Designing Personalized Medicine." *Critical Reviews in Microbiology* 47 (4). Taylor & Francis: 479–98.

Sivan, Ayelet, Leticia Corrales, Nathaniel Hubert, Jason B Williams, Keston Aquino-Michaels, Zachary M Earley, Franco W Benyamin, Yuk Man Lei, Bana Jabri, and Maria-Luisa Alegre. 2015. "Commensal *Bifidobacterium* Promotes Antitumor Immunity and Facilitates Anti-PD-L1 Efficacy." *Science* 350 (6264). American Association for the Advancement of Science: 1084–9.

Soccol, Carlos Ricardo, Luciana Porto de Souza Vandenberghe, Michele Rigon Spier, A B Pedroni Medeiros, Caroline Tiemi Yamaguishi, J De Dea Lindner, Ashok Pandey, and Vanete Thomaz-Soccol. 2010. "The Potential of Probiotics: A Review." *Food Technology and Biotechnology* 48 (4): 413–34.

Turroni, Francesca, Douwe Van Sinderen, and Marco Ventura. 2011. "Genomics and Ecological Overview of the Genus *Bifidobacterium*." *International Journal of Food Microbiology* 149 (1). Elsevier: 37–44.

Vadder, Filipe De, Petia Kovatcheva-Datchary, Carine Zitoun, Adeline Duchampt, Fredrik Bäckhed, and Gilles Mithieux. 2016. "Microbiota-Produced Succinate Improves Glucose Homeostasis via Intestinal Gluconeogenesis." *Cell Metabolism* 24 (1). Elsevier: 151–7.

Wang, Yiming, Rena Ma, Fang Liu, Seul A Lee, and Li Zhang. 2018. "Modulation of Gut Microbiota: A Novel Paradigm of Enhancing the Efficacy of Programmed Death-1 and Programmed Death Ligand-1 Blockade Therapy." *Frontiers in Immunology* 9. Frontiers Media SA: 374.

Williams, Nancy Toedter. 2010. "Probiotics." *American Journal of Health-System Pharmacy* 67 (6). Oxford University Press: 449–58.

Wrzosek, Laura, Sylvie Miquel, Marie-Louise Noordine, Stephan Bouet, Marie Joncquel Chevalier-Curt, Véronique Robert, Catherine Philippe, Chantal Bridonneau, Claire Cherbuy, and Catherine Robbe-Masselot. 2013. "Bacteroides Thetaiotaomicron and *Faecalibacterium Prausnitzii* Influence the Production of Mucus Glycans and the Development of Goblet Cells in the Colonic Epithelium of a Gnotobiotic Model Rodent." *BMC Biology* 11. Springer: 1–13.

Wu, Tsung-Ru, Chuan-Sheng Lin, Chih-Jung Chang, Tzu-Lung Lin, Jan Martel, Yun-Fei Ko, David M Ojcius, Chia-Chen Lu, John D Young, and Hsin-Chih Lai. 2019. "Gut Commensal *Parabacteroides goldsteinii* Plays a Predominant Role in the Anti-Obesity Effects of Polysaccharides Isolated from Hirsutella Sinensis." *Gut* 68 (2). BMJ Publishing Group: 248–62.

Zhang, Huanchang, Yunfeng Duan, Feng Cai, Demin Cao, Lei Wang, Zhenyi Qiao, Qing Hong, Nan Li, Yuanrong Zheng, and Miya Su. 2022. "Next-Generation Probiotics: Microflora Intervention to Human Diseases." *BioMed Research International* 2022. Hindawi: 5633403.

8 Relevance of Genetically Engineered and Specific Bioengineered Strain and Their Role in Probiotics Production

Samia S. Alkhalil, Jumoke Florence Joda, Sherifat Ozavize Enejiyon, Mordecai Gana, Oluwafemi Adebayo Oyewole, Charles Oluwaseun Adetunji, Muhammad Lawal Attanda, and S.A. Ado

8.1 INTRODUCTION

Progress in designing new delivery systems for mucosal administration has been greatly facilitated by recent developments in recombinant technology, particularly in genetic engineering, together with improved knowledge of the human immune system (Börner et al., 2019). When compared to other conventional procedures like systemic injections, the administration of medicinal drugs via mucosal pathways presents numerous noteworthy advantages. According to Charbonneau et al. (2020) and Bermúdez-Humarán et al. (2013), these benefits include fewer side effects, easier administration, and the capacity to affect mucosal and systemic immune responses.

Moreover, for health-related compounds that exert their effects on mucosal surfaces, namely within the gastrointestinal tract (GIT), direct transport to the targeted region is crucial. However, one significant obstacle in mucosal delivery is the need to provide substantial amounts of protein, as only tiny amounts are able to withstand breakdown at mucosal surfaces such as the GIT.

Various alternative methods to traditional mucosal delivery systems, including inert systems like liposomes or nanoparticles, as well as live attenuated bacterial or viral vectors, are now acknowledged as essential for the treatment of human diseases in the 21st century (Rochet et al., 2008). The innovation of new strategies that combine genetic engineering with probiotic bacteria to facilitate the targeted delivery of health-related molecules to the mucosa represents a promising alternative to conventional attenuated pathogenic vectors. As a result, the development and utilization of genetically modified probiotics (GMPs) that produce therapeutic molecules may yield significant insights into their impacts on food, nutrition, the environment, and overall health. When compared to conventional abiotic therapies, biotic systems may offer a more effective drug delivery method since they may efficiently distribute enzymes, vaccines, antimicrobials, and cytokines. Furthermore, probiotics can include biotic sensors, which would serve as noninvasive diagnostic instruments. Probiotics have mostly only been used to deliver drugs in the form of protein-based compounds that are easily produced or altered by the commensal microbiota. Thus, in order to increase the variety of probiotic-based treatments available, it is imperative to enhance the biosynthetic properties of widely used probiotics. Furthermore, it is imperative to develop probiotics that are developed to react dynamically to diverse stimuli and modify their behavior according to specific circumstances (Bermúdez-Humarán et al., 2013). Even though tailored probiotics have several advantages in medicinal applications, there are still certain obstacles to overcome (Al Hawaj et al., 2022).

 DOI: 10.1201/9781003399797-9

A major hurdle in the engineering of probiotics is the selection of the most suitable chassis. A delicate balance must be struck between ensuring survival within the host and maintaining safety. Numerous engineered strains of *Lactobacillus* do not form part of the native human microbiota, resulting in their swift elimination by more dominant microorganisms, which diminishes their therapeutic effectiveness. Engineered probiotic strains refer to microorganisms that have been modified to optimize their metabolic pathways, typically by the use of synthetic biology and omics technologies. The engineering procedure entails using bacterial strains that have a high level of proficiency in colonizing the gastrointestinal system and are capable of synthesizing therapeutic compounds in their natural environment (Börner et al., 2019). The essential objective of probiotic engineering is to optimize metabolic processes and augment the therapeutic capabilities of these microbes. Comprehensive knowledge of the processes by which probiotics exert their effects is essential for the advancement of novel probiotic strains. Probiotics engage with the host and its resident microbiome, facilitating the enhancement of microbial composition, competition-based inhibition of pathogens, and degradation of toxic substances.

8.2 MECHANISMS OF ACTION OF ENGINEERED PROBIOTICS

Prior research has shown that probiotics influence health through multiple mechanisms, such as providing protection against pathogens, enhancing immune modulation, modifying gut microbiota composition, strengthening the gut epithelial barrier, and generating short-chain fatty acids (SCFAs).

8.3 MODULATION OF THE IMMUNE SYSTEM

The functional dynamics of engineered probiotics can be enhanced by integrating the mechanistic pathways identified across various strains, leading to the creation of more effective probiotic formulations (Canale et al., 2021). Several mechanisms underpin these novel therapeutic approaches: pathogen exclusion via the synthesis of antimicrobial peptides (AMPs); biosensing for the identification of disease markers; changes in host metabolic processes; and immunomodulation, which includes the synthesis of cytokines, antigens, and allergens. These strategies have been effectively implemented in various therapeutic settings. To combat infectious disorders, for example, several *Lacticaseibacillus* species have undergone genetic modification to specifically target *Listeria monocytogenes* or to generate a recombinant fusion protein of the cholera toxin B component (Ghazanfar et al., 2022). Similarly, in order to cure cancer, *Lactococcus lactis* has been altered to express the anti-metastatic KISS1 peptide.

Additionally, Pusch et al. (2005) have effectively engineered recombinant strains of *L. lactis* and *Lactiplantibacillus plantarum* to synthesize the microbiocidal cyanovirin-N, which has shown efficacy in neutralizing HIV-1 infectivity in vitro, thus aiding in the prevention of its transmission (Muhammad et al., 2023). Probiotics exert their effects by either directly interacting with host cells or positioning themselves in proximity to elicit inflammatory responses (Lenoir et al., 2016). They influence immune markers by modulating pathways such as MAPK, interleukin (IL)-6, IL-8, TNF-α, Akt, and IFN-γ, along with other contact-dependent mechanisms. Notable probiotic actions include the secretion of TNF-α mediated by *Lactobacillus* through lipoteichoic acid, IL-10 secretion induced by *Bifidobacterium longum* via cell surface pilli, and the modulation of cytokine and helper-T cell responses by *B. longum* 36524 through exopolysaccharides (Alsalman et al., 2022). Furthermore, *Lactobacillus rhamnosus* affects TNF-α and IL-6/10/12 levels in the intestinal mucosa through its cell surface structures, alongside immune modulation and IgA secretion, which can also be mediated by *Bifidobacterium lactis* using the L-92 effector, SLP.

8.3.1 Engineered Probiotics

The conceptual framework surrounding engineered probiotics encompasses the alteration of existing probiotic strains through gene editing methodologies to develop innovative probiotics with specific desired traits (Wang et al., 2021). This engineering approach facilitates the precise evaluation of modifications in genetic sequences, protein synthesis, and the functional attributes of these newly created microorganisms. Within the human organism, enzymes play a crucial role in regulating cellular metabolism and various intricate biological functions. A lack of these enzymes can result in a range of metabolic disorders, where the accumulation of harmful substances may provoke abnormal physiological responses, occasionally leading to fatal outcomes. Consequently, the engineering of advantageous probiotic bacterial strains to incorporate particular enzymes or metabolic pathways can provide a safeguard against such disorders. Furthermore, the engineering of probiotics enhances their efficacy, allowing for more targeted actions and the ability to function across diverse host organisms. Presently, there is an increasing interest in the integration of synthetic biology with probiotic engineering to enhance the delivery of specific molecules, improve sensing abilities, and bolster anti-pathogenic functions. A variety of molecular biology techniques can be employed to engineer multiple regulatory systems within beneficial probiotic strains (Figure 8.1). By carrying out tasks that the gut microbiota normally does not, these engineered strains offer clear benefits over conventional microbiota-mediated treatments. For instance, in the case of diabetes, the pancreas may not produce enough insulin, which results in abnormal glucose levels. Recombinant probiotic strains are being actively developed by researchers with the goal of treating diabetes. The genetically altered *L. lactis* is one such example; it generates the immunomodulatory cytokine IL-10 and shows promise for oral administration in the treatment of type 1 diabetes.

Moreover, a genetically altered *Lactobacillus gasseri* strain was employed in a separate investigation. The present strain underwent genetic modification to increase the expression of the glucagon-like peptide (GLP)-1 gene, renowned for its capacity to induce insulin production in both live and laboratory settings. Oral administration of the GLP-1 strain of *L. gasseri* to diabetic rats resulted in increased insulin secretion by pancreatic beta cells and a reduction in blood glucose levels by 10–20%. In a parallel investigation, genetically modified (GM) strains of *L. lactis* were manipulated to generate the autoantigen GAD65370–575, which mitigates inflammation in pancreatic islets induced by IL-10 during hyperglycemia.

The absence of phenylalanine hydroxylase causes the metabolic disorder called phenylketonuria (PKU), which is defined by the buildup of phenylalanine in the body, potentially causing various health problems. Integration of the phenylalanine ammonia-lyase (PAL) gene from Anabaena variabilis into *Lactobacillus reuteri* shown efficacy in decreasing blood phenylalanine levels in a mouse in vivo experiment. Moreover, the GMP bacteria SYNB1618, which were particularly designed to produce both PAL and l-amino acid deaminase (LAAD), effectively reduced blood phenylalanine levels by as much as 38%. A new variation of SYNB1618 (thyA-, argR-) was recently created by incorporating argR215 DNA into its genetic architecture. The oral administration of this altered strain led to a significant decrease in ammonia levels, therefore enhancing survival rates by up to 50%. Furthermore, over the past ten years, many mutagenesis techniques, including single-crossover insertion, double deletion, homologous recombination, and inducible plasmid self-destruction (IPSD)-mediated genome engineering, have been refined for *Bifidobacterial* probiotic strains. Nevertheless, these methods have many common disadvantages, such as a poor degree of transformation effectiveness, unstable mutations, plasmid loss, and the requirement for thorough screening from big colonies.

Within the field of genome editing, the recently identified CRISPR-Cas system is a remarkable development offering a quick and efficient means of introducing genetic changes (Zhang et al., 2021). Especially useful for managing transcription and supervising genomic circuits is this innovative cutting-edge tool. From homologous recombination to the development of first-generation Zinc Finger Nucleases (ZFNs), then second-generation Transcription Activator-Like Effector Nucleases

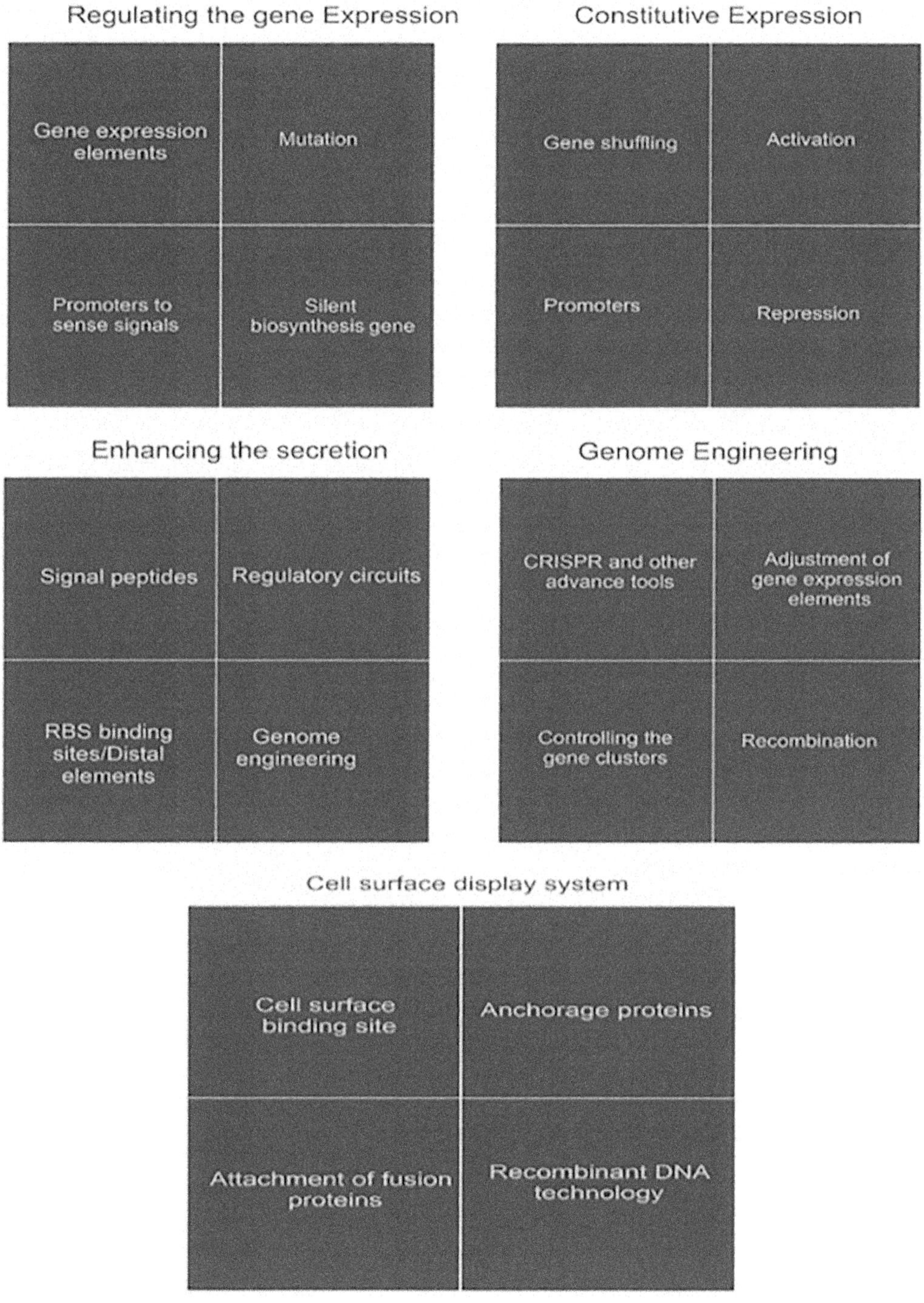

FIGURE 8.1 The functionality of probiotic bacteria can be enhanced by using molecular biology tools alongside synthetic biology approaches. Improvements can be achieved by: (a) regulating gene expression, (b) ensuring continuous gene expression using appropriate promoters, (c) boosting the secretion of target enzymes with the addition of signal peptides, (d) editing genes through genome engineering, and (e) developing cell surface display systems.

(TALENs), and lastly third-generation Clustered Regularly Interspaced Short Palindromic Repeats (CRISpen-Cas)-associated technology (CRISpen-Cas). While CRISpen gene editing was named as the best breakthrough of 2015 by Science News, culminating in the Nobel Prize in Chemistry in 2020, which marked a turning point in the life sciences, ZFNs and TALENs were honored as the methods of the year by Nature Methods. This part tries to give a quick summary of these three generations of gene editing technologies (Zhang et al., 2021). With either the more precise Homology Directed Repair (HDR) mechanism or the error-prone Non-homologous End Joining (NHEJ) pathway, these approaches enable the generation of Double Strand Breaks (DSBs) in specific genes, followed by DNA repair. Operating as designed restriction endonucleases, zinc finger or TAL-effector DNA domains specifically target and modify assigned DNA sequences (Barrangou & Doudna, 2016). These technologies do, however, have drawbacks, including ZFNs' poor selectivity and the possibility of inadvertent alterations. Moreover, labor-intensive and time-consuming procedures define the usual development of ZFNs and TALENs. Consequently, these restrictions have hampered their general acceptance, even though they have been used in genome editing across human, animal, and plant cells since 2002 and 2011, respectively.

From late 2012 onwards, there has been a significant surge in research dedicated to exploring new approaches in the field of CRISPR technology. This technology is known for its exceptional editing efficiency and flexible, user-friendly features. This breakthrough has created a novel paradigm in the field of gene editing methodologies. The CRISPR system, together with its related Cas genes, is recognized as a component of the bacterial immune defense machinery, serving as a protective mechanism against phage attacks. Through molecular mechanisms, bacteria incorporate viral genetic information from early infections into their CRISPR system (Roberts & Barrangou, 2020). Subsequent viral attacks trigger the transcription of the CRISPR locus to generate the precursor CRISPR RNA (pre-crRNA), which is then transformed into crRNA that corresponds to the genetic sequence of the virus. A critical function of this crRNA is to detect homologous regions within the viral genome, therefore facilitating the binding and cleavage of viral genes by Cas proteins. The classification of the CRISPR-Cas system is largely based on the spatial arrangement of the Cas protein gene in close proximity to the CRISPR array. The existing classifications define two primary categories, six unique types, and 33 varied subtypes of the CRISPR-Cas system. Significantly, the nucleases Cas9 and Cas12a (Cpf1) from the two categories of the Cas protein family are well recognized for their very reliable ability to specifically target specific sections of DNA. The CRISPR-Cas9 system, particularly type II, is widely used as the prevailing protein for targeted genome modification. This system's architecture consists of the Cas9 protein, crRNA, and trans-activating RNA (tracrRNA). Cas9 is distinguished among the several Cas proteins derived from *Streptococcus thermophilus* because of its potent DNA catalytic activity. This activity allows it to cleave double-stranded DNA at specific sequence locations by the process of crRNA pairing. The combination of crRNA and tracrRNA in synthetic design enables the formation of small guide RNA (sgRNA), which facilitates the precise cleavage of DNA, knockdown of genes, and insertion of cas9 at specific sites. The versatility of CRISPR technology enables precise exploration of loss-of-function scenarios, including CRISPR knockout (CRISPRKO) and CRISPR interference, along with various other genomic modifications. The use of CRISPR interference (CRISPRi) to suppress certain processes, together with the investigation of gain-of-function methods like CRISPR activation (CRISPRa), represents significant progress in the arena of genetic research.

The development of next-generation probiotics hinges on the crucial progress made in gene editing technologies, particularly CRISPR. The application of CRISPR-Cas systems for genome engineering has been conducted in various families of lactic acid bacteria (LAB), including *Lactobacillus*, *Lactococcus*, and *Streptococcus*. The application of CRISPR-Cas with IPSD is expected to improve genome editing in probiotic bacterial species. Furthermore, the advancement of genetic tools such as plasmid expression systems and genome engineering approaches would enhance the practical applications of synthetic biology in the domain of probiotic biology. The effectiveness of certain broad-host range plasmids, including pSH71, pAMβ11, pBC1, and pTB6, in encoding genes in LAB is still uncertain, and its effectiveness in other probiotic strains is similarly undetermined. The transfer of plasmid DNA to different microbial strains can be accomplished by many methods, including transformation, conjugation, electroporation, or phage transduction. However, the challenge has become more challenging due to the emergence of resistance mechanisms in bacterial hosts, including CRISPR-Cas and restriction-modification systems that hinder the uptake of foreign DNA (Oh & van Pijkeren, 2014). A successful strategy involves the transmission of plasmid DNA in a single-stranded orientation from the donor bacterium to the recipient bacteria. Viable transmission of plasmids from *Escherichia coli* to other *Bifidobacterial* strains has been achieved using a conjugative transfer method.

However, in order to ensure the effective manifestation of foreign gene products, it is imperative to consider several host characteristics. These factors encompass promoters, regulatory elements that include ribosome binding sites (RBS), cell-surface anchoring components, and other signals necessary for secretion and localization. Significantly, promoters such as hup, which encodes a histone-like protein, and gap, which encodes glyceraldehyde-3-phosphate dehydrogenase, have a crucial role in facilitating the expression of high-level genes in *Bifidobacterial* strains (Yan et al., 2021).

The Pgap promoter has been effectively used to enhance the functional expression of homologous and heterologous genes, as well as to generate various fluorescent proteins and medicinal compounds in *Bifidobacteria*. Furthermore, genetic circuits can be precisely engineered to regulate gene expression in response to particular stimuli (Barrangou et al., 2007). CR By employing bacterial one/two-component systems and quorum sensing, genetic circuits are developed with the purpose of illness diagnostics, bactericidal effects, and therapeutic agent delivery (Duan et al., 2015). Moreover, there is ongoing development of gene expression systems that are responsive to gastrointestinal signals such as pH, bile salts, and inflammation. These systems aim to establish genetic circuits that can be used for various diagnostic and therapeutic applications.

Over the past several years, there has been an increasing fascination with the advancement of targeted probiotics designed to combat bacterial infections and suppress the growth of pathogens in the GIT (Martín et al., 2013). Through processes such as antimicrobial release, immune response regulation, bacterial adhesion inhibition, and the creation of toxic proteins directed against pathogens, these novel "programmed" probiotics provide a safer alternative for attacking multidrug-resistant bacteria. Particular attention has been given to bacteria that create biofilms, which are responsible for persistent infections and can impair immune function. This makes them a key subject for probiotic engineering aimed at inhibiting biofilm formation. An example is the development of a genetically engineered strain of *E. coli* that targets *Pseudomonas aeruginosa* by modifying quorum sensing and integrating the E7 lysis protein with pyocin S5. This intervention led to a 99% drop in *P. aeruginosa* growth and a 90% reduction in biofilm formation. The results of in vivo studies showed that including dispersin B, an anti-biofilm agent, into this modified strain greatly enhanced its efficacy in infection models, including *Caenorhabditis elegans* and mice. In addition, microcin H47 was added to *E. coli* Nissle 1917 (ECN) to specifically attack Salmonella infections. This was achieved by using a tetrathionate sensing mechanism to trigger the creation of H47, which effectively suppressed Salmonella. Furthermore, ECN was GM to produce the antibacterial peptide Mccj25, resulting in a 97% decrease in Salmonella populations while preserving the original microbiome.

The presence of *Vibrio cholerae* infection leads to a severe manifestation of diarrheal illness, which, if left untreated, can progress to shock and potentially culminate in mortality. This infection's pathogenesis is controlled by a quorum-sensing system that functions in a density-dependent way. Hence, there have been efforts to interfere with this sensing process by employing modified *E. coli* probiotics. The available research suggests that the administration of this particular strain successfully reduces the production of toxins in mouse models. Moreover, the ability of *L. lactis* to suppress *V. cholerae* infection by the secretion of lactic acid has been assessed. Within the realm of engineered probiotics, researchers have endeavored to develop resilient sensing systems with the ability to identify particular chemicals or produce substantial signals, therefore augmenting

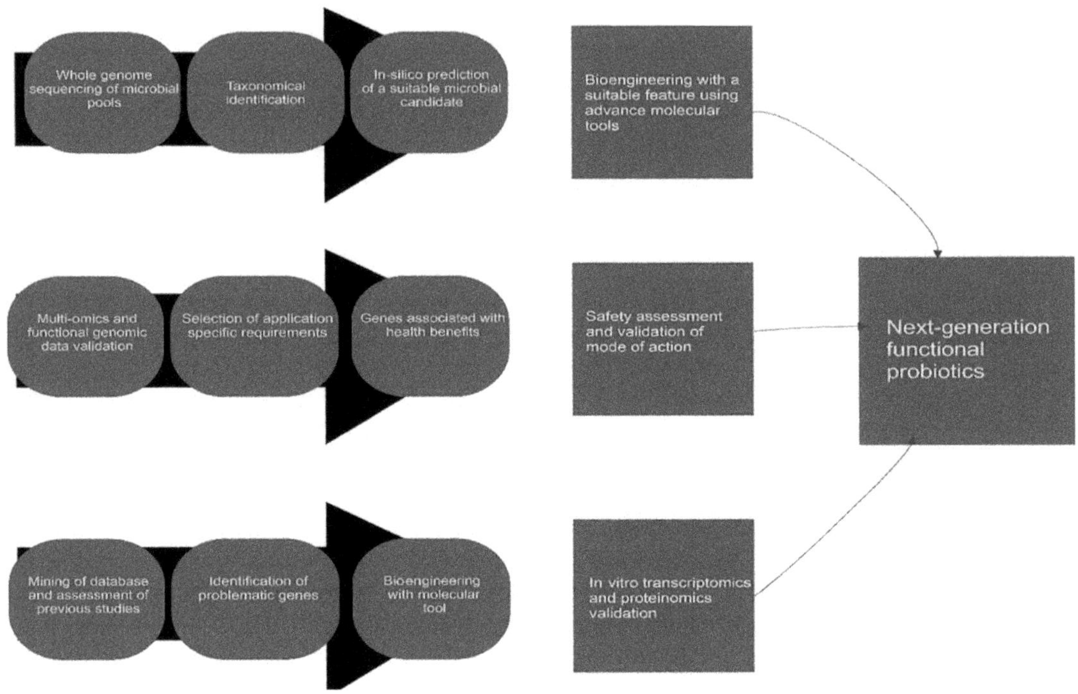

FIGURE 8.2 A schematic diagram outlines the pathways for developing next-generation probiotics. Probiotic bacteria can be engineered to improve their desirable traits using advanced molecular biology tools. Their effectiveness is then tested through in vivo evaluations, followed by safety assessments and validation of their mechanism of action.

their effectiveness as diagnostic tools. An integrated diagnostic circuit was created in *L. lactis* by combining the CqsS domain from *Vigiobacter cholerae* with the NisK domain from *L. lactis*. The introduction of this altered *L. lactis* strain to mice infected with cholera resulted in a significant stimulation of β-lactamase release, as measured by a colorimetric shift. In addition, Sedlmayer et al. (2018) integrated formyl peptide sensors with quorum-sensing devices to specifically target pathogenic infections. A GM strain of *E. coli* was developed to produce PROP-Z, which utilizes luciferase (lux CDABE) and β-galactosidase (lac Z) for the luminous identification of liver metastases. In addition, modified probiotics have been specifically designed to detect nitric oxide (NO), a prominent inflammatory indicator in inflammatory bowel disease (IBD), in order to evaluate the extent of inflammation.

Recent advances in metabolic engineering and synthetic biology have paved the way for the creation of novel probiotic strains with useful characteristics. To mitigate any potential dangers to patients, however, these tailored probiotics must first undergo comprehensive clinical testing (Figure 8.2). The genetic stability of newly introduced gene functions must be rigorously examined during the production phase. Moreover, additional research is necessary to ascertain the effective colonization of these engineered strains within the gut's ecological niche.

8.3.2 Research and Application of Engineered Probiotics

Significant advancements have been achieved in the field of gene editing technology, which has opened up new possibilities for genetically enhancing probiotic cultures. The development of GMPs has shown a substantial acceleration in recent years. Numerous studies have been conducted to explore the biological uses of these altered probiotics in the context of illness diagnosis and therapy. Significantly, a number of bioengineered probiotics have progressed to the stage of clinical trials. Moreover, scientists have created bioengineered microbes for several industrial applications. Through the utilization of the intrinsic properties of probiotics or the integration of new features, modified probiotics show significant potential for focused therapeutic treatments. The precise genetic modifications they possess allow for customized remedies to a range of health problems, including metabolic, immunological, and neurological diseases. Furthermore, apart from exhibiting therapeutic efficacy in animal models, numerous clinical studies utilizing probiotics have produced promising initial results. Recent studies have shown significant progress in using GMPs for the management of IBD, cancer, neurological diseases, and other medical problems.

Werner Kollath introduced the term "probiotic," derived from Greek and intended to signify "for life," initially in 1953. Probiotics, as defined, are live microorganisms that, when given in enough amounts, confer a health advantage to the host. The notion of probiotics emerged in the early 1900s, notably in 1906 when French physician Henry Tissier from the Pasteur Institute in Paris documented the advantages of *Bifidobacteria* in the treatment of pediatric intestinal illness. A study by Hill et al. (2014) revealed that the gut microbiome of breastfed children harbors *Bifidobacteria*, which have the ability to inhibit the development of pathogenic bacteria.

Subsequently, Elie Metchnikoff, a renowned Russian microbiologist and recipient of the Nobel Prize, suggested that the impressive lifespan seen in Bulgarians was linked to their significant consumption of fermented foods abundant in live LAB (Isolauri et al., 1995). The results of his research led to the emergence of two core concepts: the possibility of live bacteria engaging in advantageous interactions with people and the development of intricate communication mechanisms between these microbes and their hosts to enhance or reinstate health. In 1965, Lilly and Stilwell proposed the adoption of probiotics as a means to enhance intestinal well-being.

Recent studies have identified several bacteria, primarily from the *Bifidobacterium* and *Lactobacillus genera*, as probiotics. These indigenous microorganisms, present in the GIT, help maintain the equilibrium of intestinal microbiota by inhibiting harmful bacteria, safeguarding the integrity of the epithelial barrier, and aiding in food digestion. Nevertheless, the commercial exploitation of the word "probiotic" has resulted in substantial problems, as many medicines do not meet the necessary criteria. Regulatory authorities have grown more stringent in protecting customers from misleading marketing strategies. Current approval by the European Food Safety Authority (EFSA) is limited to a single probiotic claim. The enhanced lactose digestion is attributed to the yogurt cultures formed by *Lactobacillus delbrueckii* subsp. bulgaricus and *Streptococcus salivarius* subsp. thermophilus. Scientific evidence substantiates this assertion by demonstrating that these microorganisms synthesize β-galactosidase, an enzyme that facilitates the hydrolysis of lactose in yogurt and the gastrointestinal system. This is beneficial for individuals who are lactase-deficient, particularly those who often encounter symptoms such as diarrhea, abdominal pain, and gas following milk consumption. Advancements in technology, the emergence of the "omics" era (such as proteomics, metabolomics, and metagenomics), the implementation of closely regulated preclinical and clinical studies, and other contributing elements all indicate that new probiotic claims will be approved in the future that extend beyond lactose digestion.

The following sections will briefly outline some of the beneficial effects of different probiotic strains in treating human diseases, particularly gastrointestinal disorders.

8.3.3 Probiotics and COVID-19

Coronaviruses, which are distinguished by their unique crown-shaped surface features, form a heterogeneous cohort of viruses. Currently, there are seven officially recognized varieties of human coronaviruses (CoVs), which were first identified in the 1960s. Infections of the respiratory tract induced by CoVs often lead to changes in the gut microbiota. Significantly, COVID-19 has been associated with various gastrointestinal manifestations, lymphopenia, acute respiratory distress syndrome, and multi-organ dysfunction. In many cases, this disorder is accompanied by a phenomenon referred to as a "cytokine storm," characterized by the excessive synthesis of pro-inflammatory cytokines, which may cause harm to several organs such as the lungs, brain, liver, and cardiovascular system. Evidence from recent studies suggests that infection with COVID-19 is linked to a decrease in populations of beneficial bacteria, including *Bifidobacterium* spp. and *Lactobacillus* spp. The initial care of COVID-19 was based on mechanical ventilation and antiviral drugs. However, the strategy has now expanded to include vaccinations and biological agents designed to regulate the immune response. Baud et al. have identified six probiotics, namely *Bifidobacterium bifidum*, *Bifidobacterium breve*, *Lactobacillus casei*, *L. gasseri*, *L. plantarum*, and *Pediococcus pentosaceus*, as possible remedies to mitigate the impact of the COVID-19 epidemic. Furthermore, patients who experienced conditions such as antibiotic-associated diarrhea had a higher incidence of COVID-19, underscoring the significance of maintaining the equilibrium of microbiota in these individuals by the use of probiotic supplements. Infection with COVID-19 generally leads to significant hypoxia, which, when coupled with a decrease in probiotic species, might facilitate the proliferation of pathogenic bacteria such as *Actinobacteria* spp., *Corynebacterium* spp., and *Ruthenibacterium* spp. Notwithstanding the theoretical evidence suggesting that probiotics could mitigate cytokine storms linked to COVID-19, there remains an urgent need for more clinical and laboratory investigations. Experimental studies have shown that strains such as *Lactococcus paracasei* and *Lactococcus plantarum* have the ability to support the reduction of immune-inflammatory responses.

Studies have shown that probiotics from the *Lactobacillus* and *Bifidobacterium* genera can effectively treat gastrointestinal dysbiosis caused by COVID-19 disease. Moreover, probiotics that form biofilms, such as *L. reuteri* and selected species of *Lactobacillus*, are recognized for producing bioactive chemicals that demonstrate anti-inflammatory properties. The research undertaken by Bottari et al. (2020) demonstrated that probiotics have the potential to strengthen immune responses during COVID-19 by augmenting mucosal immunity via the production of IgA, boosting macrophage function, and activating regulatory T plasma cells. Furthermore, probiotics are acknowledged for their capacity to regulate specific classes of T cells, enhance the synthesis of antibacterial peptides, and impact the development of Th17 cells in the small intestine. The identification of specific *Lactobacillus* strains as peptide makers that engage with the ACE2 receptor, which is employed by SARS-CoV-2 for cellular entrance, has been reported. Recently, Minato et al. have identified *Paenibacillus* bacteria as natural reservoirs of carboxypeptidase, which shares resemblance with the ACE2 receptor.

8.3.4 Probiotics and COVID-19

Emerging studies indicate that the spike proteins of SARS-CoV-2 are binding to human angiotensin-converting enzyme 2 (ACE2). Bioengineered probiotics that express ACE2 have been developed in response to this interaction in order to neutralize the virus. Secretory human ACE2 (sACE2) is produced by the probiotic *L. paracasei* (LP), which has shown promise in containing the virus by either enclosing it or preventing the spike protein from binding to host cell receptors. The probiotic bacteria's produced sACE2 can regulate viral entrance into many organs, including the lungs, with systemic consequences. Furthermore, in SARS-CoV-2-infected people, these bioengineered probiotics improve innate immunity and help manage dysbiosis.

8.3.5 Engineered Probiotics and Inflammatory Bowel Disease

IBD is defined by persistent inflammation of the intestines caused by a disruption in the adaptive capacity of the intestinal microbiota. Effective management of IBD requires ongoing treatment measures, which have substantial medical and financial consequences. Therefore, it is crucial to create treatments that are cost-effective and easily administered, while simultaneously attenuating any adverse effects (Riglar et al., 2017). The occurrence of IBD is closely associated with the dysregulation of the microbiota. The microbiome may be enhanced by probiotics through the modification of the intestinal milieu, suppression of harmful bacteria, improvement of inflammatory disorders in the human immune system, and advantageous regulation of inflammation. The latest advancements in gene editing protocols have further established the association between probiotics and IBD. Dedicated probiotics have been specifically formulated for the sole indication of diagnosing and managing IBD. In 2017, researchers used genetically engineered bacteria as recognition components in biosensors specifically developed to identify IBD by detecting biomarkers such as thiosulphate, tetrasulphate, and NO. The primary focus of current research has been on the application of GMPs in the management of IBD. Achieving genetic alteration of ECN has enabled scientists to produce a matrix of curly fibers, therefore improving the structural integrity of the intestinal epithelium. Pichet Praveschotinunt and his team at Harvard University are recognized as outstanding researchers in this particular sector. The presence of trefoil factor (TFF) in this matrix highly enhances the structural integrity of the intestinal barrier and promotes the regeneration of the epithelium. Empirical evidence suggests that the modified extracellular matrix (ECM) has the ability to produce protective proteins at its initial site, therefore providing defense against colitis in mouse models induced by dextran sodium sulphate (DSS). This article proposes a prospective therapy strategy for IBD. Furthermore, some researchers have used modified extracellular symbiotic networks (ECN) to synthesise the Schistosoma immunomodulatory protein Sj16, as well as the ketone body (R)-3-hydroxybutyrate (3HB) and IL-10. These measures have been implemented to alleviate intestinal discomfort. Through the stimulation of probiotic growth, inhibition of pathogenic bacteria, augmentation of the intestinal microbiota, and inhibition of cells or proteins linked to inflammation, these compounds help to safeguard the intestinal mucosa. Moreover, they efficiently and precisely target and relieve the symptoms linked to intestinal inflammation. However, it is crucial to consider that these research studies have mostly demonstrated outcomes in mice models. Therefore, more clinical trials are required to assess the practicality of these modified extracellular vesicles for treating IBD in people.

Benjamin M. Scott from the University of Maryland and Cristina Gutierrez-Vazquez from Harvard Medical School have developed a probiotic yeast that autonomously regulates and expresses authentic human P2Y2 purinergic receptors. Through a self-regulatory mechanism and subsequent protein release, this innovative probiotic connects the activation of the P2Y2 receptor to the release of enzymes that degrade ATP, allowing it to recognize and counteract substances that promote inflammation. This modified yeast probiotic has demonstrated the capacity to decrease intestinal fibrosis and dysbiosis, as well as suppress intestinal inflammation, in preclinical investigations utilizing mice models of IBD. These findings indicate a potentially effective novel strategy for managing IBD and maybe other disorders by using GMPs.

Liu Jinyao and his research team at Shanghai Jiaotong University School of Medicine in China produced a polydopamine nano-immunosuppressant specifically designed to coat ECN. Within animal models of colitis, this approach effectively inhibits excessive immune responses and assists in the control of the gut flora to decrease inflammation. Studies indicate that administering immunosuppressants to mice exposed to beneficial microorganisms can enhance the progress of colitis therapy. This discovery has enabled the production of probiotics capable of expressing immunosuppressants for the treatment of inflammatory diseases.

Since the late 20th century, Gram-positive LAB have become potential carriers for delivering medicinal compounds orally, intranasally, vaginally, and rectally due to the need for novel modes of transporting these molecules to mucosal surfaces. The capacity of these bacteria to act as live mucosal delivery systems, particularly targeting food-grade strains of *L. lactis*, has been the subject of intense investigation over the past two decades. Notably, preclinical investigations using mouse models have demonstrated that genetically engineered *Lactococci* are highly efficient carriers for delivering functional proteins at the mucosal level. Through genetic modification, these *Lactococci* have effectively reproduced many human disorders, including irritable bowel syndrome (IBS), cancer, obesity, and diabetes.

An essential aim of probiotic therapy is to modulate the immunological response of the body and enhance its defense mechanisms. Hence, it can be inferred that GMPs have great potential in delivering therapeutically important constituents to mucosal surfaces, so aiding in the prevention and treatment of certain human diseases. Crucially, the adoption of good manufacturing practices (GMPs) has the capacity to lead to a substantial decrease in treatment costs, alleviate negative consequences, streamline administration, and enable the control of both systemic and mucosal immune responses. The most prevalent and convenient method for drug delivery is oral administration, which is characterized by its simplicity, noninvasiveness, and minimal discomfort for patients (McKay et al., 2018). Furthermore, this methodology has demonstrated effective therapy outcomes for specific gastrointestinal disorders, such as IBD and IBS. Thus far, *L. lactis* has been the most extensively modified lactic acid bacterium for the production and delivery of pharmaceutical medications. Furthermore, the advancements achieved in phase I and II clinical trials involving GM strains of *L. lactis* that generate IL-10 and tumor necrosis factor (TNF) for the management of Crohn's disease and mucositis, respectively, have created new opportunities for the application of gene regulatory programs (GMPs) in human medicine. Notwithstanding the significant advancements achieved with *L. lactis* as a live delivery carrier, this bacterium has many limitations, including a very short lifespan in the GIT and a lack of intrinsic immunomodulatory properties. In contrast, *Lactobacilli* possess the capacity to persist for an extended duration in the GIT and exhibit notable immunomodulatory characteristics, such as anti-inflammatory actions. The aforementioned characteristics possess the capacity to enhance the therapeutic effectiveness of GM strains in the treatment of IBD. Moreover, the pro-inflammatory properties can be advantageous in the context of immunization against infections, serving as an extra augmentation.

Contemporary research has focused on two other genera, *Bifidobacterium* and *Lactobacillus* spp., in order to produce proteins that have therapeutic importance by means of heterologous expression. Utilizing GM *Lactobacilli* or *Bifidobacteria* for the production and delivery of recombinant proteins is an extremely promising and expanding area of research. When utilized as live mucosal carriers, these taxa provide certain advantages in comparison to *L. lactis*. The advantages encompass enhanced survivability within the GIT and the immunomodulatory characteristics of particular strains. IBD is often linked to the presence of oxidative stress and cell membrane injury. Therefore, the distribution of antioxidant enzymes by GMPs has the potential to serve as valuable preventive and therapeutic agents (Binda et al., 2020). Therefore, the use of GMPs for the conveyance of antioxidant enzymes like catalase, glutathione peroxidase, glutathione reductase, glutathione-S-transferase, or superoxide dismutase (SOD) can substantially reduce the concentrations of reactive oxidative species (ROS) in the GIT. This strategy gives a feasible option for reducing and managing intestinal inflammation. In a mouse colitis model without IL-10, a genetically altered strain of *L. gasseri* that synthesizes SOD has shown anti-inflammatory effects. Experimental investigations on mice have demonstrated that GM strains of *L. casei* BL23, which generate either manganese-dependent catalase or SOD, offer significant defense against inflammation in the intestines and reactive oxygen species. An intricately designed study has shown that in a murine colitis model, two GM strains of *S. thermophilus* CRL807, a bacterium previously used as a yogurt starter and having anti-inflammatory and anticancer properties, as well as the ability to produce either catalase or SOD enzymes, have better anti-inflammatory effects than the wild-type strain. Evidence from recent studies has shown that proteases and the corresponding endogenous inhibitors of these enzymes have a substantial impact on the development of IBD.

8.3.6 *Bifidobacterium* as a Genetically Modified Probiotic for Drug Delivery

The genus *Bifidobacterium* has been recently utilized as a GMP for the specific application of medication delivery. Preliminary studies on the application of GM *Bifidobacterium* species for drug delivery primarily examined intravenous or systemic routes, rather than oral approaches, to treat solid tumors in different animal models (Chen et al., 2020). This strategy used the genus's ability to move from areas with limited blood supply to cancerous growths, which is associated with its strong attraction to anaerobic environments, such as the low oxygen conditions seen in solid tumors. Previous studies evaluated the administration of a genetically altered strain of *B. longum* orally to convey an anti-inflammatory molecule, namely the IL-10 cytokine, in mice with inflammation. A recent research study demonstrated that orally administering a GM *B. longum* strain expressing alpha-melanocyte-stimulating hormone (α-MSH) successfully relieved colitis in mice. These findings suggest a potentially effective alternative approach for treating IBD by using GMPs as living carriers for delivering α-MSH.

8.3.7 Engineered Probiotics and Bacterial Infections

Bacterial infections pose a considerable threat to public health, resulting in nearly one million fatalities each year globally. The challenge of combating these infections is exacerbated by the absence of effective bacterial resistance mechanisms and the limited availability of new antibiotics. In response, researchers have investigated innovative strategies to mitigate this problem, including the development of engineered probiotics.

An American research team, led by Arun K. Bhunia, has successfully adapted a strain of *L. casei* to produce Listeria adhesion protein exclusively. This modified strain effectively populates the intestines of mice, so competitively inhibiting Listeria colonization and systemic spread and offering protection against lethal infections. Moreover, it increases the abundance of mucosal regulatory T cells, CD11c+ dendritic cells, and natural killer cells in the intestinal milieu, therefore enhancing the regulation of the intestinal immune system. An especially impressive accomplishment is the intentional modification of *L. casei* ATCC334 to efficiently combat Listeria infections and safeguard the gastrointestinal system.

Simultaneously, other researchers have modified ECN to selectively target and eliminate the two primary vancomycin-resistant *Enterococcus* species, leading to a significant reduction in *Enterococcus faecalis* and *Enterococcus faecium* colonies in the feces of experimental mice. Moreover, Saccharomyces boulardii has undergone genetic modification to produce a fusion protein (ABAB) that efficiently neutralizes four distinct *Clostridium difficile* toxins. Prophylactic administration of these GM bacteria has been shown to substantially reduce inflammation and tissue damage in the intestines of mice suffering from *C. difficile*, resulting in a decrease in mortality rates.

Considerable advancements have been made by scientists in the domain of engineered probiotics with the explicit objective of enhancing the prevention, diagnosis, and therapy of bacterial illnesses. Probiotics capable of detecting the cholera autoinducer 1 (CAI-1) emitted by *V. cholerae*, resulting in noticeable changes in the color of fecal material, have been generated using an engineered version of *L. lactis* (Schultz et al., 2002). A further GM strain of LAB has been specifically engineered to identify and monitor real-time variations in the autoinducer peptide-I (AIP-I) produced by *Staphylococcus aureus.* In addition, researchers have modified ECN to establish interactions with the autoinducer N-acyl homoserine lactone (AHL) generated by *P. aeruginosa.* This contact induces the release of antibiotics and antibiofilm enzymes engineered specifically to attack these bacteria. The present study has the capacity to support the advancement of innovative strategies in the control and reduction of drug-resistant bacterial diseases. Additional validation of the safety and efficacy of these molecularly modified probiotics may accelerate their application in clinical environments in the foreseeable future.

8.3.8 Engineered Probiotics and Tumors

The utilization of bacteria as a therapeutic approach for cancer has been a subject of investigation for several decades. Certain probiotics have been shown to exert anticancer effects by promoting apoptosis in malignant cells and mitigating oxidative stress. Recent advancements in gene editing technologies have enabled the design of probiotics that actively participate in cancer treatment

(Leventhal et al., 2020). Ongoing research is focused on the development of novel probiotics aimed at targeting tumors effectively.

Candice R. Gurbatri and her team at Columbia University utilized synthetic biology methods to purposefully modify a particular strain of *E. coli*, known as "SLIC." Upon reaching a critical population density, this GM strain is capable of colonizing cancer cells and undergoing spontaneous lysis to produce nanoantibodies that target PD-L1 and CTLA-4. By promoting T cell activation, increasing the number of memory T cells, and improving the overall anti-tumor immune response, this strategy successfully suppresses tumor growth, triggers tumor regression, and prevents metastasis (Huizhuo et al., 2021). The findings of this study have significant ramifications, especially in improving methodologies for treating tumors and reducing the growth of cancer cells (Praveschotinunt et al., 2019).

In a parallel study, Chinese scientists adapted ECN to selectively transport the angiogenesis inhibitor TUM-5 and the tumor suppressor p53 to anaerobic tracts inside tumors. This novel methodology led to a substantial decrease in the proliferation of transplanted tumors in murine models, emphasizing the possibility of using bacterial tumor-targeting abilities for the administration of anti-tumor treatments. Furthermore, an alternative GM form of ECN was created with the purpose of establishing itself in tumors and transforming metabolic waste products into L-arginine so as to augment the immune response against tumors. The aforementioned strain significantly enhanced the therapeutic effectiveness of inhibitors that specifically target programmed cell death protein-1 and its ligand in models of mouse tumors. Moreover, a meticulously designed ECN strain with the ability to destroy tumors by inducing color changes in response to light has been effectively developed.

Recent research has shown significant advancements in the therapy of malignancies. Currently, there are two ongoing clinical trials (NCT04167137 and NCT04025307) focused on the use of genetically engineered probiotics for cancer treatment. Of specific significance is the SYNB1891 strain, which has been derived from ECN by the Synlogic Corporation. The current strain exhibits the ability to produce cyclic di-AMP, therefore initiating the interferon gene pathway and consequently augmenting innate immunity by inducing antigen-presenting cells to see tumor antigens. The ongoing Phase 1 clinical trials for this strain indicate that this GMP is anticipated to be available for clinical application in the near future (He et al., 2019).

8.3.9 Engineered Probiotics and Metabolic Diseases

Metabolic diseases arise from either insufficient or excessive accumulation of particular metabolites, which are caused by disturbed metabolic pathways. These conditions pose difficult treatment obstacles and typically need long-term changes in lifestyle and food, which impose substantial financial and personal hardships on those affected (Canale et al., 2021). Bioengineered probiotics have the potential to offer a practical treatment strategy by breaking down or transforming stored metabolites to relieve various disorders. PKU is a condition resulting from mutations in the phenylalanine hydroxylase gene, which generates increased concentrations of phenylalanine in the circulation, ultimately posing a significant risk to health. The phenylalanine lyase gene has been effectively inserted into the genetic modification of *L. reuteri* 100-23C, resulting in the development of a probiotic strain that can produce phenylalanine lyase. In a mouse model of PKU, this genetic alteration has shown the capacity to decrease blood phenylalanine levels (Gurbatri et al., 2020). The present work offers promising insights into the potential use of genetically engineered probiotics for the therapy of PKU. However, the prolonged colonization of the host by this strain presents an unresolved enigma. Furthermore, Synlogic has undertaken a comparable study by integrating genes that encode phenylalanine ammonia lyase and LAAD into the genetic composition of ECN. This has led to the creation of a hybrid strain known as SYNB1618. The present GMP has been tailored to efficiently degrade phenylalanine within the GIT while avoiding the formation of a durable colony. Evidence of clinical safety has been established (NCT03516487). Furthermore, the scientists improved their prognostic model for the therapeutic efficacy of SYNB1618 by integrating in vitro studies with an understanding of human physiology to quantify the assumed reduction in plasma phenylalanine levels induced by this probiotic. This breakthrough enables the utilization of the biological SYNB1618 enzyme.

The clinical trials for SYNB1618 (NCT04534842 and NCT04984525) are now in progress to explore the possible advantages of this bioactive medication for individuals suffering from PKU disorder. Mutations in the alanine glyoxylate aminotransferase gene can cause hyperoxaluria, a physiological disorder characterized by the accumulation of glyoxylic acid and subsequent increase in oxalate levels, resulting in the formation of urinary stones. In order to reduce urinary oxalic acid levels and alleviate kidney damage linked to hyperoxaluria, Synlogic has created an orally modified strain of *E. coli* called SYNB8802, which is capable of metabolizing oxalic acid within the GIT. An initial analysis of a Phase 1a study has shown a positive safety profile for this strain, and further trials are now under progress (NCT04629170). Furthermore, Novome has launched a therapeutic medication called NOV-001, specifically designed to target intestinal hyperoxaluria. Currently, clinical trials are being conducted to evaluate the efficacy of this formulation, which consists of the GM strain NB1000S and the polysaccharide NB2000P obtained from plants (NCT04909723). We believe that the combination of tailored probiotics with additional pharmaceutical therapies may offer a very promising strategy for illness management. Impaired clearance of ammonia, a highly neurotoxic byproduct, caused by deficits in urea cycle enzymes or liver disease, presents substantial hazards. Inadequate treatment of hyperammonemia can result in grave consequences, such as coma or mortality. Genetic modification of the oral probiotic ECN has resulted in the development of a modified strain called SYNB1020. This strain has shown the capacity to transform ammonia into L-arginine, leading to a decrease in intestinal ammonia levels in a mouse model of hyperammonemia and an improvement in survival rates. Nevertheless, the administration of SYNB1020 was discontinued in individuals with cirrhosis because it showed inadequate effectiveness (as assessed by plasma ammonia AUC) as compared to a placebo (NCT03447730).

Current clinical trials are underway for patients diagnosed with urea cycle abnormalities (NCT03179878), with the objective of attaining the best possible results. Concurrently, scientists have been creating bioengineered probiotics specifically designed to address certain metabolic problems. Significantly, a GM variant of *L. gasseri* has been developed to produce glucagon-like peptide-1 (GLP-1), which has demonstrated potential in improving the control of diabetes. The GLP-1 synthesized by this strain possesses the capacity to reprogram intestinal cells into cells that make insulin, hence correcting hyperglycemia in mice models. Further investigations have documented the action of a GM strain, EcN-GM, in enhancing the production of GLP-1 and potentially exerting beneficial effects on diseases such as obesity, hyperglycemia, and liver steatosis. Additionally, Precigen ActoBio has successfully created a genetically altered variant of *L. lactis*, which has been termed AG019. The purpose of this GM bacterium is to transport proinsulin and IL-10 to the gastrointestinal mucosa, with the aim of reducing harm to pancreatic β cells and improving or maintaining indigenous insulin synthesis. Furthermore, this study is now in progress (NCT03751007).

8.4 ENGINEERED BACTERIA AND INDUSTRIAL PRODUCTION

In addition to its application in illness diagnosis and treatment, the utilization of modified probiotics has significant implications for industrial production. By employing the CRISPRi technique, scientists, including Jaewoo Son and his colleagues at the Korea Academy of Science and Technology, have created a modified strain of Leuconostoc citrate that decreases the expression of two specific genes and incorporates a co-expression operon, leading to enhanced synthesis of riboflavin. A series of concurrent investigations have been conducted on *E. coli* LS31T to augment the synthesis of riboflavin, suggesting a promising prospect for its application as a dietary supplement (Jijon et al., 2004). Furthermore, photosynthetic cyanobacteria have the ability to transform carbon dioxide into monosaccharides. Substantial research has focused on the manipulation of *E. coli* to transform these sugars into 2,3-butanediol, a chemical with potential applications as rocket fuel or for interstellar exploration. In addition, a different GM strain of *E. coli* has been documented to synthesize benzoic acid from glucose derived from plants, providing a safer alternative for producing this often-used preservative. Empirical investigations on altered *Clostridium ljungdahlii* have demonstrated its capacity to produce isopropanol, 3-hydroxybutyric acid, and ethanol, emphasizing the potential of using single-carbon raw materials for the synthesis of novel chemicals. Prior research has focused on optimizing *E. coli*, LAB, and yeast to synthesize economically important compounds such as L-citrulline, flavan-3-ols, and β-carotene.

By acting as a physical barrier that prevents harmful bacteria from colonizing the GIT, probiotics serve a vital function in protecting against pathogenic organisms (O'Mahony et al., 2005). Specific strains of *Lactobacilli* and *Bifidobacteria* have demonstrated exceptional effectiveness in controlling diarrhea in newborns. Moreover, probiotics have been shown to regulate the immunological responsiveness of the host and enhance its defense mechanisms. In mouse models of intestinal inflammation, VSL#3, a formulation consisting of eight distinct bacterial strains, including *Lactobacilli* and *Bifidobacteria*, has demonstrated anti-inflammatory properties and protective advantages. Numerous clinical studies have emphasized the therapeutic benefits of particular probiotic strains in the setting of inflammatory bowel disorders (IBDs). Exemplary instances are VSL#3 for patients with pouchitis, ECN for people with ulcerative colitis, and *Lactobacillus salivarius* UCC4331 in conjunction with *Bifidobacterium* infantis 35624 for persons afflicted with irritable bowel syndromeIBS (Kruis et al., 2004). Experimental investigations using IL-10 mutant mice shown that the oral delivery of *Lactobacillus plantarum* effectively reduced the intensity of spontaneous colitis. Moreover, preclinical studies conducted on mouse models have shown that *L. casei* BL23 has the ability to augment systemic immunity and effectively defend against intestinal inflammation and colorectal cancer. *Bifidobacterium* animalis subsp. Lactis CNCM I-2494, which is abundantly present in fermented dairy products, has been shown to effectively repair the integrity of the gut barrier in instances of gut inflammation. Furthermore, *L. casei* has demonstrated the ability to stimulate an immunological response in children who are administered an oral vaccination for rotavirus, which is a common cause of sudden intestinal illness in newborns and young children in underdeveloped countries (Gionchetti et al., 2000).

8.4.1 Probiotics in Reducing Oxidative Stress and Treating IBD

In the context of IBD, probiotics have shown promise in reducing oxidative stress, a condition often characterized by an uncontrolled increase of ROS in the GIT. A multitude of studies have shown the effectiveness of different probiotic strains in the treatment of IBD, especially when using animal models. For instance, *L. rhamnosus* CNCM I-3690, known for its antioxidant activities in laboratory settings, has demonstrated anti-inflammatory characteristics in animal models of colitis and has showed protective effects against induced barrier hyperpermeability in experimental mice. These results suggest that probiotics might influence the adaptive immune response of the host and highlight the complex interactions between the host and its related bacterial populations.

8.4.2 Safety Concerns Regarding Engineered Probiotics

A comprehensive investigation is necessary to address the significant safety issues associated with the therapeutic potential of bioengineered probiotics. Engineered probiotics, being genetically modified organisms (GMOs), are subject to stringent regulatory frameworks, especially when they are designed for human use. Evaluation of the environmental safety and monitoring of the behavior of recombinant probiotics in laboratory and living organisms pose significant difficulties. Integration of these factors is crucial during the early design phases of probiotic engineering projects. One notable unresolved concern in biosafety is the development of bacterial hosts that lack the ability to survive in natural environments. Government regulations on GMOs emphasize the need of including biocontainment mechanisms into the genetic structures of probiotics to reduce and control their possible spread in the environment. Possible methods for biocontainment encompass conditional plasmid replication, auxotrophy, selective control of crucial gene expression, and the implementation of toxin-antitoxin systems. In order to guarantee efficient biocontainment of engineered probiotics, Gallagher et al. have suggested the implementation of multi-layered protective measures. Successful integration and expression of foreign DNA within designed probiotics present another significant obstacle. Introduction of foreign DNA may cause unintentional alterations and the disturbance of essential activities, perhaps resulting in the production of toxic compounds. Hence, it is imperative to carefully regulate alterations to the genetic makeup of modified probiotics in order to maintain the beneficial traits of the organism and avoid the generation of harmful byproducts in generally recognized as safe (GRAS) species.

8.5 PROSPECTS

To fully harness the potential of probiotic engineering, it is essential to ensure the safety of bioengineered probiotics. This necessitates addressing the regulatory, environmental, and technical hurdles that accompany their development and application. The advent of novel gene-editing technologies for the creation of synthetic engineered probiotics presents considerable promise for the diagnosis and treatment of a variety of human diseases. These engineered probiotics may provide an innovative strategy for tackling conditions such as cancer, inflammation, and infections. Researchers have optimized these probiotics to enhance their efficacy while minimizing side effects and reducing costs in comparison to conventional treatments or wild-type strains, thus extending benefits to a broader patient population and their families. The merits of engineered probiotics, including stability, specificity, preference, affordability, and relative safety, position them as a promising alternative for the treatment of numerous diseases and potentially more. Nevertheless, the deployment of engineered probiotics faces several obstacles. For example, many microorganisms do not possess adequate genetic tools for manipulation, and the strains currently utilized in research represent only a limited subset of the vast microbial diversity, constraining probiotic modifications to a select few strains. Furthermore, very few engineered probiotics have progressed from laboratory research to clinical application, and even those that enter clinical trials do not always yield the anticipated results, complicating the endeavors of researchers in this field. Additionally, ongoing concerns regarding GMOs continue to impede the effective integration of engineered probiotic therapies.

8.5.1 Challenges in the Application of Engineered Probiotics

Although engineered probiotics hold considerable promise, their implementation faces numerous obstacles. A primary issue is the scarcity of genetic tools for a wide range of microorganisms, which limits the ability to modify probiotics to a select few strains. Moreover, only a minor portion of the extensive microbial diversity is actively utilized in research initiatives. The shift from experimental studies to practical clinical applications presents a significant challenge, and even when engineered probiotics progress to clinical trials, there is no assurance that they will produce the intended outcomes. Additionally, public apprehensions regarding the safety of GMOs further complicate the introduction of engineered probiotic treatments.

8.6 CONCLUSIONS

While acknowledging the existing limitations, it is important to recognize the significant potential that engineered probiotics hold for enhancing human health. The journey of research and development in this domain will inevitably involve both achievements and setbacks. It is crucial to pursue further investigations to uncover avenues for success, which entails not only the creation of diverse probiotic modification techniques but also a comprehensive understanding of disease mechanisms. Such insights will aid in the formulation of effective probiotic solutions.

The effectiveness of engineered probiotics warrants thorough evaluation; however, prioritizing their safety is essential prior to any clinical application. Compliance with the standards set forth by health regulatory bodies is vital in the design and development of safe and efficacious probiotics. As our comprehension of the human microbiome and disease processes expands, coupled with the swift progress in gene editing technologies, it is anticipated that safe, stable, and effective engineered probiotics will play a pivotal role across various fields.

REFERENCES

Al Hawaj, M. A., Farid, A., Al Mohaini, M., Alsalman, A. J., Muzammal, M., Khan, M. H., … Ismail, S. (2022). Biosurfactant screening and antibiotic analysis of *Bacillus salmalaya. International Journal of Current Research and Review, 14*(12), 56–64.

Alsalman, A. J., Farid, A., Al Mohaini, M., Al Hawaj, M. A., Muzammal, M., Khan, M. H., … Ismail, S. (2022). Analysis and characterization of chitinase in *Bacillus salmalaya* strain 139SI. *International Journal of Current Research and Review, 14*, 31–36.

Barrangou, R., & Doudna, J. A. (2016). Applications of CRISPR technologies in research and beyond. *Nature Biotechnology, 34*(9), 933–941. doi:10.1038/nbt.3659.

Barrangou, R., Fremaux, C., Deveau, H., Richards, M., Boyaval, P., & Moineau, S.. (2007). CRISPR provides acquired resistance against viruses in prokaryotes. *Science (New York, NY), 315*(5819), 1709–1712. doi: 10.1126/science.1138140.

Bermúdez-Humarán, L. G., Aubry, C., Motta, J. P., Deraison, C., Steidler, L., Vergnolle, N., Chatel, J. M., & Langella, P. (2013). Engineering *Lactococci* and *Lactobacilli* for human health. *Current Opinion in Microbiology, 16*, 278–283.

Binda, S., Hill, C., Johansen, E., Obis, D., Pot, B., & Sanders, M. E. (2020). Criteria to qualify microorganisms as "probiotic" in foods and dietary supplements.

Börner, R. A., Kandasamy, V., Axelsen, A. M., Nielsen, A. T., & Bosma, E. F. (2019). Genome editing of lactic acid bacteria: Opportunities for food, feed, pharma, and biotech. *FEMS Microbiology Letters, 366*, 291. https://doi.org/10.1093/femsle/fny291

Bottari, B., Castellone, V., & Neviani, E. (2020). Probiotics and COVID-19. *International Journal of Food Science and Nutrition, 12*, 1–7.

Canale, F. P., Basso, C., Antonini, G., Perotti, M., Li, N., & Sokolovska, A. (2021). Metabolic modulation of tumours with engineered bacteria for immunotherapy. *Nature, 598*(7882), 662–666. https://doi.org/10.1038/s41586-021-04003-2

Charbonneau, M. R., Isabella, V. M., Li, N., & Kurtz, C. B. (2020). Developing a new class of engineered live bacterial therapeutics to treat human diseases. *Nature Communications, 11*, 1738. https://doi.org/10.1038/s41467-020-15508-1

Chen, K., Zhu, Y., Zhang, Y., Hamza, T., Yu, H., & Saint Fleur, A. (2020). A probiotic yeast-based immunotherapy against *Clostridioides difficile* infection. *Science Translational Medicine, 12*, 567. https://doi.org/10.1126/scitranslmed.abb5674

Ghazanfar, S., Ali, G. M., Abid, R., Farid, A., Akhtar, N., Batool, N. A., … Hameed, Y. (2022). *An overview of functional food.* IntechOpen.

Gionchetti, P., Rizzello, F., Venturi, A., Brigidi, P., Matteuzzi, D., Bazzocchi, G., Poggioli, G., Miglioli, M., & Campieri, M. (2000). Oral bacteriotherapy as maintenance treatment in patients with chronic pouchitis: A double-blind, placebo-controlled trial. *Gastroenterology, 119*, 305–309.

Gurbatri, C. R., Lia, I., Vincent, R., Coker, C., Castro, S., & Treuting, P. M. (2020). Engineered probiotics for local tumor delivery of checkpoint blockade nanobodies. *Science Translational Medicine, 12*, 530. https://doi.org/10.1126/scitranslmed.aba0437

He, L., Yang, H., Tang, J., Liu, Z., Chen, Y., & Lu, B. (2019). Intestinal probiotics *E. coli* Nissle as a targeted vehicle for delivery of p53 and tum-5 to solid tumors for cancer therapy. *Journal of Biological Engineering, 13*, 58. https://doi.org/10.1186/s13036-019-0171-4

Hill, C., Guarner, F., Reid, G., Gibson, G. R., Merenstein, D. J., Pot, B., Morelli, L., Canani, R. B., Flint, H. J., Salminen, S., Calder, P. C., & Sanders, M. E. (2014). Expert consensus, document. The International Scientific Association for probiotics and prebiotics consensus statement on the scope and appropriate use of the term probiotic. *Nature Reviews Gastroenterology & Hepatology*, *11*(8), 506–514.

Huizhuo, P., Lianyue, L., Gaoju, P., Chunli, H., Baona, L., & Yingying, Z. (2021). Engineered NIR light-responsive bacteria as anti-tumor agents for targeted and precise cancer therapy. *Chemical Engineering Journal*, *426*, 130842. https://doi.org/10.1016/j.cej.2021.130842

Isolauri, E., Joensuu, J., Suomalainen, H., Luomala, M., & Vesikari, T. (1995). Improved immunogenicity of oral d x RRV reassortant rotavirus vaccine by *Lactobacillus casei* GG. *Vaccine*, *13*, 310–312.

Jijon, H., Backer, J., Diaz, H., Yeung, H., Thiel, D., McKaigney, C., De Simone, C., & Madsen, K. (2004). DNA from probiotic bacteria modulates murine and human epithelial and immune function. *Gastroenterology*, *126*, 1358–1373.

Kruis, W., Fric, P., Pokrotnieks, J., Lukás, M., Fixa, B., Kascák, M., Kamm, M. A., Weismueller, J., Beglinger, C., Stolte, M., Wolff, C., & Schulze, J. (2004). Maintaining remission of ulcerative colitis with the probiotic *Escherichia coli* Nissle 1917 is as effective as with standard mesalazine. *Gut*, *53*(11), 1617–1623.

Lenoir, M., Del Carmen, S., Cortes-Perez, N. G., Lozano-Ojalvo, D., Munoz-Provencio, D., Chain, F., Langella, P., de Moreno de LeBlanc, A., LeBlanc, J. G., & Bermudez-Humaran, L. G. (2016). *Lactobacillus casei* BL23 regulates Treg and Th17 t-cell populations and reduces DMH-associated colorectal cancer. *Journal of Gastroenterology*, *51*(9), 862–873. https://doi.org/10.1007/s00535-016-1186-3

Leventhal, D. S., Sokolovska, A., Li, N., Plescia, C., Kolodziej, S. A., & Gallant, C. W. (2020). Immunotherapy with engineered bacteria by targeting the STING pathway for anti-tumor immunity. *Nature Communications*, *11*(1), 2739. https://doi.org/10.1038/s41467-020-16541-4

Martín, R., Miquel, S., Ulmer, J., Kechaou, N., Langella, P., & Bermúdez-Humarán, L. G.. 2013. Role of commensal and probiotic bacteria in human health: A focus on inflammatory bowel disease. *Microb Cell Fact, 12*, 71.

McKay, R., Ghodasra, M., Schardt, J., Quan, D., Pottash, A. E., & Shang, W. (2018). A platform of genetically engineered bacteria as vehicles for localized delivery of therapeutics: Toward applications for Crohn's disease. *Bioeng Transl Med*, *3*(3), 209–21.

Muhammad, Z., Anjum, M. Z., Akhter, S., Irfan, M., Amin, S., Jamal, Y., … Ghazanfar, S.. (2023). Effect of *Lactobacillus plantarum* and *Pediococcus pentosaceus* on the growth performance and morphometry of the genetically improved farmed tilapia (*Oreochromis niloticus*). *Pakistan Journal of Zoology*, *56*(1), 9–16.

O'Mahony, L., McCarthy, J., Kelly, P., Hurley, G., Luo, F., Chen, K., O'Sullivan, G. C., Kiely, B., Collins, J. K., Shanahan, F., & Quigley, E. M. M. (2005). *Lactobacillus* and *Bifidobacterium* in irritable bowel syndrome: Symptom responses and relationship to cytokine profiles. *Gastroenterology*, *128*, 541–551. https://doi.org/10.1053/j.gastro.2004.11.050.

Oh, J.-H., & van Pijkeren, J.-P. (2014). CRISPR–Cas9-assisted recombineering in *Lactobacillus reuteri*. *Nucleic Acids Research*, *42*(17), e131. https://doi.org/10.1093/nar/gku623

Praveschotinunt, P., Duraj-Thatte, A. M., Gelfat, I., Bahl, F., Chou, D. B., & Joshi, N. S. (2019). Engineered *E. coli* Nissle 1917 for the delivery of matrix-tethered therapeutic domains to the gut. *Nature Communications*, *10*(1), 5580.

Pusch, O., Boden, D., Hannify, S., Lee, F., Tucker, L. D., & Boyd, M. R. (2005). Bioengineering lactic acid bacteria to secrete the HIV-1 virucide cyanovirin. *Journal of Acquired Immune Deficiency Syndromes*, *40*, 512–520.

Riglar, D. T., Giessen, T. W., Baym, M., Kerns, S. J., Niederhuber, M. J., & Bronson, R. T. (2017). Engineered bacteria can function in the mammalian gut long-term as live diagnostics of infammation. *Nature Biotechnology*, *35*(7), 653–658.

Roberts, A., & Barrangou, R. (2020). Applications of CRISPR-cas systems in lactic acid bacteria. *FEMS Microbiology Reviews*, *44*(5), 523–537. doi: 10.1093/femsre/fuaa016.

Rochet, V., Rigottier-Gois, L., Ledaire, A., Andrieux, C., Sutren, M., Rabot, S., Mogenet, A., Bresson, J. L., Cools, S., Picard, C., Goupil-Feuillerat, N., & Doré, J.. 2008. Survival of *Bifidobacterium* animalis DN-173 010 in the faecal microbiota after administration.

Schultz, M., Veltkamp, C., Dieleman, L. A., Grenther, W. B., Wyrick, P. B., Tonkonogy, S. L., & Sartor, R. B. (2002). *Lactobacillus plantarum* 299V in the treatment and prevention of spontaneous colitis in interleukin-10-deficient mice. *Inflammatory Bowel Disease*, *8*, 71–80. https://doi.org/10.1097/00054725-200203000-00001.

Sedlmayer, F., Hell, D., Müller, M., Ausländer, D., & Fussenegger, M. (2018). Designer cells programming quorum-sensing interference with microbes. *Nature Communications*, *9*, 1822.

Wang, L., Liao, Y., Yang, R., Zhu, Z., Zhang, L., & Wu, Z. (2021). An engineered probiotic secreting Sj16 ameliorates colitis via *Ruminococcaceae*/butyrate retinoic acid axis. *Bioengineering & Translational Medicine*, *6*(3), e10219.

Yan, X., Liu, X. Y., Zhang, D., Zhang, Y. D., Li, Z. H., & Liu, X. (2021). Construction of a sustainable 3-hydroxybutyrate-producing probiotic *Escherichia coli* for treatment of colitis. *Cellular & Molecular Immunology*, *18*(10), 2344–2357.

Zhang, M., Eshraghian, E. A., Jammal, O. A., Zhang, Z., & Zhu, X. (2021). CRISPR technology: The engine that drives cancer therapy. *Biomedicine & Pharmacotherapy*, *133*, e45.

9 Microbiota-Targeted Foods in Relation to Probiotics

Rabia Mehmood, Maliha Rashid, Saiqa Noureen, Hafiz Muhammad Bilal Akhtar, Muhammad Zia ur Rehman, Aatka Jamil, Nazia Rehman, Shakira Ghazanfar, Haseeb Raza, and Amina Najam

9.1 INTRODUCTION

Microbiota in our body is shaped by the food we eat and tends to affect our health. Our immune system is closely related to the microbiota. An imbalance in the microbiota called dysbiosis is caused by a disturbance in the diversity of the microbiota. Dysbiosis causes many health issues, such as inflammation and chronic diseases. Taking prebiotics and probiotics in our food can shape and diversify the gut microbiota and prevent many diseases in our body (Valdes et al. 2018). The term symbiosis means a product that combines probiotics and prebiotics. Probiotics can be used to solve various health issues, such as diarrhea caused by antibiotics, infant sepsis, infant colic, and ulcerative colitis (Markowiak and Śliżewska 2017; Sekhon and Jairath 2010). An imbalance in the gut microbiota can lead to many gastrointestinal problems, such as gas, bloating, constipation, and diarrhea. Taking probiotics with antibiotics can prevent these issues and also help with the immune system. Probiotics associated with the foods are the easiest ways to balance microbiota for a healthy gut, such as yogurt, kefir, sauerkraut, which is fermented vegetables or pickles, and kimchi. However, there is a need for more scientific studies to determine whether they are safe to take for adults (Akbar et al. 2024; Dahiya and Nigam 2023; Éliás et al. 2023). Probiotics can create a balance in the healthy digestive system and can prevent many gastrointestinal issues that are either due to taking antibiotics, due to some illness, or due to a lack of nutrients in the food we are eating. However, probiotics should be prescribed by the physician because gastrointestinal problems and other illnesses are related to the specific strain. Some individuals reported serious infections while taking probiotics. Some foods that contain rich probiotics include kombucha, miso, pickles, tempeh, and Greek yogurt (Peng et al. 2020; Rau et al. 2024; Reyes-Castillo et al. 2021). Some countries, such as Japan, China, and Korea, have a tradition of eating these probiotic-rich foods regularly. There have been some cases of infants having serious infections who were given the probiotics. The probiotic we should take must be FDA-approved and prescribed by the physician.

Due to the development of scientific techniques such as majorly next-generation sequencing, it is possible to understand complex gastrointestinal microbiota. Our diet is directly related to the gut microbiome balance. Gut microbiota coverts the food components into the nutrients that affect health directly. In short, a healthy and balanced microbiome leads to a healthy person.

Gut microbiota is closely affecting the innate and adaptive immune systems. The flagellin production by microorganisms in our gut is regulated by the immune system, and this flagellin helps in homeostasis. Due to the influence of microbiota on the adaptive immune system, it can resist many diseases. Thus, our immune system is shaped by the gut microbiota. Our central nervous system, skin, and sleep cycle are also affected by the immune system, which is directly linked to the gastrointestinal microbial communities. Taking probiotics in our diet can improve our immune system (Wang, Zhang, and Zhang 2021). In this chapter, we focused on how our diet can impact on the gut microbiota, impact on our health, and food that can be taken as probiotics.

DOI: 10.1201/9781003399797-10

9.2 DIET INFLUENCING MICROBIOTA COMPOSITION

There are several factors that impact the microbial composition, such as genetics of the host, diet, age, medication, antibiotics, and stress. Diet can significantly influence the balance of microbiota in our gut. A healthy diet can help to colonize the beneficial microbial communities, such as fiber in our diet, playing an important role in the colonization of the beneficial microbiota (Conlon and Bird 2015).

Actinobacteria, Proteobacteria, Firmicutes, and *Bacteroidetes* members make up the majority of the gut microbiota in infants, and the quantity of each phylum is found in accordance with the sequence in which they occur. On the other hand, in adults, the abundance changes to *Firmicutes,* then *Bacteroidetes*, and then *Actinobacteria.* While adults do contain individuals of *Proteobacteria, Fusobacteria, Cyanobacteria,* and *Verrucomicrobia*, their numbers are quite low (Ghosh and Pramanik 2021). Only some of the species make up 99% of the total biomass. Just two phyla, *Firmicutes* and *Bacteroidetes*, account for nearly all (~90%) of the bacteria found in the human gut. Most people have comparable proportions of each of the two prominent phyla, while the proportional amounts of each can fluctuate and be impacted by a variety of events ("Categorization of the Gut Microbiota: Enterotypes or Gradients? | Nature Reviews Microbiology" 2023).

Dietary patterns can influence the microbial population on genus and species levels, but do not influence microbial communities on the phylum level ("Linking Long-Term Dietary Patterns with Gut Microbial Enterotypes | Science" 2023). Different components of the diet affect human gut microbiota differently; these components include protein, carbohydrates, fats, vitamins, and other micronutrients, as shown in Figure 9.1.

Arrows in Figure 9.1 showed the dominancy and decreasing the microbial communities. Growth rate of microbial species increases as well as decreases based on the type of diet one is taking, such as the *Bacteroids* species growth rate, which increases as a result of animal-based protein and decreases while taking plant-based protein. The relative diversity of microbiota becomes dominant based on protein types such as whey protein, plant extract protein, and animal protein. A high-protein diet can increase the insulin-like growth factor 1 (IGF-1) as a result of a high mortality rate,

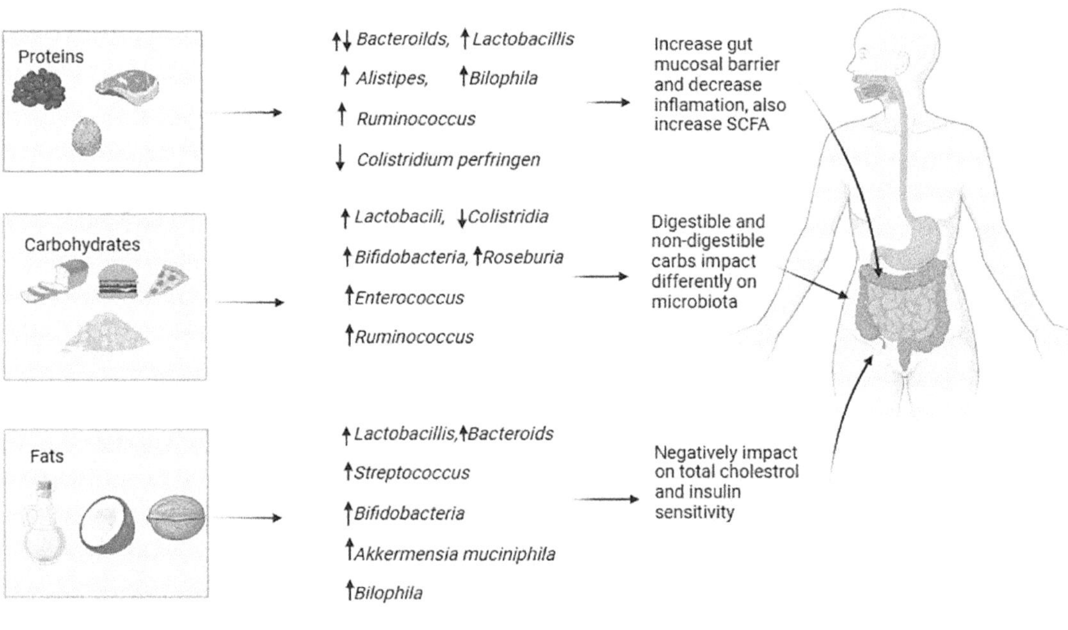

FIGURE 9.1 Effect of different nutrients in diet on the gut microbiota.

but on the other hand, a diet based on plant protein can decrease the mortality rate and lower the risk of some diseases, such as cancer and diabetes (Levine et al. 2014).

Complex carbohydrates in dietary fiber reduce microbial growth, alter community composition, and reduce sensitivity to perturbations in synthetic human gut communities. Fermentation kinetics and end-products of non-digestible carbohydrates in broiler chickens' caecal contents vary significantly, potentially impacting gut microflora and health (Lan et al. 2005; Loss et al. 2022). Short-chain fatty acids (SCFAs) produced by fermentation of resistant starches and dietary fiber may reduce the risk of developing gastrointestinal disorders, cancer, and cardiovascular disease. SCFAs produced by gut microbiome fermentation of dietary fiber have numerous health benefits, including anti-inflammatory, immunoregulatory, and anti-obesity properties. SCFAs in prebiotics may enhance colonic health and contribute to overall health benefits, but human study evidence is inconsistent (Roy et al. 2006; Wong et al. 2006; Xiong et al. 2022).

High-fat diets with high levels of lipid oxidation products (LOPs) cause selective alteration of the gut microbiome and mycobiome compared to high-fat diets alone, and studies showed that gut microbiome plays a significant role in modulating fatness in pigs, with specific microbes producing SCFAs that may regulate energy homeostasis, protect against inflammation, and inhibit fat mass development (Arowolo et al. 2020; He et al. 2016).

Selected diets (Figure 9.2), such as ketogenic diet, impact on the composition of the gut microbiota. Ketogenic diets alter gut microbiome structure and function, leading to decreased intestinal proinflammatory Th17 cells, unlike high-fat diets (Ang et al. 2020). This is a very low-calorie diet that reduces the amount of SCFAs in the gut. It may restore the health of the gut but may diminish the diversity of the commensal microbial communities, such as *Bifidobacteriace*, may decrease *E. coli* and Shigella. Keto-diet alter the composition of the microbial community, which develops disease symptoms, such as an increase in *Bacteroids* to *Firmicutes* ratio and a decrease in *Proteobacteria* (Gutiérrez-Repiso et al. 2019; Kong et al. 2021; Rondanelli et al. 2021).

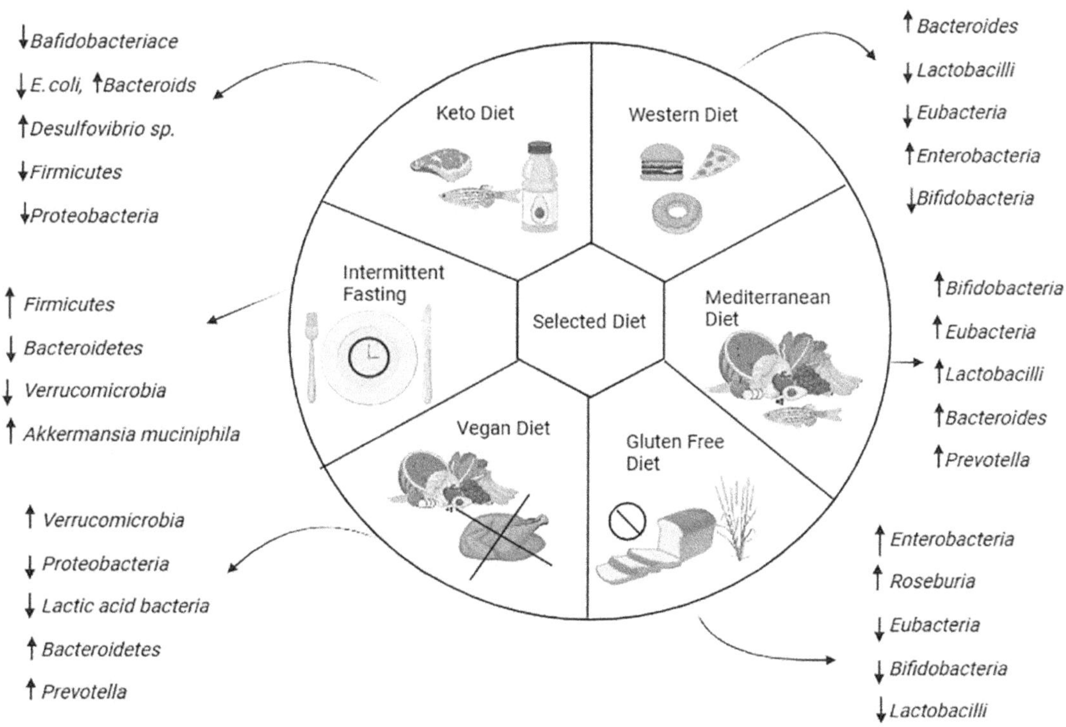

FIGURE 9.2 Effect of selected diets on the microbial composition of gut.

Intermittent fasting is a dietary pattern in which an individual takes meals alternatively after specific hours. Intermittent fasting effectively reduces body weight, improves insulin sensitivity, blood lipids, and lowers blood pressure, effectively preventing and treating metabolic diseases like obesity and type-2 diabetes (Liu and Zhu 2018). Plant-based diets may promote a more favorable gut microbial profile, potentially reducing cardiometabolic risk factors and diabetes incidence. Well-planned vegan diets positively impact gut microbiota diversity and metabolism, with potential benefits for medical disorders but also limitations, including nutritional deficiencies (Sakkas et al. 2020; Wong 2014).

9.3 PROBIOTIC

Probiotics may help reduce gut inflammation by improving gut microbial population, increasing mucus secretion, and preventing the destruction of tight junction proteins (Cristofori et al. 2021). It can also improve gut health, prevent diarrhea, and modulate immunity, potentially preventing cancer and lowering cholesterol levels, such as *Lactobacilli* and *Bifidobacteria* (Kaur, Chopra, and Saini 2002). Probiotic products tailored to specific human populations, such as extreme age groups and patients with diseases, can provide nutritional benefits and enhance health (Arboleya et al. 2012). Probiotics can be safely and effectively encapsulated in food, beverages, and supplements, offering health benefits and enhancing the immune system (Vandamme et al. 2016). Fermented foods and beverages, such as kefir, kombucha, sauerkraut, tempeh, natto, miso, kimchi, and sourdough bread, can reduce the risk of type 2 diabetes, metabolic syndrome, and heart disease and improve weight management (Farooqui 2021). *Lactobacillus plantarum* LHB55, an antibacterial bacterium from Kimchi, has high antibacterial activity without affecting microbial or physicochemical properties (Lee et al. 2010). Other fermented foods, particularly kefir, have limited clinical evidence for their effectiveness in gastrointestinal health and disease, with potential benefits in lactose malabsorption and Helicobacter pylori eradication (Dimidi et al. 2019). Kombucha consumption reduces oxidative stress, inflammation, and intestinal dysbiosis, improving obesity control and promoting gut microbiota modulation (Costa et al. 2021).

Yogurt consumption with probiotics, particularly *Lactobacillus amylovorus* and *Lactobacillus fermentum*, reduces body fat mass and alters gut microflora composition, potentially contributing to altered energy metabolism and body composition (Omar et al. 2013). High consumption of yogurt, rich in lactic acid-producing bacteria, improves gastrointestinal health and longevity by promoting gut microflora, bowel transit, and enhancing gastrointestinal immune responses. It also stimulates the activity of colon bacteria such as *Bifidobacteria* and *Lactobacilli* (Mazahreh and Ershidat 2009). Consumption of fermented food is the most convenient way to take probiotics.

9.4 RULES AND REGULATIONS FOR CONSUMPTION OF PROBIOTICS

Probiotics and prebiotics currently lack positive recommendations from international medical societies for treating or preventing food allergy, allergic rhinitis, and asthma, with varying recommendations for eczema prevention in high-risk infants (Koletzko 2016). Autoprobiotics, derived from healthy individuals' intestines, show potential in treating and preventing various diseases, with a recommended daily dose of 10^8–10^9 CFU (Okrugina et al. 2022). Higher doses of probiotics, above 10^{10} CFU/day, are required for strains to survive, persist, and be efficacious in the gut, with treatment duration not affecting fecal recovery (Morelli and Pellegrino 2021). The USDA meal plan has the highest total amount of microbes ingested per day, followed by the VEGAN and AMERICAN meal plans, with no significant difference in diversity among the three dietary patterns (Lang, Eisen, and Zivkovic 2014).

9.5 CONCLUSION

This chapter delves into the intricate relationship between diet, gut microbiota, and health outcomes, particularly focusing on the impact of microbiota-targeted foods like probiotics. The symbiotic relationship between our dietary choices and the composition of the gut microbiome significantly influences our health and immune system. It's clear that dysbiosis, an imbalance in gut microbiota, can lead to various health issues, including inflammation and chronic diseases.

However, caution is advised, especially with self-prescription, as specific strains of probiotics should be tailored to individual needs and prescribed by a physician due to potential adverse effects reported by some individuals. The chapter underscores that diet profoundly impacts microbial populations, notably at the genus and species levels, while certain factors like genetics, age, medication, antibiotics, and stress also influence the microbiota. Various nutrients, such as protein, carbohydrates, fats, vitamins, and other micronutrients, exhibit differential effects on gut microbial communities. For instance, high-protein diets have contrasting effects based on the protein source, influencing microbial diversity and potentially impacting disease risk. Different dietary patterns, including high-fat diets, ketogenic diets, intermittent fasting, and plant-based diets, have unique implications for gut microbiota composition and overall health. Probiotics emerge as potential aids in reducing gut inflammation, improving gut health, preventing diarrhea, modulating immunity, and impacting disease prevention. Various fermented foods and beverages, when consumed regularly, offer potential health benefits like reducing the risk of type 2 diabetes, heart disease, and obesity. However, caution is advised regarding their clinical effectiveness, especially in specific health conditions, as evidence for some fermented foods remains limited. Regulations surrounding probiotics highlight the need for cautious consumption, particularly emphasizing the lack of unanimous positive recommendations from medical societies for treating or preventing certain conditions. Dosage and strain specificity are crucial factors to consider for probiotic efficacy and survival in the gut.

In summary, the chapter elucidates the profound impact of diet on gut microbiota composition, emphasizing the potential benefits and cautionary measures associated with microbiota-targeted foods like probiotics. While these foods hold promise in promoting health and preventing diseases, their usage should be approached judiciously under medical guidance due to varying individual responses and potential risks associated with certain dietary interventions.

REFERENCES

"Categorization of the Gut Microbiota: Enterotypes or Gradients? | Nature Reviews Microbiology." 2023. Accessed November 8. https://www.nature.com/articles/nrmicro2859.

"Linking Long-Term Dietary Patterns with Gut Microbial Enterotypes | Science." 2023. Accessed November 8. https://www.science.org/doi/abs/10.1126/science.1208344.

Akbar, Muhammad Yasir, Humera Azad, Maliha Rashid, Wajya Ajmal, Adnan Ahmad Ansari, Mohamed M. Salem, Ram Kumar Sahu, Mounir M. Salem-Bekhit, and Shakira Ghazanfar. 2024. "ES-PredHSP: Improved Prediction of Heat Shock Proteins Using Machine Learning by Enhanced Sampling Technique." *Journal of Biological Regulators and Homeostatic Agents* 38 (1): 665–673.

Ang, Qi Yan, Margaret Alexander, John C. Newman, Yuan Tian, Jingwei Cai, V. Upadhyay, and Jessie A. Turnbaugh et al. 2020. "Ketogenic Diets Alter the Gut Microbiome Resulting in Decreased Intestinal Th17 Cells." *Cell* 181: 1263–1275. doi: 10.1016/j.cell.2020.04.027.

Arboleya, S., Sonia González, N. Salazar, P. Ruas-Madiedo, C. G. de los Reyes-Gavilán, and M. Gueimonde. 2012. "Development of Probiotic Products for Nutritional Requirements of Specific Human Populations." *Engineering in Life Sciences* 12. doi: 10.1002/elsc.201100129.

Arowolo, Folagbayi K., Joseph F. Pierre, Morgan Blaser, and D. Shanmuganayagam. 2020. "Longitudinal Effects of Dietary Oxidized Lipids on the Gut Microbiome and Mycobiome in Pigs." *The FASEB Journal* 34. doi: 10.1096/fasebj.2020.34.s1.09676.

Conlon, Michael A., and Anthony R. Bird. 2015. "The Impact of Diet and Lifestyle on Gut Microbiota and Human Health." *Nutrients* 7 (1): 17–44. Multidisciplinary Digital Publishing Institute. doi: 10.3390/nu7010017.

Costa, M. a. C., D. L. S. Vilela, G. M. Fraiz, Isabelle Lima Lopes, Ana Íris Mendes Coelho, Luiza Carla Vidigal Castro, and J. G. P. Martin. 2021. "Effect of Kombucha Intake on the Gut Microbiota and Obesity-Related Comorbidities: A Systematic Review." *Critical Reviews in Food Science and Nutrition*: 1–16. doi: 10.1080/10408398.2021.1995321.

Cristofori, F., V. Dargenio, C. Dargenio, V. Miniello, M. Barone, and R. Francavilla. 2021. "Anti-Inflammatory and Immunomodulatory Effects of Probiotics in Gut Inflammation: A Door to the Body." *Frontiers in Immunology* 12. doi: 10.3389/fimmu.2021.578386.

Dahiya, Divakar, and Poonam Singh Nigam. 2023. "Antibiotic-Therapy-Induced Gut Dysbiosis Affecting Gut Microbiota—Brain Axis and Cognition: Restoration by Intake of Probiotics and Synbiotics." *International Journal of Molecular Sciences* 24 (4): 3074. Multidisciplinary Digital Publishing Institute. doi: 10.3390/ijms24043074.

Dimidi, E., S. Cox, M. Rossi, and K. Whelan. 2019. "Fermented Foods: Definitions and Characteristics, Impact on the Gut Microbiota and Effects on Gastrointestinal Health and Disease." *Nutrients* 11. doi: 10.3390/nu11081806.

Éliás, Anna Júlia, Viktória Barna, Cristina Patoni, Dóra Demeter, Dániel Sándor Veres, Stefania Bunduc, Bálint Erőss, Péter Hegyi, László Földvári-Nagy, and Katalin Lenti. 2023. "Probiotic Supplementation during Antibiotic Treatment Is Unjustified in Maintaining the Gut Microbiome Diversity: A Systematic Review and Meta-Analysis." *BMC Medicine* 21 (1): 262. doi: 10.1186/s12916-023-02961-0.

Farooqui, A. A. 2021. "Importance of Fermented Foods on Human Health," 69–86. doi: 10.1016/B978-0-12-821039-0.00012-5.

Ghosh, Soma, and Sreemanta Pramanik. 2021. "Structural Diversity, Functional Aspects and Future Therapeutic Applications of Human Gut Microbiome." *Archives of Microbiology* 203 (9): 5281–5308. doi: 10.1007/s00203-021-02516-y.

Gutiérrez-Repiso, C., C. Hernández-García, J. M. García-Almeida, D. Bellido, G. Martín-Núñez, L. Sánchez-Alcoholado, J. Alcaide-Torres, I. Sajoux, F. Tinahones, and I. Moreno-Indias. 2019. "Effect of Synbiotic Supplementation in a Very-Low-Calorie Ketogenic Diet on Weight Loss Achievement and Gut Microbiota: A Randomized Controlled Pilot Study." *Molecular Nutrition & Food Research*. doi: 10.1002/mnfr.201900167.

He, Maozhang, S. Fang, Xiaochang Huang, Yuanzhang Zhao, S. Ke, Hui Yang, Zhuo-min Li, Jun Gao, Congying Chen, and Lusheng Huang. 2016. "Evaluating the Contribution of Gut Microbiota to the Variation of Porcine Fatness with the Cecum and Fecal Samples." *Frontiers in Microbiology* 7. doi: 10.3389/fmicb.2016.02108.

Kaur, I., K. Chopra, and A. Saini. 2002. "Probiotics: Potential Pharmaceutical Applications." *European Journal of Pharmaceutical Sciences : Official Journal of the European Federation for Pharmaceutical Sciences* 15 (1): 1–9. doi: 10.1016/S0928-0987(01)00209-3.

Koletzko, S. 2016. "Probiotics and Prebiotics for Prevention of Food Allergy: Indications and Recommendations by Societies and Institutions." *Journal of Pediatric Gastroenterology and Nutrition* 63: 9–10. doi: 10.1097/MPG.0000000000001220.

Kong, C., Xuebing Yan, Yongqiang Liu, Linsheng Huang, Yefei Zhu, J. He, and R. Gao et al. 2021. "Ketogenic Diet Alleviates Colitis by Reduction of Colonic Group 3 Innate Lymphoid Cells through Altering Gut Microbiome." *Signal Transduction and Targeted Therapy* 6. doi: 10.1038/s41392-021-00549-9.

Lan, Y., B. Williams, S. Tamminga, H. Boer, A. Akkermans, G. Erdi, and M. Verstegen. 2005. "In Vitro Fermentation Kinetics of Some Non-Digestible Carbohydrates by the Caecal Microbial Community of Broilers." *Animal Feed Science and Technology* 123: 687–702. doi: 10.1016/J.ANIFEEDSCI.2005.04.027.

Lang, Jenna M., J. Eisen, and A. Zivkovic. 2014. "The Microbes We Eat." *PeerJ*. https://consensus.app/papers/microbes-lang/9c2714e383e35200a5dd218e5809e629/.

Lee, Seung-Gyu, Yeon-Jung Lee, Min-kyung Kim, Ki-Sung Han, Seok-Geun Jeong, M. Oh, A. Jang, In-Hyu Donghoon Kim, J. Bae, and Ham. 2010. "A Study on the Yogurt Manufacture Suitability and Antimicrobial Activity of Lactobacillus Plantarum LHB55 Isolated from Kimchi." *Journal of Animal Science and Technology* 52: 141–148. doi: 10.5187/JAST.2010.52.2.141.

Levine, Morgan E., Jorge A. Suarez, Sebastian Brandhorst, Priya Balasubramanian, Federica Chia-Wei Cheng, Luigi Madia, and Fontana et al. 2014. "Low Protein Intake Is Associated with a Major Reduction in IGF-1, Cancer, and Overall Mortality in the 65 and Younger but Not Older Population." *Cell Metabolism* 19 (3): 407–417. Elsevier. doi: 10.1016/j.cmet.2014.02.006.

Liu, Meijuan, and Huijuan Zhu. 2018. "Effects of Intermittent Fasting in Improving Metabolic Diseases and Its Mechanism" 38: 415–18. doi: 10.3760/CMA.J.ISSN.1673-4157.2018.06.014.

Loss, Erin Ostrem, Jaron Thompson, Kevin Cheung, Y. Qian, and Ophelia S. Venturelli. 2022. "Growth Limitation via Chemically Complex Carbohydrates Reduces Sensitivity of Synthetic Human Gut Communities to Perturbations." *bioRxiv*. doi: 10.1101/2022.04.08.487701.

Markowiak, Paulina, and Katarzyna Śliżewska. 2017. "Effects of Probiotics, Prebiotics, and Synbiotics on Human Health." *Nutrients* 9 (9): 1021. Multidisciplinary Digital Publishing Institute. doi: 10.3390/nu9091021.

Mazahreh, A., and O. Ershidat. 2009. "The Benefits of Lactic Acid Bacteria in Yogurt on the Gastrointestinal Function and Health." *Pakistan Journal of Nutrition* 8: 1404–1410. doi: 10.3923/PJN.2009.1404.1410.

Morelli, L., and P. Pellegrino. 2021. "A Critical Evaluation of the Factors Affecting the Survival and Persistence of Beneficial Bacteria in Healthy Adults." *Beneficial Microbes*: 1–12. doi: 10.3920/BM2021.0017.

Okrugina, Svetlana Alekseevna Okrugina Svetlana Alekseevna, Anastasiya Vitalievna Naumova, Vladimir Leonidovich Ishchenko, Sergey Bronislavovich Stavitskiy, I. Eremin, and I. Nadelyaeva. 2022. "Selection of an Effective Dose of Autologous Probiotics for Patients in the Rehabilitation Period." *Fizioterapevt (Physiotherapist)*. doi: 10.33920/med-14-2212-08.

Omar, J. M., Yen-Ming Chan, M. Jones, S. Prakash, and P Jones. 2013. "*Lactobacillus fermentum* and *Lactobacillus amylovorus* as Probiotics Alter Body Adiposity and Gut Microflora in Healthy Persons." *Journal of Functional Foods* 5: 116–123. doi: 10.1016/J.JFF.2012.09.001.

Peng, Mengfei, Nana Frekua Kennedy, Andy Truong, Blair Arriola, and Ahlam Akmel. 2020. "Probiotics and Prebiotics on Intestinal Flora and Gut Health." In *Gut Microbiome and Its Impact on Health and Diseases*, edited by Debabrata Biswas and Shaik O. Rahaman, 85–103. Cham: Springer International Publishing. doi: 10.1007/978-3-030-47384-6_4.

Rau, Sameeha, Andrew Gregg, Shelby Yaceczko, and Berkeley Limketkai. 2024. "Prebiotics and Probiotics for Gastrointestinal Disorders." *Nutrients* 16 (6): 778. Multidisciplinary Digital Publishing Institute. doi:10.3390/nu16060778.

Reyes-Castillo, Pedro A., Raquel González-Vázquez, Edgar Torres-Maravilla, Mario Tello, Luis G. Bermúdez-Humarán, and Lino Mayorga-Reyes. 2021. "Probiotics against Viral Infections: Current Clinical Trials and Future Perspectives." *Immuno* 1 (4): 468–498. Multidisciplinary Digital Publishing Institute. doi: 10.3390/immuno1040034.

Rondanelli, M., C. Gasparri, G. Peroni, M. Faliva, M. Naso, S. Perna, Philip Bazire, I. Sajoux, R. Maugeri, and C. Rigon. 2021. "The Potential Roles of Very Low Calorie, Very Low Calorie Ketogenic Diets and Very Low Carbohydrate Diets on the Gut Microbiota Composition." *Frontiers in Endocrinology* 12. doi: 10.3389/fendo.2021.662591.

Roy, C., C. L. Kien, L. Bouthillier, and E. Levy. 2006. "Short-Chain Fatty Acids: Ready for Prime Time." *Nutrition in Clinical Practice : Official Publication of the American Society for Parenteral and Enteral Nutrition* 21 (4): 351–366. doi: 10.1177/0115426506021004351.

Sakkas, H., Petros Bozidis, Christos Touzios, Damianos Kolios, Georgia Athanasiou, E. Athanasopoulou, Ioanna Gerou, and C. Gartzonika. 2020. "Nutritional Status and the Influence of the Vegan Diet on the Gut Microbiota and Human Health." *Medicina* 56. doi: 10.3390/medicina56020088.

Sekhon, Bhupinder Singh, and Saloni Jairath. 2010. "Prebiotics, Probiotics and Synbiotics: An Overview. | Journal of Pharmaceutical Education & Research | EBSCOhost." December 1. https://openurl.ebsco.com/contentitem/gcd:59543341?sid=ebsco:plink:crawler&id=ebsco:gcd:59543341.

Valdes, Ana M., Jens Walter, Eran Segal, and Tim D. Spector. 2018. "Role of the Gut Microbiota in Nutrition and Health." *BMJ* 361. British Medical Journal Publishing Group: k2179. doi: 10.1136/bmj.k2179.

Vandamme, T., G. Gbassi, T. Nguyen, and Xiang Li. 2016. "Microencapsulation of Probiotics," 97–128. doi: 10.1002/9781118946893.CH5.

Wang, Xinzhou, Peng Zhang, and Xin Zhang. 2021. "Probiotics Regulate Gut Microbiota: An Effective Method to Improve Immunity." *Molecules* 26 (19): 6076. Multidisciplinary Digital Publishing Institute. doi: 10.3390/molecules26196076.

Wong, J. M. 2014. "Gut Microbiota and Cardiometabolic Outcomes: Influence of Dietary Patterns and Their Associated Components." *The American Journal of Clinical Nutrition* 100: 369–377. doi: 10.3945/ajcn.113.071639.

Wong, J. M., R. D. de Souza, C. Kendall, A. Emam, and D. Jenkins. 2006. "Colonic Health: Fermentation and Short Chain Fatty Acids." *Journal of Clinical Gastroenterology* 40: 235–243. doi: 10.1097/00004836-200603000-00015.

Xiong, Ruoxuan, Dan-Dan Zhou, Silan Wu, Si-Yu Huang, Adila Saimaiti, Zhijing Yang, Ao Shang, Cai-Ning Zhao, R. Gan, and Huabin Li. 2022. "Health Benefits and Side Effects of Short-Chain Fatty Acids." *Foods* 11. doi: 10.3390/foods11182863.

10 Probiotics Use and Applications in Early Life as Well as in Old Age

Zahid Ullah, Shamsa Kanwal, Shakira Ghazanfar, and Saiqa Noureen

10.1 INTRODUCTION

The term "probiotics" was first used by Lilley and Stillwell to denote chemical compounds or chemicals that facilitate the growth and proliferation of other microorganisms. The definition of this phrase has been revised and limited to "a viable microbial agent(s) that, when administered to animals or humans, has a positive impact on the host, potentially by enhancing the natural balance of the native microflora" (Moradi et al. 2020; Sánchez et al. 2017). Several terms, including "friendly," "beneficial," or "healthy" bacteria, are also often associated with probiotics based on their interpretation. Metchnikoff conducted the first research relevant to the beneficial effects of probiotics in the early 1900s (Sarao and Arora 2017). During his discussion of the advantages of sour milk for humans, he made the hypothesis that ingesting *Lactobacilli* or other live microbes found in such fermented milk would help restore the gut microbiota's normal balance. Microbial probiotics have since gained success, and their usage is now widely acknowledged (Sarao and Arora 2017). The gastrointestinal system of humans serves as a complex and dynamic environment where a mutually beneficial interaction exists between the host and the gut bacteria (Idrees et al. 2022). The host's immune system often plays a key role in the mutualistic relationship between the gut microbiota and many physiological functions in the host (Honda and Littman 2016). Furthermore, the components of the gut microbiota generate a range of chemical substances that may be useful to the host organism as well as other microbes. Host-microbe and microbial community interactions are the terms used to describe these relationships (Fischbach and Segre 2016). Throughout an individual's lifetime, these relationships are essential to the development of stable communities that preserve health and demonstrate resilience to crises. Additionally, they play a crucial role in the development of microbial organisms and hosts into symbiotic interactions (Sommer et al. 2017). While more study is needed to identify the particular functions that the gut microbiota plays in many aspects of health and illness, the composition of the gut microbiota is strongly linked to many diseases and has become a prominent target for nutrition and medicine (Skelly et al. 2019; Zmora, Suez, and Elinav 2019). Probiotics may have a direct impact on the immune system's reaction as well as the gut microbiota's composition and activity (Wegh et al. 2019). This study aims to present the most recent findings about the use of probiotics in both children and adults, as well as the effects of these interventions on people's general health and quality of life (Abdelnour et al. 2023).

10.2 GUT MICROBIOTA AND ITS IMPORTANCE

Microbial probiotics should ideally benefit the host and not harm it in any way (Kanwal et al. 2021). As a result, all strains are designated Generally Regarded as Safe (GRAS) after conducting extensive research before being used on people or animals. The evaluation of the safety of a probiotic strain or strains typically includes several factors. These include the cellular capacity to produce metabolites and enzymes, the properties related to colonization or adhesion, the influences on the strain's survival, and the interactions with the host, particularly concerning pathogenicity. As previously

 DOI: 10.1201/9781003399797-11

mentioned, the predominant selection of species used as probiotics consists of organisms belonging to the lactic acid bacteria (LAB) classification. The majority of LAB species are regarded as commensal microbes with no pathogenic potential because of their lengthy history of safe usage in food. Numerous species belonging to the genera *Lactobacillus*, *Pediococcus*, *Lactococcus*, *Aerococcus*, *Leuconostoc*, *Bifidobacterium*, and *Weissella* make up the LAB, a group of Gram-positive bacteria. The genus *Lactobacillus* is the most often found probiotic among the LAB group (Havenaar 1992; Reid 1999); these include *Leptospharyngebus acidophilus*, *Leptosphaericus*, *Leptosphaericus fermentum*, *Leptosphaericus*, *Leptosphaericus reuteri*, *Leptosphaericus*, and *Leptosphaericus* are some of them. These *Lactobacillus* species are useful probiotic organisms because they have several significant characteristics (Salvetti, Torriani, and Felis 2012), including effective adhesion to intestinal epithelial cells to lessen or stop pathogen colonization, competitive growth, and metabolite synthesis to limit or eliminate pathogen and nonpathogen (Peng et al. 2019). In recent times, the use of genetically modified LAB, also known as GM-LAB, has been employed to enhance the sensory characteristics such as flavor, smell, texture, and overall quality of many food products (Naiel et al. 2023), including yogurt and buttermilk (Beena Divya et al. 2012). Moreover, it has been shown and acknowledged that probiotics may be derived from several bacterial species, including *Propionibacterium* species, *Bifidobacterium,* and *Bacillus* (Sanders et al. 2010). For novel probiotic possibilities like *Vibrio alginolyticus* and *Carnobacterium* spp. that are resistant to acid, this approach might be ideal (Pérez-Sánchez et al. 2014).

10.3 BENEFICIAL EFFECTS OF PROBIOTICS

The exact mechanism of action of microbial probiotics, despite their long history of usage as additional foods to improve health, is still unknown. First proposed by Metchnikoff in 1908, the idea that certain bacteria or substances included in fermented foods, including yogurt, might alter the normal equilibrium of intestinal microflora. Since Metchnikoff's discovery, several studies have been carried out in this area, and further information on the advantages of using probiotics is being acquired. Probiotics are being employed to enhance both human and animal health in several different ways (Chukeatirote 2003).

10.3.1 Role of Gut Microbiota in the Maintenance of Health

The intestinal tract of humans is thought to include over 100 distinct types of bacteria, with a total population estimated to be between 1010 and 1012 (Simon and Gorbach 1984). The types of microbes that live in the gut (GI) tract vary from person to person and over time within the same person. There are both "friendly" and pathogenic organisms in the GI system, which makes the link between them complex. This state of balance can be disrupted by a variety of factors, such as nutrition, environment, aging, medicine (especially antibiotic usage), disease, stress, and lifestyle (Sanderson and Walker 1993), resulting in mucosal irritation, diarrhea, or other severe illnesses. To establish a state of optimal gut flora balance, often referred to as "eubiosis," it is essential for the predominant presence of beneficial bacteria, including Gram-positive *Lactobacilli* and *Bifidobacterium*, to exceed 85% of the total bacterial population. These friendly bacteria play a crucial role in acting as a protective barrier against the proliferation of harmful microbes inside the gut. Therefore, taking probiotics is probably the safest and most natural way to maintain this state of balance. Predominant "friendly" bacteria will compete with bacterial pathogens for vital nutrients or attachment sites, preventing the pathogens from colonizing. One example is the effective colonization of intestinal epithelial cells by *Lactobacillus casei* Shirota found in Japanese soured milk, Yakult (Spanhaak, Havenaar, and Schaafsma 1998), and the prevention of the growth of other harmful bacteria in the GI region. Furthermore, several *Lactobacilli* strains included in probiotics prevent the development of enteric pathogens, including *Salmonella* species, by generating antimicrobial chemicals called bacteriocins (Gill et al. 2001). One example is the administration of

Lactobacillus strain GG at higher levels (about 10^9 CFU/day) orally, which has been shown to exhibit intestinal adherence properties. Its use has been documented to provide relief from diarrhea symptoms caused by *Enterobacteria*. (Hatakka and Saxelin 2008).

10.3.2 Enhancement of the Immune System

Microbial probiotics, particularly those derived from LAB, can have diverse effects on the systemic immune systems (Mokoena, Mutanda, and Olaniran 2016). Firstly, it is possible to augment the defensive characteristics of the intestinal mucosa, which serves as a protective barrier against antigens, such as bacterial infections (Wu et al. 2022). For instance, adding probiotic supplements to a patient's diet may help treat Crohn's disease, which is linked to a reduction in the function of the barrier (Chukeatirote 2003). It has also been suggested that *L. casei* Shirota may be used as a probiotic to boost immunity and guard against enterobacterial infections (Matsuzaki 1998). An efficient oral rotavirus vaccination is also available using *Lactobacillus* GG (Chukeatirote 2003).

10.3.3 Protection against Cancer

According to in vitro studies, *Lactobacillus* bacteria (LAB) in dairy products may suppress the development of tumor cells and reduce the mutagenicity of substances like 1,2-dimethylhydrazine (Kumar, Singh, and Sinha 2010). LAB may indirectly lower the ability of bacterial enzymes that activate carcinogens in the digestive system to cause cancer. The *Lactobacillus acidophilus* NCFM strain can reduce the concentration of unbound amines inside the gastrointestinal tract, hence decreasing the risk of developing colon cancer (Chukeatirote 2003). Ling et al. (Crowley et al. 1994) indicated that eating yogurt containing *Lactobacillus* strain GG, which is believed to be involved in the etiology of colon cancer, reduced the enzyme activity in human volunteers. The anti-carcinogenic effect of probiotics is shown in Figure 10.1.

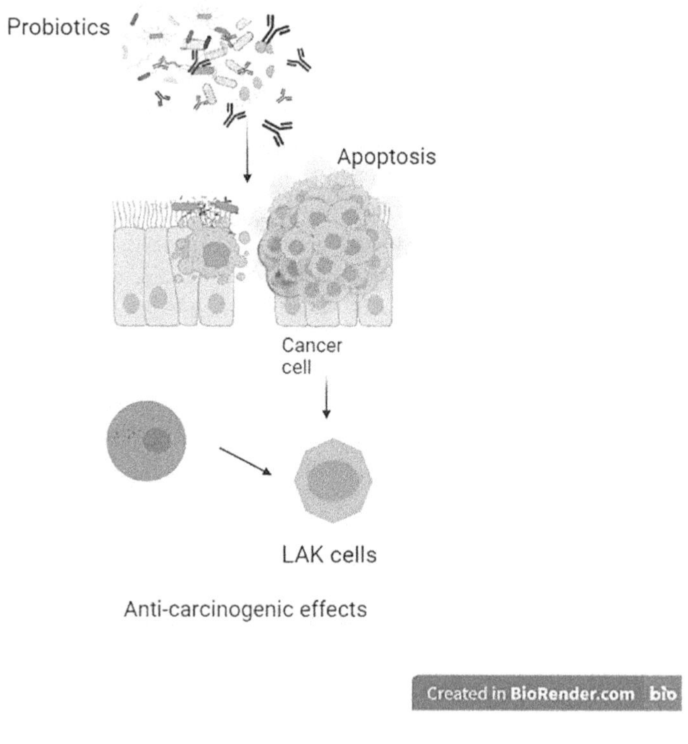

FIGURE 10.1 A schematic diagram showing the anti-carcinogenic effects of probiotics.

10.4 MECHANISM OF ACTION

Bacterial-epithelial cross-talk, which involves bacteria communicating with gastrointestinal mucosa, has been studied at the cellular level. Toll-like receptors (TLRs) on eukaryotic cells interact with pathogen and commensal bacterial molecular patterns (Walker 2008). TLR-4, the most studied TLR, interacts with Gram-negative lipopolysaccharide (LPS) and activates signaling molecules, releasing NF-κB and inflammatory cytokines. Ten more TLRs have been cloned since their discovery, and other microbial patterns have been found as ligands for these receptors (Walker 2008). The characteristics and composition of colonizing bacteria have a significant role in the establishment and enhancement of gastrointestinal host immune responses to combat infections and mitigate allergic reactions (Pickard et al. 2017). A fully developed newborn is capable of generating a sufficient defense against its host. But colonizing bacteria need to activate gut lymphoid tissue for the efferent part of that defense to work. Follicle epithelial cells are one example of this, since they express themselves and help bacteria and lymphoid cells communicate (Gutzeit, Magri, and Cerutti 2014). Polymeric IgA (pIgA) is secreted by mesenteric lymph nodes, and protective cytokines are produced by activated intraepithelial and lamina propria sites. In the lamina propria, dendritic cells (DCs) can grow extensions between epithelial cells, collect molecular patterns of commensal bacteria, and release cytokines. This enables them to control the maturation process of immature T-helper cells into a complete and fully developed immune response (Walker 2008). Commensal bacteria can cross microfold cells, hence facilitating interactions with antigen-presenting cells situated in the mesenteric lymph nodes. This interaction results in the stimulation of naïve plasma cells, which subsequently undergo differentiation into *B lymphocytes* that produce polymeric immunoglobulin A (pIgA). The previously discussed cellular mechanisms are of essential significance in the regulation of microbial and antigenic penetration. Moreover, these processes play a crucial role in preserving the integrity of the intestinal host defense system and mitigating the onset of chronic disorders inside the gastrointestinal tract (Walker 2008). The mechanism of action of probiotics is explained in detail in Figure 10.2.

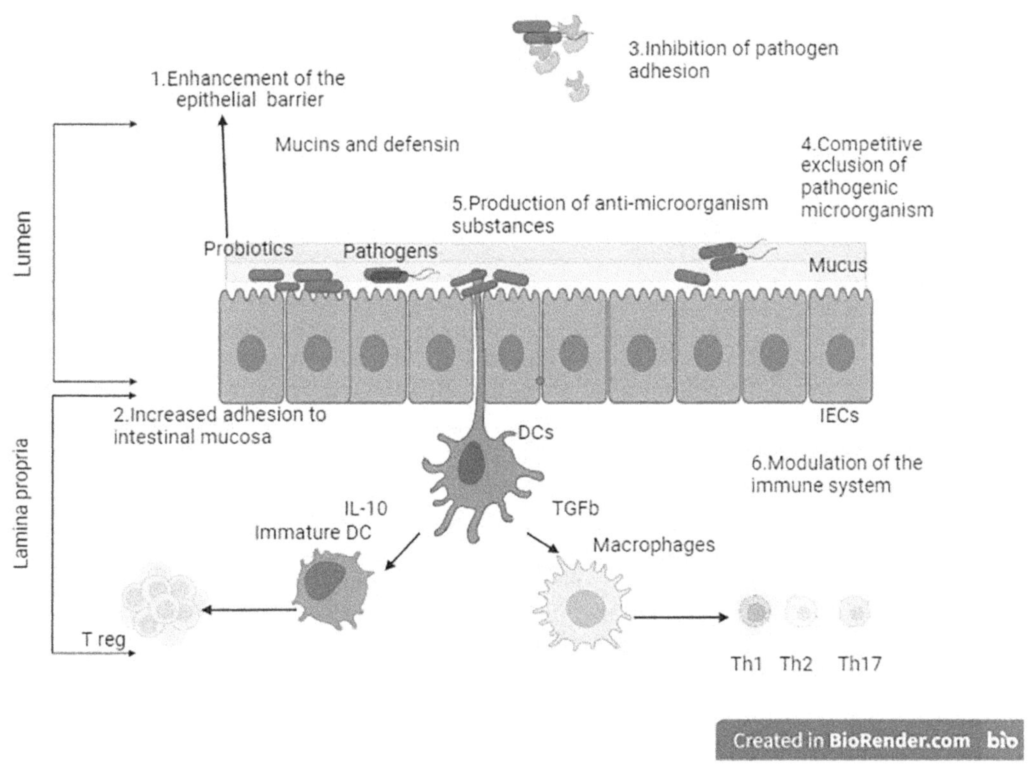

FIGURE 10.2 Mechanism of action of probiotics.

10.4.1 Improving Intestinal Inflammation with Probiotics

Studies on mice models and clinical trials involving humans both suggest that luminal contents serve as an ongoing source of antigenic stimulation for intestinal inflammation. The study of the dysbiosis of intestinal microbiota in pouchitis, Crohn's disease, and ulcerative colitis has resulted in the use of probiotics for the treatment of inflammatory bowel disease (IBD) (Guslandi 2015). According to the dysbiosis idea, persistent inflammation results from an imbalance between beneficial and detrimental gut microorganisms. Modifications in the abundance of certain commensal bacteria have been identified in individuals diagnosed with IBD. More specifically, there is a notable decline in the abundance of *Lactobacillus* and *Bifidobacterium* species, but there is a notable rise in the occurrence of Bacteroides, adherent or invasive, *Enterococci*, and *Escherichia coli* (Reiff and Kelly 2010). The study conducted by Gordon's team offered evidence supporting the idea that the modulation of the microbiota by probiotic intervention has promise for influencing the host organism. The study demonstrated that the probiotic *Bifidobacterium* longum can inhibit the synthesis of antimicrobial proteins, potentially leading to improved viability of the probiotic and influencing the composition of the microbial community (Ng et al. 2009). A single microbial compound generated from bacterial fragilities protects its host against an inflammatory state caused by *Helicobacter hepaticus*, according to research conducted using an animal model of experimental colitis. Studies in metagenomics have shown that probiotics may alter the composition of the gut microbiota and the metabolism of short-chain fatty acids, bile acids, amino acids, and plasma lipoproteins (Chung et al. 2018). Probiotic bacteria have a variety of impacts on the host, including changes in the mucosal immune system, the function of the epithelial and mucosal barriers, and the general environment within the gut lumen. These agents have an impact on a variety of cellular populations involved in both innate and adaptive immune responses, including B cells, regulatory T cells, T cells, epithelial cells, DCs, monocytes/macrophages, and NK cells (Blum et al. 2002).

10.4.2 In Vitro Studies

Probiotics have positive effects on intestinal inflammation, and bacterial colonization is necessary for animal models of IBD to generate an inflammatory phenotype. Animal colitis models provide proof of concept for both therapeutics and mechanisms. But not every commensal or probiotic bacterium affects the gut immune system in the same way. *Lactobacillus* and *Bifidobacterium* species, for example, diminish established intestinal inflammation in the IL-10 deletion animal model, whereas *Lactobacillus* GG inhibits colitis recurrence in HLA-B27 transgenic mice (Ng et al. 2009).

10.4.3 Gut Microbiota Modification

Probiotic bacteria can counteract the growth and activity of pathogenic bacteria by many mechanisms, including lowering the pH in the surrounding environment, preventing the attachment and movement of bacteria, and generating antibacterial compounds and defensins. The gut microbiota exhibits resistance to colonization by establishing a physiologically constrained environment characterized by redox potential, pH, and hydrogen sulfide generation (Ullah et al. 2023). Probiotic bacteria are capable of synthesizing antimicrobial chemicals known as bacteriocins, which exhibit diverse levels of inhibitory action (Verschuere et al. 2000). Certain strains of *Lactobacilli* or closely similar Gram-positive bacteria can stop other *Lactobacilli* from growing. However, other strains show antimicrobial action against a wider range of bacteria, including molds and yeasts, in addition to both Gram-positive and Gram-negative bacteria. The probiotic *L. salivarius* subsp. *Salivarius* UCC118 is one example that might be used. It can produce a peptide that has inhibitory effects on a variety of pathogens, including *Salmonella*, *Bacillus*, *Staphylococcus*, *Enterococcus*, and *Listeria* (Harzallah and Belhadj 2013). Probiotics have the potential to mitigate the adverse impacts generated on the epithelium by *E. coli* strains O157:H7 and O127:H6. The use of lactic acid-producing bacteria as a pretreatment for intestinal cells resulted in a decrease in the capacity of pathogenic *E. coli* to deliver virulence factors into cells and cause damage to tight junctions inside cells. The adhesion and invasion of intestinal epithelial cell lines were significantly reduced via co- or preincubation with the probiotic strain *E. coli* Nissle 1917 (Boudeau et al. 2003). A family of antimicrobial peptides known as defensins is essential to innate defense mechanisms. Human beta-defensin 2 (hBD-2) was up-regulated in Caco-2 intestinal epithelial cells after exposure to the probiotic *E. coli* Nissle strain. This phenomenon helped to strengthen the mucosal barrier, which prevented intestinal bacteria from penetrating (Wehkamp et al. 2004).

10.4.4 Development in the Integrity of the Barrier

The use of probiotics has been seen to prevent the proliferation of conventional organisms and possible pathogens. Additionally, probiotics can impact the contacts between mucosal cells and enhance cellular stability. The augmentation of the intestinal barrier function is accomplished by regulating the phosphorylation of cytoskeletal and tight junctional proteins. The preservation of intestinal barrier integrity is maintained by a network of interrelated mechanisms. These mechanisms include the secretion of water, mucus, and chloride as well as the tight junction protein-supported adhesion of epithelial cells at their apical junctions (Lodemann 2010). The deterioration of the epithelial barrier function has been associated with several illnesses, including enteric infections, celiac disease, IBD, and various autoimmune disorders, including Type 1 diabetes. In these situations, the use of probiotic bacteria may help the host organism by improving the mucosal barrier's performance (Ohland and MacNaughton 2010). This effect has been shown in research, including the whole body as well as in vitro experimental models. Barrier integrity was normalized by the probiotic combo known as VSL#3. The results of the in vitro study demonstrated that VSL#3 revealed a significant reduction in monolayer permeability and permeability when applied to T84 epithelial cells. However, this effect was not seen with the other probiotic bacteria in the study. It was found that at least one of the bacteria in VSL#3 was the only one to exhibit this decrease in resistance (Mennigen et al. 2009). The findings of many studies have shown that probiotic bacteria can improve barrier integrity in both healthy animals and animal models of colitis. The precise methods through which probiotic bacteria increase the operation of the gut mucosal barrier remain uncertain. However, alterations in the production of chloride or mucus, or changes in the expression of tight junction proteins by epithelial cells, could account for these improvements. Specific varieties of probiotic bacteria are capable of activating MUC gene expression regulation and augmenting mucus secretion. Other probiotic bacterial strains, on the other hand, are capable of inhibiting the secretion of chloride and water (Ng et al. 2009). Tight junction proteins have dynamic characteristics, undergoing structural modifications that determine their functional state. The potential relevance of probiotic bacteria in avoiding the redistribution of ZO-1 was shown by the observation of maintained cytoskeleton architecture and maintained barrier function in epithelial cells when co-cultured with VSL#3 probiotic bacteria and *Salmonella dublin* (Ng et al. 2009).

10.5 THE IMPACT OF BENEFICIAL BACTERIA ON EPITHELIAL CELLS

Epithelial cells can manifest distinct reactions in the presence of commensal or probiotic microbes as compared to pathogenic germs. The probiotic bacterial combination included in VSL#3 was shown to have a lower capacity to stimulate the release of IL-8 by epithelial cells in comparison to intestinal pathogens such as *Listeria monocytogenes*, *S. dublin*, *E. coli*, and *Shigella dysenteriae* (Madsen et al. 2001). Probiotic microorganisms can reduce the effects caused by pathogenic bacteria. Commensal bacteria may stimulate the production of protective cytokines, which can then interact with TLRs or pattern recognition molecules to promote epithelial cell regeneration and prevent programmed cell

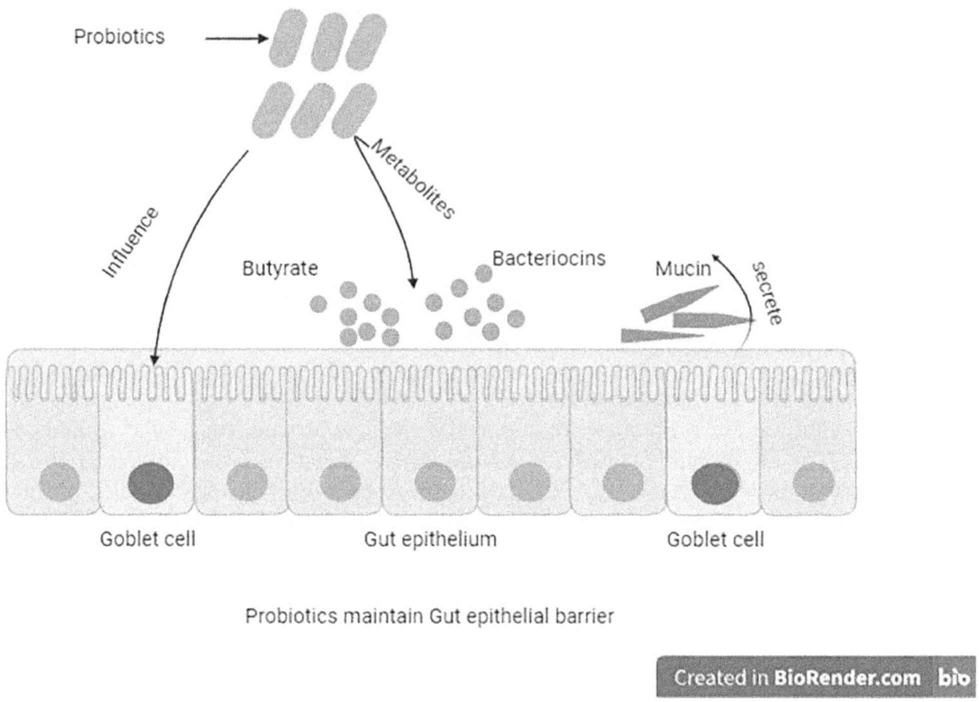

FIGURE 10.3 Impact of probiotics on epithelial cells.

death. Various signaling pathways serve to distinguish pathogenic bacteria from commensal or probiotic species (Gómez-Llorente, Munoz, and Gil 2010). By inhibiting the breakdown of the counterregulatory factor IB, nonpathogenic species can reduce the proinflammatory responses caused by pathogenic bacteria's activation of the transcription factor NF-κB. Studies analyzing how probiotic bacteria affect intestinal epithelial cells subjected to cytokine-induced apoptosis have shown that these bacteria may either promote epithelial regeneration or prevent apoptosis (Jin et al. 2012). The way in which probiotics maintain the epithelial barrier is shown in Figure 10.3.

10.6 PROBIOTICS IN PRIMARY CARE PEDIATRICS

10.6.1 Pregnancy

Research has shown that using probiotics in the early stages of pregnancy and throughout the perinatal period has beneficial effects on newborn health, effectively reducing the occurrence of illnesses both before and after delivery, as well as mitigating the development of atopy in children. However, further research and randomized controlled trials are necessary to ascertain the impact of probiotics on the maturation of the neonatal neurological system (Lenfestey and Neu 2017). Although probiotics are often used throughout pregnancy and the perinatal period, there is a limited amount of research investigating their safety in this specific context. A clinical study using a randomized, double-blind, placebo-controlled methodology demonstrated that the administration of probiotic supplements exhibited no adverse effects (Lenfestey and Neu 2017). Baldassarre et al. (2019) verified that there were no negative effects on moms or babies from the early use of probiotics during pregnancy. Figure 10.4 explains the effects of probiotics on pregnant women.

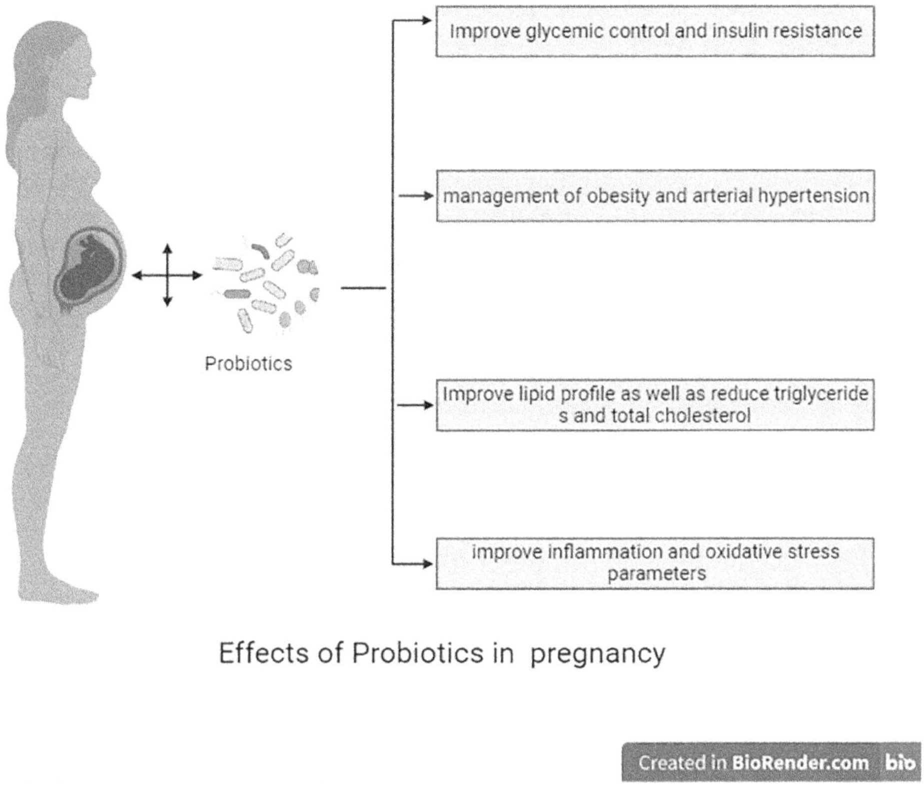

FIGURE 10.4 Effects of probiotic intake during pregnancy.

10.6.2 Maternal Gut Microbiota and Development of the Fetus

According to the concept, the presence of bacteria in the uterus might result in the colonization of fetal tissues by maternal germs far before the start of the birthing process. The potential transfer of microorganisms from the mother's gastrointestinal system to the fetus during prenatal development is a theoretical concept that requires scientific confirmation in subsequent studies. The reported increase in the predominance of the phylum *Proteobacteria* during pregnancy is thought to be related to proinflammatory changes, which may result in greater mucosal permeability and microbiota transfer (Navarro-Tapia et al. 2020). Previously, the development of neonatal gut microbiota was thought to be sterile, but emerging results from non-cultivational, PCR, and DNA sequencing-based studies support novel hypotheses concerning maternal-fetal transfer of microorganisms in pregnancy (Miko et al. 2022). There have been raised concerns about the interpretation of these findings, including potential contamination of samples characterized by low bacterial density as well as the selection of the best suitable analytical platform. Even beneficial microbiota components cannot pass through the intestinal epithelial barrier, which acts as a physiological barrier. Internalizing bacteria from the intestinal lumen and moving them to neighboring lymph nodes is a capability of DCs. This mechanism accelerates the spread of microorganisms into the bloodstream, ultimately arriving in the developing fetal gut (Miko et al. 2022). Once reaching the fetal location, bacteria are most likely recognized by TLRs, conserved innate immune receptors. Despite this, the TLR pathway is significantly less activated than in adults, suggesting immature innate immunity and ongoing immunological development throughout the perinatal period (Maródi 2006). To better understand the possibilities of maternal-fetal microbiota interaction, further research is required.

10.7 APPLICATIONS OF PROBIOTICS IN THE TREATMENT OF DIFFERENT PEDIATRIC DISEASES

10.7.1 Infectious Diarrhea

Infectious diarrhea is a major problem in clinical pediatrics, and studies on probiotics in children have shown their effectiveness in treating and preventing it (Szajewska and Mrukowicz 2001). Gastrointestinal problems such *as Clostridium difficile* diarrhea and infectious diarrhea, which are caused by antibiotics, may be relieved by probiotics (Marteau et al. 2001). Children with rotaviral infections, creche participants, and pediatric inpatients may all benefit from probiotics by shortening the length of phases of infectious diarrhea (Weizman, Asli, and Alsheikh 2005). Increased frequency and reduced consistency of stools with antibiotic treatment are symptoms of antibiotic-associated diarrhea (AAD), a common problem in clinical pediatrics (Johnston, Supina, and Vohra 2006). By the tenth day of oral antibiotic treatment, *Lactobacillus* GG has been shown, in comparison to a placebo group, to dramatically decrease stool frequency and enhance stool consistency (Arvola et al. 1999). Probiotics may also help with gastrointestinal distress, such as cramps and nausea (Pyne et al. 2015). Children in hospitals who need to be fed intravenously or have their enteral intake limited are more likely to have disruptions in their gut flora, which may be triggered by receiving broad-spectrum antibiotic treatment (Btaiche et al. 2010). The majority of research about the use of probiotics in the treatment of *C. difficile*-associated diarrhea comes from adult populations since *C. difficile* exacerbates AAD more often in adults than in children (McFarland et al. 2016).

10.7.2 Necrotizing Enterocolitis

The gastrointestinal tract of healthy newborns undergoes colonization by nonpathogenic bacteria, such *as Bifidobacteria*, consistently and predictably (Boehm et al. 2004). Premature newborns have a distinct intestinal colonization pattern than healthy infants, which may lead to the development of necrotizing enterocolitis (NEC) (Shane 2008). Accordingly, there is a theory that suggests probiotic supplementation might lower the risk of NEC in premature babies. This study evaluated the impact of a probiotic supplement that contains *L. acidophilus* and *Bifidobacterium infantis* on the incidence of NEC in premature infants using a randomized controlled trial. The results indicated that infants who received the probiotic supplement experienced a lower frequency and severity of NEC (Deshpande, Rao, and Patole 2007). Although probiotics seem to be promising, further research in other cases is required to validate these results. The role of probiotics in alleviating NEC is shown in Figure 10.5.

10.7.3 Inflammatory Bowel Disease

A theory suggests that defects in the mucosal barrier's functionality might trigger the aberrant immune responses associated with IBDs, such as Crohn's disease and ulcerative colitis. A mixed probiotic formulation has been utilized in the adult population to help people with ulcerative colitis stay in remission (Figure 10.6). Furthermore, it has been used as a prophylactic against the development of acute pouchitis after ileal pouch-anal anastomosis (Cabana et al. 2006; Rioux and Fedorak 2006). Moreover, there has been a possibility that altering the intestinal flora may be helpful in treating Crohn's disease (Schultz, Schölmerich, and Rath 2003). Additional research is required to investigate the use of probiotics in the context of children's IBD since the existing body of evidence is currently limited to exploratory studies (Kerman and Deshpande 2014).

10.7.4 Urogenital Infections

Certain probiotic bacteria can colonize not just the gut but also the urogenital tract. Therefore, it is hypothesized that probiotic supplementation may be able to stop uropathogenic organisms from colonizing urinary tracts, thereby reducing infections (MacPhee et al. 2010). While there are several accounts of probiotics being used successfully in this way (Gerasimov 2004).

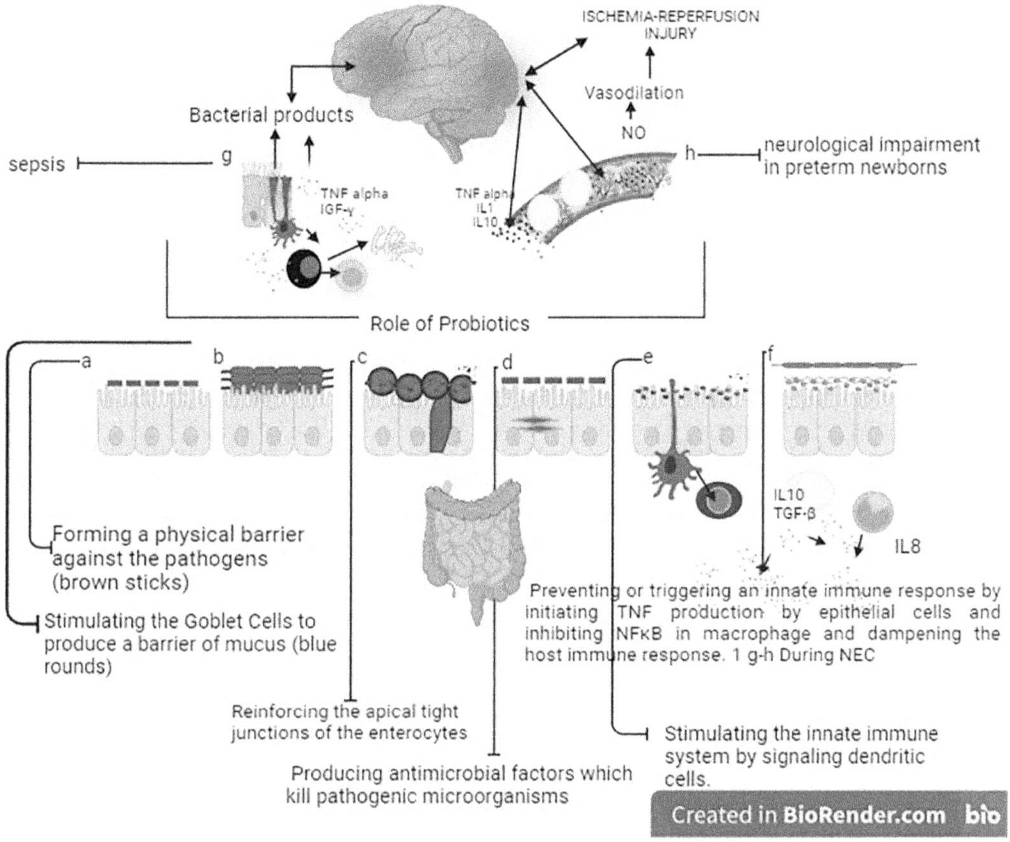

FIGURE 10.5 Role of probiotics in control of necrotizing enterocolitis.

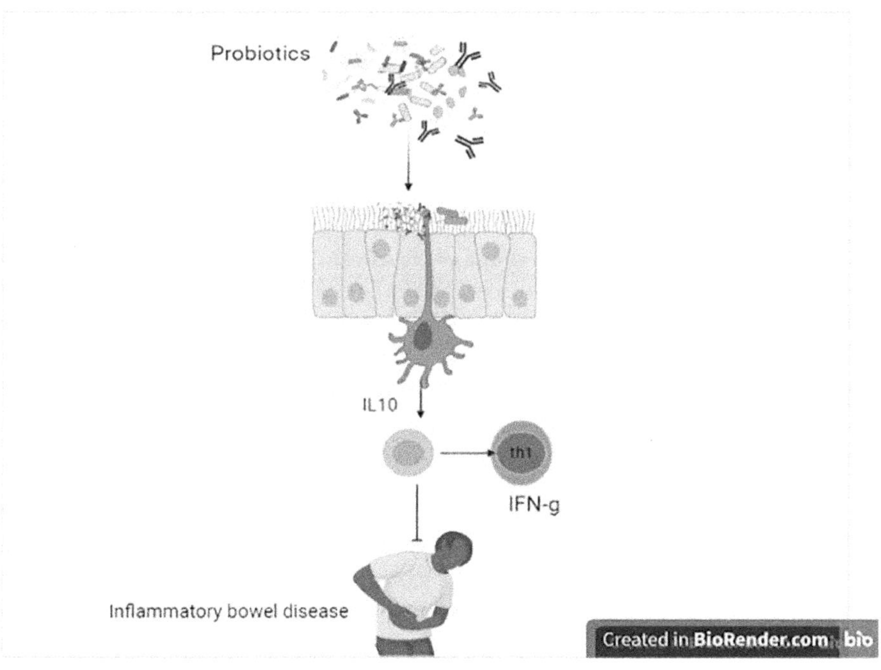

FIGURE 10.6 Pathway of probiotics in inflammatory bowel disease.

There is a limited number of randomized controlled studies available, and there is an even smaller number of papers documenting the use of this application specifically in the pediatric population.

10.8 PROBIOTICS AND THE ELDERLY POPULATION

The findings indicate that the microbiota is notably impacted by the aging process, a phenomenon that may not be exclusively attributed to aging itself but may also be affected by many variables, including deteriorating overall health, malnutrition, and increased use of prescription medications, including nonsteroidal anti-inflammatory drugs (NSAIDs) and antibiotics. Variations in microbial composition have been documented between older individuals in good health and those who are hospitalized or living in institutional environments (Ticinesi et al. 2017). Large bowel function depends heavily on colonic transit, with longer transit periods being associated with lower bacterial cell mass in the feces. Due to prolonged bed rest and a lack of activity, elderly individuals in hospitals often have these transit durations markedly prolonged. Variations in eating patterns during hospital stays, different food options, and parenteral and enteral artificial feeding may all be factors. Normal gut microbiota functions as a crucial natural defensive system against pathogen invasion and stops opportunistic bacteria from growing out of control (colonization resistance) (Rondanelli et al. 2015). Age-related changes in the microbiota composition may have negative consequences, including decreased immune system function and an increased risk of gastrointestinal infections. Aging-related inflammatory alterations include decreased B and T cell responses and increased circulating antibodies against commensal gut microorganisms (Rondanelli et al. 2015). The aged population has a lower variety of microbiota, with more interindividual variability. The researchers investigated the intestinal microbiota of old people with and without long-term care facilities, finding that the composition of the microbiome was connected to where the elderly stayed. *Firmicutes* and *Bacteroidetes* account for more than 93% of the total bacteria in centenarians' gut microbiota. The reduction of *Faecalibacterium prausnitzii* in centenarians is of importance since it is known to alter intestinal inflammation. The gut microbiota of centenarians is richer in *Proteobacteria*, with numerous "pathobiont" bacteria that may proliferate and cause illness (Salazar et al. 2017).

10.8.1 Probiotic Efficacy in the Elderly

The study examines how probiotics affect the microbiota of the elderly population, with particular importance on how probiotic intake affects the composition of the microbiota (McKean et al. 2017) Taking into account how effectively it addresses the signs and symptoms of a variety of gastrointestinal disorders, including iatrogenic ones like diarrhea brought on by antibiotics and constipation (Leong and Chan 2006). A range of observational studies was conducted to investigate changes in the microbiota composition in elderly adults after the administration of probiotics (Aljumaah et al. 2022).

10.8.2 Microbiota Composition in Elderly

Three double-blind studies were conducted on the effects of probiotics on the microbiota composition in the elderly population (Bartosch et al. 2004). The first investigation had a six-month duration and focused on the examination of a fermented oat-based beverage that contained *B. longum* 46 and *B. longum* 2C (Lahtinen et al. 2009). In the second trial, individuals were given a beverage containing reconstituted skim milk that included various concentrations of *Bifidobacterium lactis* HN019 over four weeks. The third trial used inulin in conjunction with *Bifidobacterium bifidum* and *B. lactis*. The results showed that the experimental group's microbiota composition differed significantly from that of the control group (Ahmed et al. 2007).

10.8.3 Effect of Probiotics on Elderly Patients

In a comprehensive study involving a sample size of 360 individuals aged 60 years and above, it was demonstrated that a three-week intervention incorporating the consumption of fermented milk enriched with yogurt cultures and probiotic casei DN-114001 led to a significant decrease in winter illnesses in comparison to the control group receiving placebo (Baker, Tran, and Thomas 2009). A further double-blind clinical experiment was undertaken, including a cohort of 24 elderly individuals who had undergone artificial enteral nourishment. The findings of the experiment indicated that the group that received the probiotic had a statistically significant decrease in the duration of infection days after the end of the therapy (Rondanelli et al. 2015). The use of probiotics in conjunction with antibiotics has been utilized as a treatment strategy for addressing *C. difficile*, a pathogenic microorganism accountable for around 20–25% of occurrences of AAD and more than 95% of cases of *Pseudomembranous colitis* (Rondanelli et al. 2015). A research conducted by Hickson et al. (Hickson et al. 2007). The evaluation of 135 hospitalized patients took place before and after they consumed a drink containing *L. casei*, *L. bulgaricus*, and *Streptococcus thermophilus* for a week. The mixture was consumed twice a day along with an antibiotic therapy. The findings indicated that 12 individuals out of a total of 57 participants (12%) who consumed the probiotic drink had diarrhea in relation to antibiotic use. In contrast, 19 out of 56 participants (34%) from the placebo group reported the same symptom.

10.8.4 Effect of Probiotics on Elderly Gastrointestinal Track

Beausoleil researched the utilization of probiotics as supplementary agents to antibiotic therapy to mitigate intestinal diseases (Beausoleil et al. 2007). The research population included 89 elderly people, with an average age of 72 years. The used mixture consisted of a fermented dairy product containing a minimum of 50×10^9 colony-forming units of *L. casei* and *L. acidophilus* CL1285. The experimental group that was administered probiotics had a median hospitalization duration of eight days, while the placebo group displayed a median length of stay of ten days. Constipation is a commonly seen illness that is frequent among the elderly population. It reveals a variety of symptoms, as described by the "Rome III criteria." A 2010 research provided evidence supporting the efficacy of providing *B. lactis* DN-173010, *E. coli* Nissle, and *L. casei* 1917 as therapeutic interventions for the regulation of defecation frequency and enhancement of stool consistency. The observed improvement may be ascribed to the decrease in colonic pH after the ingestion of probiotics, which can be related to the production of short-chain fatty acids like butyric acid, propionic acid, and lactic acid (Chmielewska and Szajewska 2010). Ouwehand et al. performed research comparing the efficacy of a double-blind treatment with a placebo (Ouwehand et al. 2008). They investigated the impact of a symbiotic combination of lactitol and *L. acidophilus* NCFM on a cohort of older individuals. The results exhibited advantageous impacts on the health of the intestinal mucosa, including a higher rate of bowel movements in the group treated with probiotics, as well as significantly elevated levels of PGE2, IgA, and spermidine.

10.8.5 Augmentation of Elderly Health with Probiotics

Studies have shown that probiotics improve oral health in the elderly by reducing the prevalence of Candida and hyposalivation, two conditions that are common in this population. In addition, probiotic use has been linked to beneficial outcomes for respiratory tract infections, such as shortening their duration or rapid healing process in those suffering from acute distal radius fractures. In mouse models, probiotics have also demonstrated possible impacts on bone metabolism; however, more clinical trials are required to evaluate these effects in older adults. The use of probiotics in the elderly population has shown a reduction in mental health issues such as anxiety and depression, particularly when combined with a 12-week resistance-training schedule. This finding highlights the significance of the brain-gut axis. Furthermore, probiotics have been shown to have a positive impact on stress reduction and cognitive flexibility. Treating problems such as neuroinflammation and silent systemic inflammation, which often manifest in the first phases of Alzheimer's disease, may be particularly advantageous. Increased blood vitamin levels have been linked to additional effects of probiotics on the elderly. For instance, Fischer et al. (2015) found that subjects who took a commercial probiotic supplementation had lower plasma homocysteine levels and higher serum concentrations of vitamin B12 and folate. Researchers hypothesized that an increase in *F. praunitzii* may reduce the build-up of LPSs and β-amyloid in the brain, which are overproduced during the development of Alzheimer's disease. The gut microbiota has also shown similar positive benefits from symbiotics, which are a combination of probiotics and prebiotics. Symbiotics improved mucosal integrity and stool frequency, reduced pro-inflammatory response, increased butyrate synthesis, and improved lipid metabolism. They also raised the quantity of *Lactobacilli* and *Bifidobacteria.* The frequency of metabolic syndrome, several cardiovascular risk factors, and signs of insulin resistance were significantly decreased in older individuals. Additionally, prebiotics have demonstrated benefits for immunological function and the overall composition of the gut flora (Ale and Binetti 2021).

10.9 POTENTIAL OF PROBIOTICS IN LONGEVITY

The gut microbiota inhibits the growth of toxins and the colonization of pathogens, contributing significantly to immunosenescence and inflammation. Gut microbiota homeostasis may have an impact on the preservation of host longevity and fitness. It may be possible to extend life by using microbiota-targeted treatments, like pro/prebiotics, for extended periods and at an earlier age (Ghazanfar et al. 2023). However, it is not feasible to do longitudinal research on humans over their entire existence. Recent research using mouse models has shed light on the role gut bacteria play in longevity (Biagi et al. 2013). According to a study, giving mice an 11-month supply of *B. lactis* extended their lifespan, presumably because it reduced intestinal inflammation. The *B. lactis*-supplemented group's colonic mucosa showed improved tight junction maintenance and higher mucus output. The treated mice had high fecal contents of polyamines (PA), and their gut microbiota composition was similar to that of younger mice. It also shows that using probiotics or prebiotics to raise the level of luminal PA may be a positive first step in encouraging a longer and healthier life (Matsumoto et al. 2011).

10.10 APPROPRIATE SELECTION METHODS OF PROBIOTICS

The procedure of choosing probiotic microbes is systematic and adheres to a "step-by-step" blueprint. The objective of this group of trials is to reduce the number of possible probiotic choices. The strains selected have the finest practical characteristics and no undesirable qualities. To facilitate the selection process of probiotic bacteria, it is important to maintain pH levels within the range of 2–5 and bile salt concentrations between 0.3% and 2%. To evaluate the resistance qualities of the target strain, a variety of enzymes such as lysozyme and amylase, as well as substances including pancreatic enzymes, taurodeoxycholic acid, phenol, NaCl, oxgall, and porcine gastric juice, may be used for cultivation purposes. This cultivation process involves subjecting the target strain to various pH levels. Depending on the time, one may use either absorbance or colony count to quantify drug resistance. Using this approach can help you choose the best probiotic bacteria for a range of uses (de Melo Pereira et al. 2018).

10.10.1 Ability of Probiotics to Adhere to Surfaces

The process of microbial adhesion to epithelial cells is a multifaceted phenomenon that includes the interaction between two distinct membranes, namely those of microbial and human cells. The probiotic strain's cell surface chemical and physiochemical composition affects this adhesion process. The relationship between Van der Waals and electrostatic interactions may affect a target surface's adhesion. The capacity of microorganisms to self-aggregate and the hydrophobic properties of their cell surface both affect how well they adhere to epithelial cells. Auto-aggregation can be accomplished using a straightforward technique, but the quantification of cell surface hydrophobicity can be conducted using the MATH approach. Evaluating microbial adhesion qualities can be achieved by directly evaluating the capacity of microorganisms to adhere to mammalian epithelial cells, including Caco-2, HT-29, and fetal I-407 cells (Monteagudo-Mera et al. 2019).

10.10.2 Effects of Probiotics That Are Antagonistic to Harmful Organisms

Through their metabolism of proteins, carbohydrates, and other substances in the gastrointestinal system, probiotics can produce antimicrobial compounds that aid in the removal of pathogenic microbes. The antagonist activities exhibit variability based on the microbial strains and can be assessed using the inhibition zone experiment conducted on agar plates. The quantification of competition activity can be achieved by evaluating the ability of probiotics to block the adherence of pathogens to cell lines. Coaggregation is an additional antagonistic probiotic activity that involves the agglomeration of pathogens with probiotic cells, hence facilitating their removal by fecal excretion (Rabetafika et al. 2023).

10.10.3 Safety Evaluation of Probiotics

It's essential to assess the safety of probiotics rather than assuming they are harmless. To establish guidelines for the safety assessment of probiotics designed for human consumption, the US, Canada, and the EU have started projects. Documentation of the isolation history, taxonomic identification, and the absence of virulence, infectivity, toxicity, and genes that confer transferable antibiotic resistance are some of these requirements. Most probiotics on the market today were isolated from healthy persons to increase probiotic compatibility and survival in the gut flora (Sanders et al. 2010). Probiotic use has raised concerns about antimicrobial resistance because of indications of horizontal gene transfer from beneficial to pathogenic bacteria in the host's gut. PCR's great sensitivity and ease of use have made it the gold standard for identifying resistance genes in probiotic bacteria. The presence of the risk of gene transfers makes antibiotic resistance a safety concern (Sanders et al. 2010).

10.11 CONCLUSION

Microbial probiotics have become increasingly well-liked on account of their well-established benefits in disease prevention and health maintenance. Traditional probiotic dairy strains, including LAB, are considered substantial contributors to the probiotic industry due to their established and secure usage. However, current research and speculation advocate for the potential alternative applications of numerous probiotic strains. Recent preclinical and clinical investigations have explored the effects of probiotics administered by mothers on the health of their infants. Several academic searches have contributed to the current corpus of knowledge concerning development and gastrointestinal health, in addition to the implications of metabolic syndrome and behavior. Existing research has demonstrated that probiotics may offer potential advantages that become apparent during the phases of puberty and maturity. An increasing number of scientists are focusing on the therapeutic uses of probiotics in the context of pediatric primary care. Probiotics have been the subject of extensive research due to their purported effectiveness in treating AAD and infection, among other conditions. Based on the available evidence, probiotic organisms may possess the capacity to alleviate three common concerns among the elderly: constipation, immune system decline, and malnutrition. In this regard, yogurts or supplements might be viable alternatives. Further research is necessary to determine the probiotic strains suitable for preserving the long-term effects.

REFERENCES

Abdelnour, S A, S Ghazanfar, M Abdel-Hamid, H M Abdel-Latif, Z Zhang, and M A Naiel. 2023. "Therapeutic uses and applications of bovine lactoferrin in aquatic animal medicine: An overview." *Veterinary Research Communications* 47 (3): 1015–29.

Ahmed, M, J Prasad, H Gill, L Stevenson, and P Gopal. 2007. "Impact of consumption of different levels of *Bifidobacterium lactis* HN019 on the intestinal microflora of elderly human subjects." *Journal of Nutrition Health and Aging* 11 (1): 26.

Ale, Elisa C, and Ana G Binetti. 2021. "Role of probiotics, prebiotics, and synbiotics in the elderly: Insights into their applications." *Frontiers in Microbiology* 12: 631254.

Aljumaah, Mashael R, Urja Bhatia, Jeffery Roach, John Gunstad, and M Andrea Azcarate Peril. 2022. "The gut microbiome, mild cognitive impairment, and probiotics: A randomized clinical trial in middle-aged and older adults." *Clinical Nutrition* 41 (11): 2565–76.

Arvola, Taina, Kirsi Laiho, Sari Torkkeli, Hannu Mykkänen, Seppo Salminen, Leena Maunula, and Erika Isolauri. 1999. "Prophylactic *Lactobacillus* GG reduces antibiotic-associated diarrhea in children with respiratory infections: A randomized study." *Pediatrics* 104 (5): e64–e64.

Baker, Hannah Claire, Dang Ngoc Tran, and Linda Valerie Thomas. 2009. "Health benefits of probiotics for the elderly: A review." *Journal of Foodservice* 20 (5): 250–62.

Baldassarre, Maria Elisabetta, Antonio Di Mauro, Manuela Capozza, Valentina Rizzo, Federico Schettini, Raffaella Panza, and Nicola Laforgia. 2019. "Dysbiosis and prematurity: Is there a role for probiotics." *Nutrients* 11 (6): 1273.

Bartosch, Sabine, Alemu Fite, George T Macfarlane, and Marion ET McMurdo. 2004. "Characterization of bacterial communities in feces from healthy elderly volunteers and hospitalized elderly patients by using real-time PCR and effects of antibiotic treatment on the fecal microbiota." *Applied and Environmental Microbiology* 70 (6): 3575–81.

Beausoleil, Mélanie, Nadia Fortier, Stéphanie Guénette, Amélie L'Ecuyer, Michel Savoie, Martin Franco, Jean Lachaîne, and Karl Weiss. 2007. "Effect of A fermented milk combining *Lactobacillus acidophilus* Cl1285 and *Lactobacillus casei* in the prevention of antibiotic-associated diarrhea: A randomized, double-blind, placebo-controlled trial." *Canadian Journal of Gastroenterology and Hepatology* 21: 732–6.

Beena Divya, Jayakumar, Kontham Kulangara Varsha, Kesavan Madhavan Nampoothiri, Bindhumol Ismail, and Ashok Pandey. 2012. "Probiotic fermented foods for health benefits." *Engineering in Life Sciences* 12 (4): 377–90.

Biagi, Elena, Marco Candela, Silvia Turroni, Paolo Garagnani, Claudio Franceschi, and Patrizia Brigidi. 2013. "Ageing and gut microbes: Perspectives for health maintenance and longevity." *Pharmacological Research* 69 (1): 11–20.

Blum, Stephanie, Dirk Haller, Andrea Pfeifer, and Eduardo J Schiffrin. 2002. "Probiotics and immune response." *Clinical Reviews in Allergy & Immunology* 22: 287–309.

Boehm, G, J Jelinek, B Stahl, Koen Van Laere, J Knol, S Fanaro, G Moro, and V Vigi. 2004. "Prebiotics in infant formulas." *Journal of Clinical Gastroenterology* 38: S76–9.

Boudeau, J, A-L Glasser, S Julien, J-F Colombel, and A Darfeuille-Michaud. 2003. "Inhibitory effect of probiotic *Escherichia coli* strain Nissle 1917 on adhesion to and invasion of intestinal epithelial cells by adherent–invasive *E. coli* strains isolated from patients with Crohn's disease." *Alimentary Pharmacology & Therapeutics* 18 (1): 45–56.

Btaiche, Imad F, Lingtak-Neander Chan, Melissa Pleva, and Michael D Kraft. 2010. "Critical illness, gastrointestinal complications, and medication therapy during enteral feeding in critically ill adult patients." *Nutrition in Clinical Practice* 25 (1): 32–49.

Cabana, Michael D, Andi L Shane, Cewin Chao, and Maria Oliva-Hemker. 2006. "Probiotics in primary care pediatrics." *Clinical Pediatrics* 45 (5): 405–10.

Chmielewska, Anna, and Hania Szajewska. 2010. "Systematic review of randomised controlled trials: Probiotics for functional constipation." *World Journal of Gastroenterology: WJG* 16 (1): 69.

Chukeatirote, Ekachai. 2003. "Potential use of probiotics." *Songklanakarin Journal of Science and Technology* 25 (2): 275–82.

Chung, Hyuk-Jin, Jae-Hun Sim, Tae-Sun Min, and Hyung-Kyoon Choi. 2018. "Metabolomics and lipidomics approaches in the science of probiotics: A review." *Journal of Medicinal Food* 21 (11): 1086–95.

Crowley, Craig, Susan D Spencer, Merry C Nishimura, Karen S Chen, Sharon Pitts-Meek, Mark P Armaninl, Lanway H Ling, Stephen B McMahon, David L Shelton, and Arthur D Levinson. 1994. "Mice lacking nerve growth factor display perinatal loss of sensory and sympathetic neurons yet develop basal forebrain cholinergic neurons." *Cell* 76 (6): 1001–11.

de Melo Pereira, Gilberto Vinícius, Bruna de Oliveira Coelho, Antonio Irineudo Magalhães Júnior, Vanete Thomaz-Soccol, and Carlos Ricardo Soccol. 2018. "How to select a probiotic? A review and update of methods and criteria." *Biotechnology Advances* 36 (8): 2060–76.

Deshpande, Girish, Shripada Rao, and Sanjay Patole. 2007. "Probiotics for prevention of necrotising enterocolitis in preterm neonates with very low birthweight: A systematic review of randomised controlled trials." *The Lancet* 369 (9573): 1614–20.

Fischbach, Michael A, and Julia A Segre. 2016. "Signaling in host-associated microbial communities." *Cell* 164 (6): 1288–300.

Fischer, André, Christèle Blanc-Bisson, Lorenzo Maria Donini, Luzia Valentini, Alessandro Pinto, Isabelle Bourdel-Marchasson, Rita Ostan, Patrizia Brigidi, Silvia Turroni, and Silvana Hrelia. 2015. "Impact of personalized diet and probiotic supplementation on inflammation, nutritional parameters and intestinal microbiota e The "RISTOMED project": Randomized controlled trial in healthy older people." *Clinical Nutrition* 34 (4): 593–602.

Gerasimov, Sergei V. 2004. "Probiotic prophylaxis in pediatric recurrent urinary tract infections." *Clinical Pediatrics* 43 (1): 95–8.

Ghazanfar, S., R. Abid, A. Khan, M. Y. Akbar, I. Asif, R. Nadeem, and A Ansari. 2023. "Machine Learning to Explore the Psychrophilic Soil Microbiome." In *Soil Microbiome of the Cold Habitats*, 213–22. CRC Press.

Gill, Harsharnjit S, Quan Shu, Hai Lin, Kay J Rutherfurd, and Martin L Cross. 2001. "Protection against translocating *Salmonella typhimurium* infection in mice by feeding the immuno-enhancing probiotic *Lactobacillus rhamnosus* strain HN001." *Medical Microbiology and Immunology* 190: 97–104.

Gómez-Llorente, Carolina, Sergio Munoz, and Angel Gil. 2010. "Role of toll-like receptors in the development of immunotolerance mediated by probiotics." *Proceedings of the Nutrition Society* 69 (3): 381–9.

Guslandi, Mario. 2015. "Role of probiotics in Crohn's disease and in pouchitis." *Journal of Clinical Gastroenterology* 49: S46–9.

Gutzeit, Cindy, Giuliana Magri, and Andrea Cerutti. 2014. "Intestinal IgA production and its role in host-microbe interaction." *Immunological Reviews* 260 (1): 76–85.

Harzallah, Daoud, and Hani Belhadj. 2013. "Lactic Acid Bacteria as Probiotics: Characteristics, Selection Criteria and Role in Immunomodulation of Human GI Muccosal Barrier." In *Lactic Acid Bacteria—R & D for Food, Health and Livestock Purposes*, edited by M. Kongo, 197–216.

Hatakka, K, and M Saxelin. 2008. "Probiotics in intestinal and non-intestinal infectious diseases-clinical evidence." *Current Pharmaceutical Design* 14 (14): 1351–67.

Havenaar, Robert. 1992. "Selection of strains for probiotics use." *Probiotics: the Scientific Basis*: 209–24.

Hickson, Mary, Aloysius L D'Souza, Nirmala Muthu, Thomas R Rogers, Susan Want, Chakravarthi Rajkumar, and Christopher J Bulpitt. 2007. "Use of probiotic *Lactobacillus* preparation to prevent diarrhoea associated with antibiotics: Randomised double blind placebo controlled trial." *BMJ* 335 (7610): 80.

Honda, Kenya, and Dan R Littman. 2016. "The microbiota in adaptive immune homeostasis and disease." *Nature* 535 (7610): 75–84.

Idrees, M, M Imran, N Atiq, R Zahra, R Abid, M Alreshidi, and S Ghazanfar. 2022. "Probiotics, their action modality and the use of multi-omics in metamorphosis of commensal microbiota into target-based probiotics." *Frontiers in Nutrition* 9: 959941.

Jin, Bo, Tao Sun, Xiao-Hong Yu, Ying-Xiang Yang, and Anthony ET Yeo. 2012. "The effects of TLR activation on t-cell development and differentiation." *Clinical and Developmental Immunology* 2012.

Johnston, Bradley C, Alison L Supina, and Sunita Vohra. 2006. "Probiotics for pediatric antibiotic-associated diarrhea: A meta-analysis of randomized placebo-controlled trials." *CMAJ* 175 (4): 377–83.

Kanwal, H, A Di Cerbo, F Zulfiqar, C Sabia, A Nawaz, F M Siddiqui, and S Ghazanfar. 2021. "Probiotic characterization and population diversity analysis of gut-associated pediococcus acidilactici for its potential use in the dairy industry." *Applied Sciences* 11 (20): 9586.

Kerman, David H, and Amar R Deshpande. 2014. "Gut microbiota and inflammatory bowel disease: The role of antibiotics in disease management." *Postgraduate Medicine* 126 (4): 7–19.

Kumar, Arvind, Nikhlesh Kumar Singh, and Pushpalata Rabindra Sinha. 2010. "Inhibition of 1, 2-dimethylhydrazine induced colon genotoxicity in rats by the administration of probiotic curd." *Molecular Biology Reports* 37: 1373–6.

Lahtinen, Sampo J, Liisa Tammela, Jaakko Korpela, Riikka Parhiala, Henri Ahokoski, Hannu Mykkänen, and Seppo J Salminen. 2009. "Probiotics modulate the *Bifidobacterium microbiota* of elderly nursing home residents." *Age* 31: 59–66.

Lenfestey, Mary W, and Josef Neu. 2017. "Probiotics in newborns and children." *Pediatric Clinics* 64 (6): 1271–89.

Leong, Rupert W L, and Francis K L Chan. 2006. "Drug-induced side effects affecting the gastrointestinal tract." *Expert Opinion on Drug Safety* 5 (4): 585–92.

Lodemann, Ulrike. 2010. "Effects of Probiotics on Intestinal Transport and Epithelial Barrier Function." In *Bioactive Foods in Promoting Health*, 303–33. Elsevier.

MacPhee, Roderick A, Ruben Hummelen, Jordan E Bisanz, Wayne L Miller, and Gregor Reid. 2010. "Probiotic strategies for the treatment and prevention of bacterial vaginosis." *Expert Opinion on Pharmacotherapy* 11 (18): 2985–95.

Madsen, Karen, Anthony Cornish, Paul Soper, Conor McKaigney, Humberto Jijon, Christine Yachimec, Jason Doyle, Lawrence Jewell, and Claudio De Simone. 2001. "Probiotic bacteria enhance murine and human intestinal epithelial barrier function." *Gastroenterology* 121 (3): 580–91.

Maródi, László. 2006. "Innate cellular immune responses in newborns." *Clinical Immunology* 118 (2–3): 137–44.

Marteau, Philippe R, Michael de Vrese, Christophe J Cellier, and Jürgen Schrezenmeir. 2001. "Protection from gastrointestinal diseases with the use of probiotics." *The American Journal of Clinical Nutrition* 73 (2): 430s–6s.

Matsumoto, Mitsuharu, Shin Kurihara, Ryoko Kibe, Hisashi Ashida, and Yoshimi Benno. 2011. "Longevity in mice is promoted by probiotic-induced suppression of colonic senescence dependent on upregulation of gut bacterial polyamine production." *PloS One* 6 (8): e23652.

Matsuzaki, Takeshi. 1998. "Immunomodulation by treatment with *Lactobacillus casei* strain shirota." *International Journal of Food Microbiology* 41 (2): 133–40.

McFarland, Lynne Vernice, Metehan Ozen, Ener Cagri Dinleyici, and Shan Goh. 2016. "Comparison of pediatric and adult antibiotic-associated diarrhea and *Clostridium difficile* infections." *World Journal of Gastroenterology* 22 (11): 3078.

McKean, Jennifer, Helen Naug, Elham Nikbakht, Bianca Amiet, and Natalie Colson. 2017. "Probiotics and subclinical psychological symptoms in healthy participants: A systematic review and meta-analysis." *The Journal of Alternative and Complementary Medicine* 23 (4): 249–58.

Mennigen, Rudolf, Kerstin Nolte, Emile Rijcken, Markus Utech, Bettina Loeffler, Norbert Senninger, and Matthias Bruewer. 2009. "Probiotic mixture VSL# 3 protects the epithelial barrier by maintaining tight junction protein expression and preventing apoptosis in a murine model of colitis." *American Journal of Physiology-Gastrointestinal and Liver Physiology* 296 (5): G1140–9.

Miko, Eva, Andras Csaszar, Jozsef Bodis, and Kalman Kovacs. 2022. "The maternal–fetal gut microbiota axis: Physiological changes, dietary influence, and modulation possibilities." *Life* 12 (3): 424.

Mokoena, Mduduzi Paul, Taurai Mutanda, and Ademola O Olaniran. 2016. "Perspectives on the probiotic potential of lactic acid bacteria from African traditional fermented foods and beverages." *Food & Nutrition Research* 60 (1): 29630.

Monteagudo-Mera, Andrea, Robert A Rastall, Glenn R Gibson, Dimitris Charalampopoulos, and Afroditi Chatzifragkou. 2019. "Adhesion mechanisms mediated by probiotics and prebiotics and their potential impact on human health." *Applied Microbiology and Biotechnology* 103: 6463–72.

Moradi, Mehran, Seyedeh Alaleh Kousheh, Hadi Almasi, Arash Alizadeh, Jonas T Guimarães, Nurten Yılmaz, and Anita Lotfi. 2020. "Postbiotics produced by lactic acid bacteria: The next frontier in food safety." *Comprehensive Reviews in Food Science and Food Safety* 19 (6): 3390–415.

Naiel, M. A., A. I. El-Kholy, S. S. Negm, S. Ghazanfar, M. Shukry, Z. Zhang, and H. M Abdel-Latif. 2023. "A mini-review on plant-derived phenolic compounds with particular emphasis on their possible applications and beneficial uses in aquaculture." *Annals of Animal Science* 23 (4): 971–7.

Navarro-Tapia, Elisabet, Giorgia Sebastiani, Sebastian Sailer, Laura Almeida Toledano, Mariona Serra-Delgado, Óscar García-Algar, and Vicente Andreu-Fernández. 2020. "Probiotic supplementation during the perinatal and infant period: Effects on gut dysbiosis and disease." *Nutrients* 12 (8): 2243.

Ng, SC, AL Hart, MA Kamm, AJ Stagg, and Stella C Knight. 2009. "Mechanisms of action of probiotics: Recent advances." *Inflammatory Bowel Diseases* 15 (2): 300–10.

Ohland, Christina L, and Wallace K MacNaughton. 2010. "Probiotic bacteria and intestinal epithelial barrier function." *American Journal of Physiology-Gastrointestinal and Liver Physiology* 298 (6): G807–19.

Ouwehand, Arthur C, Kirsti Tiihonen, Markku Saarinen, Heli Putaala, and Nina Rautonen. 2008. "Influence of a combination of *Lactobacillus acidophilus* NCFM and lactitol on healthy elderly: Intestinal and immune parameters." *British Journal of Nutrition* 101 (3): 367–75.

Peng, Mengfei, Puja Patel, Vinod Nagarajan, Cassandra Bernhardt, Michael Carrion, and Debabrata Biswas. 2019. "Feasible Options to Control Colonization of Enteric Pathogens with Designed Synbiotics." In *Dietary Interventions in Gastrointestinal Diseases*, 135–49. Elsevier.

Pérez-Sánchez, Tania, Imanol Ruiz-Zarzuela, Ignacio de Blas, and José L Balcázar. 2014. "Probiotics in aquaculture: A current assessment." *Reviews in Aquaculture* 6 (3): 133–46.

Pickard, Joseph M, Melody Y Zeng, Roberta Caruso, and Gabriel Núñez. 2017. "Gut microbiota: Role in pathogen colonization, immune responses, and inflammatory disease." *Immunological Reviews* 279 (1): 70–89.

Pyne, David B, Nicholas P West, Amanda J Cox, and Allan W Cripps. 2015. "Probiotics supplementation for athletes–clinical and physiological effects." *European Journal of Sport Science* 15 (1): 63–72.

Rabetafika, Holy N, Aurélie Razafindralambo, Bassey Ebenso, and Hary L Razafindralambo. 2023. "Probiotics as antibiotic alternatives for human and animal applications." *Encyclopedia* 3 (2): 561–81.

Reid, Gregor. 1999. "The scientific basis for probiotic strains of *Lactobacillus*." *Applied and Environmental Microbiology* 65 (9): 3763–6.

Reiff, Caroline, and Denise Kelly. 2010. "Inflammatory bowel disease, gut bacteria and probiotic therapy." *International Journal of Medical Microbiology* 300 (1): 25–33.

Rioux, Kevin P, and Richard N Fedorak. 2006. "Probiotics in the treatment of inflammatory bowel disease." *Journal of Clinical Gastroenterology* 40 (3): 260–3.

Rondanelli, Mariangela, Attilio Giacosa, Milena Anna Faliva, Simone Perna, Francesca Allieri, and Anna Maria Castellazzi. 2015. "Review on microbiota and effectiveness of probiotics use in older." *World Journal of Clinical Cases: WJCC* 3 (2): 156.

Salazar, Nuria, Lorena Valdés-Varela, Sonia González, Miguel Gueimonde, and Clara G De Los Reyes-Gavilán. 2017. "Nutrition and the gut microbiome in the elderly." *Gut Microbes* 8 (2): 82–97.

Salvetti, Elisa, Sandra Torriani, and Giovanna E Felis. 2012. "The genus *Lactobacillus*: A taxonomic update." *Probiotics and Antimicrobial Proteins* 4: 217–26.

Sánchez, Borja, Susana Delgado, Aitor Blanco-Míguez, Anália Lourenço, Miguel Gueimonde, and Abelardo Margolles. 2017. "Probiotics, gut microbiota, and their influence on host health and disease." *Molecular Nutrition & Food Research* 61 (1): 1600240.

Sanders, Mary Ellen, Louis MA Akkermans, Dirk Haller, Cathy Hammerman, James T Heimbach, Gabriele Hörmannsperger, and Geert Huys. 2010. "Safety assessment of probiotics for human use." *Gut Microbes* 1 (3): 164–85.

Sanderson, Ian R, and W Allan Walker. 1993. "Uptake and transport of macromolecules by the intestine: Possible role in clinical disorders (an update)." *Gastroenterology* 104 (2): 622–39.

Sarao, Loveleen Kaur, and M Arora. 2017. "Probiotics, prebiotics, and microencapsulation: A review." *Critical Reviews in Food Science and Nutrition* 57 (2): 344–71.

Schultz, Michael, Jürgen Schölmerich, and Heiko C Rath. 2003. "Rationale for probiotic and antibiotic treatment strategies in inflammatory bowel diseases." *Digestive Diseases* 21 (2): 105–28.

Shane, Andi L. 2008. "Applications of probiotics for neonatal enteric diseases." *The Journal of Perinatal & Neonatal Nursing* 22 (3): 238–43.

Simon, Gary L, and Sherwood L Gorbach. 1984. "Intestinal flora in health and disease." *Gastroenterology* 86 (1): 174–93.

Skelly, Ashwin N, Yuko Sato, Sean Kearney, and Kenya Honda. 2019. "Mining the microbiota for microbial and metabolite-based immunotherapies." *Nature Reviews Immunology* 19 (5): 305–23.

Sommer, Felix, Jacqueline Moltzau Anderson, Richa Bharti, Jeroen Raes, and Philip Rosenstiel. 2017. "The resilience of the intestinal microbiota influences health and disease." *Nature Reviews Microbiology* 15 (10): 630–8.

Spanhaak, Steven, Robert Havenaar, and Gertjan Schaafsma. 1998. "The effect of consumption of milk fermented by *Lactobacillus casei* strain shirota on the intestinal microflora and immune parameters in humans." *European Journal of Clinical Nutrition* 52 (12): 899–907.

Szajewska, Hania, and Jacek Z Mrukowicz. 2001. "Probiotics in the treatment and prevention of acute infectious diarrhea in infants and children: A systematic review of published randomized, double-blind, placebo-controlled trials." *Journal of Pediatric Gastroenterology and Nutrition* 33: S17–25.

Ticinesi, Andrea, Christian Milani, Fulvio Lauretani, Antonio Nouvenne, Leonardo Mancabelli, Gabriele Andrea Lugli, Francesca Turroni, Sabrina Duranti, Marta Mangifesta, and Alice Viappiani. 2017. "Gut microbiota composition is associated with polypharmacy in elderly hospitalized patients." *Scientific Reports* 7 (1):11102.

Ullah, R., I. Touseef, R. Abid, A. Farid, S. Ahmad, H. A. El Enshasy, and S Ghazanfar. 2023. "Exploitation of selected plant extracts as bio-control against fungal contaminants in animal feed." *Journal of King Saud University-Science* 35 (5): 102685.

Verschuere, Laurent, Geert Rombaut, Patrick Sorgeloos, and Willy Verstraete. 2000. "Probiotic bacteria as biological control agents in aquaculture." *Microbiology and Molecular Biology Reviews* 64 (4): 655–71.

Walker, W Allan. 2008. "Mechanisms of action of probiotics." *Clinical Infectious Diseases* 46 (Supplement_2): S87–91.

Wegh, Carrie AM, Sharon Y Geerlings, Jan Knol, Guus Roeselers, and Clara Belzer. 2019. "Postbiotics and their potential applications in early life nutrition and beyond." *International Journal of Molecular Sciences* 20 (19): 4673.

Wehkamp, Jan, Jürgen Harder, Kai Wehkamp, Birte Wehkamp-von Meissner, Miriam Schlee, Corinne Enders, Ulrich Sonnenborn, Sabine Nuding, Stig Bengmark, and Klaus Fellermann. 2004. "NF-κB-and AP-1-mediated induction of human beta defensin-2 in intestinal epithelial cells by *Escherichia coli* Nissle 1917: A novel effect of a probiotic bacterium." *Infection and Immunity* 72 (10): 5750–8.

Weizman, Zvi, Ghaleb Asli, and Ahmed Alsheikh. 2005. "Effect of a probiotic infant formula on infections in child care centers: Comparison of two probiotic agents." *Pediatrics* 115 (1): 5–9.

Wu, Jianmin, Ying Zhao, Xian Wang, Lingchang Kong, Lee J Johnston, Lin Lu, and Xi Ma. 2022. "Dietary nutrients shape gut microbes and intestinal mucosa via epigenetic modifications." *Critical Reviews in Food Science and Nutrition* 62 (3): 783–97.

Zmora, Niv, Jotham Suez, and Eran Elinav. 2019. "You are what you eat: Diet, health and the gut microbiota." *Nature Reviews Gastroenterology & Hepatology* 16 (1): 35–56.

11 Gut-Brain Axis Modulation through Probiotic Interventions

Huma Ikram, Saiqa Noureen, Evelyn Saba, Rimsha Munawar, Shakira Ghazanfar, Ya Sakina, and Heba A.M. Abdalla

11.1 INTRODUCTION

The gut-brain axis (GBA) represents one of the most compelling frontiers in neuroscience and microbiome research, revealing intricate bidirectional communication between the gastrointestinal tract and the central nervous system (Petrut et al., 2025). This complex signaling network operates through multiple interconnected pathways, including direct neural connections via the vagus nerve, endocrine signaling through the hypothalamic-pituitary-adrenal (HPA) axis, and immune-mediated communication via cytokine production and circulation. Emerging evidence demonstrates that gut microbiota composition directly influences neurodevelopment, neuroinflammation, and behavior through these pathways, with particular impact during critical periods of brain development and plasticity across the lifespan (Figure 11.1).

This microbial influence has been implicated in various neurological and psychiatric conditions, ranging from mood disorders like depression and anxiety to progressive neurodegenerative disorders such as Parkinson's and Alzheimer's disease, where distinct alterations in gut microbiome composition and diversity have been consistently observed (Abou Izzeddine et al., 2025). Probiotics, defined as live microorganisms that confer health benefits when administered in adequate amounts, have emerged as promising modulators of this axis, capable of restoring microbial balance and improving brain function through their ability to compete with pathogenic bacteria and modulate host metabolic pathways (Abdul Manan, 2025). These beneficial bacteria exert their effects through diverse mechanisms, including production of neuroactive compounds like serotonin and gamma-aminobutyric acid (GABA), regulation of intestinal barrier integrity to prevent microbial translocation, and modulation of systemic inflammation that can affect central nervous system function (Chandrasekaran et al., 2024; Dacaya et al., 2025). Recent advances in microbiome research have identified specific bacterial strains that can influence neurotransmitter levels, stress response systems through glucocorticoid regulation, and even neurogenesis in key brain regions like the hippocampus, offering novel therapeutic avenues for conditions previously considered to be solely brain-related (Mehta et al., 2025). However, the translation of these findings into clinical practice faces significant challenges, including strain-specific variability in efficacy, substantial individual differences in baseline microbiome composition, and lack of standardized protocols for administration, dosing, and outcome measurement that hinder clinical application (Table 11.1).

The therapeutic potential of probiotics in modulating the GBA has gained significant attention due to their excellent safety profile, noninvasive administration route, and broad-spectrum effects on multiple physiological systems that make them attractive alternatives or adjuncts to conventional treatments (Beikmohammadi et al., 2025). Clinical studies have shown that certain probiotic strains can reduce symptoms of depression and anxiety by clinically meaningful margins, improve various domains of cognitive function, including memory and executive function, and even mitigate neuroinflammation in animal models of neurodegeneration through mechanisms that are only beginning to be elucidated (O'Riordan et al., 2025). These beneficial effects appear to be mediated through multiple interconnected mechanisms, including the production of short-chain fatty acids (SCFAs) like butyrate that have direct neuroprotective properties, regulation of the HPA axis to normalize

DOI: 10.1201/9781003399797-12

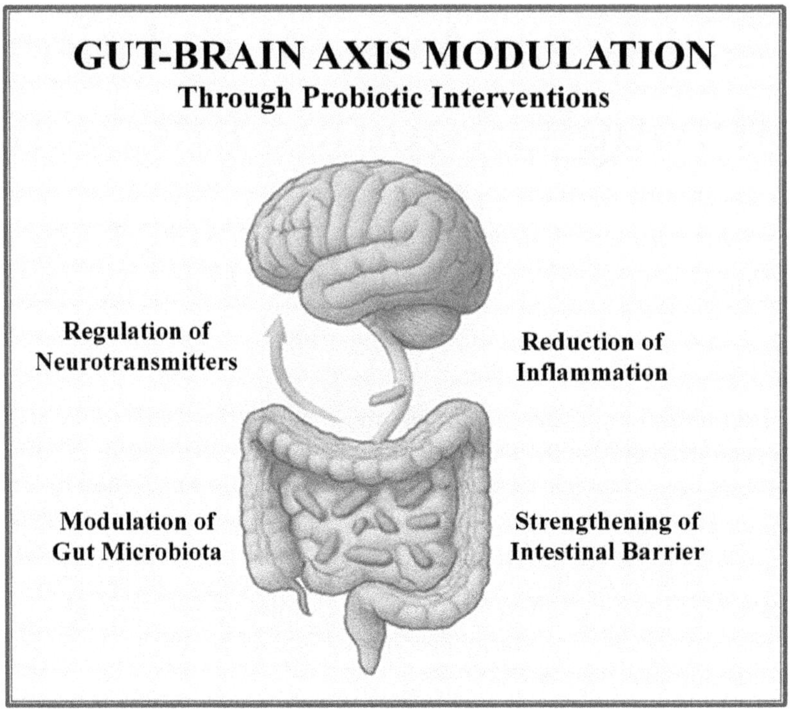

FIGURE 11.1 Gut-brain axis modulation through probiotic interventions.

TABLE 11.1
Examples of Probiotics and Their Effects on Gut-Brain Axis

Probiotic	Effects	Reference
Lactobacillus rhamnosus	• ↑ GABA receptor expression • ↓ Anxiety & stress (via vagus nerve) • ↓ Cortisol (HPA axis modulation) • Strengthens gut barrier	Feng et al. (2025)
Bifidobacterium longum	• ↑ BDNF (brain neurogenesis) • Produces butyrate (anti-inflammatory) • Modulates serotonin (5-HT) synthesis • Supports cognitive function	Xiao et al. (2025)
Lactobacillus helveticus	• ↓ Proinflammatory cytokines (IL-6, TNF-α) • Balances GABA/glutamate • Improves sleep quality • Supports mood regulation	Alipour et al. (2025)
Bifidobacterium breve	• Critical for infant neurodevelopment • Supports microglial maturation (CNS immunity) • Produces acetate (appetite regulation) • Prevents stress-induced cognitive deficits	Grønbæk et al. (2025)
Lactobacillus plantarum	• ↑ Dopamine & serotonin precursors • ↓ Oxidative stress (neuroprotection) • Reduces gut permeability (↓ LPS) • Supports antioxidant defenses	Duarte et al. (2025)

stress responses, and interaction with the enteric nervous system that forms a direct neural connection to the central nervous system (Chen et al., 2021). Despite these promising findings, the field currently suffers from substantial heterogeneity in research methodologies, with wide variations in probiotic formulations ranging from single strains to complex multi-strain combinations, dosages that vary by several orders of magnitude between studies, and treatment durations that range from weeks to months without clear rationale. Furthermore, the influence of important host factors such as dietary patterns that shape the gut environment, genetic polymorphisms affecting microbial metabolism, and baseline microbiome composition that determines ecological niche availability remains poorly characterized in most clinical trials (Cho et al., 2025). This significant knowledge gap underscores the urgent need for more rigorous, standardized research approaches that incorporate systems biology perspectives to move beyond correlations and establish causal relationships between specific probiotic interventions and neurological outcomes.

As research in this field progresses toward personalized medicine approaches, the integration of cutting-edge multi-omics technologies, including metagenomics for strain-level identification, metabolomics for functional pathway analysis, and advanced neuroimaging techniques, will be essential for fully unraveling the complex, individualized interactions between probiotics, the gut microbiome, and brain health. Sophisticated computational approaches, including machine learning algorithms applied to large microbiome datasets, may help identify reliable biomarkers of treatment response and enable precision matching of probiotic formulations to individual patient characteristics, moving beyond the current trial-and-error approach. The development of next-generation probiotics with enhanced neuroactive properties through bioengineering approaches and improved gut colonization potential represents another promising direction that could address current limitations in strain efficacy and persistence. Concurrently, regulatory frameworks must evolve to establish evidence-based standards for probiotic characterization, manufacturing consistency, and clinically validated health claims that are currently lacking for many commercial products. The current chapter aims to comprehensively synthesize existing knowledge from both basic science and clinical trials, critically address important research gaps in mechanistic understanding and translational validity, and provide a clear roadmap for future investigations that bridges the current disconnect between promising preclinical findings and inconsistent clinical outcomes. By adopting a more systematic, mechanistic approach to probiotic research that properly accounts for the complexity of host-microbe-environment interactions, the field can progress from empirical observations to truly targeted, effective interventions for brain disorders rooted in GBA dysfunction, ultimately revolutionizing our approach to neurological and psychiatric care.

11.2 OBJECTIVES

The primary objectives of this chapter are to elucidate the molecular and physiological mechanisms through which probiotics influence the GBA, assess the clinical evidence supporting their use in neurological disorders, and identify limitations in current research methodologies. We aim to analyze variations in probiotic efficacy based on strain specificity, dosage, and treatment duration and to evaluate the role of individual microbiome composition in determining therapeutic outcomes. Additionally, this chapter seeks to propose standardized protocols for future studies to enhance reproducibility and clinical relevance, ultimately facilitating the development of targeted probiotic interventions for GBA-related conditions.

11.2.1 Literature Review

The GBA represents a complex bidirectional communication network that integrates neural, endocrine, and immune signaling pathways between the gastrointestinal tract and central nervous system. Recent advances in microbiome research have revealed that gut microbiota composition significantly influences brain function and behavior through multiple mechanistic pathways, including the production of neuroactive metabolites and modulation of systemic inflammation (Yassin et al., 2025). Numerous animal studies demonstrate that germ-free mice exhibit altered stress responses and social behaviors that can be normalized through microbial colonization, providing compelling evidence for microbiota-GBA interactions (Tofani et al., 2025). Clinical observations reveal distinct gut microbiome signatures in patients with neuropsychiatric disorders compared to healthy controls, particularly in conditions like major depressive disorder and autism spectrum disorder. Probiotic interventions have emerged as a promising therapeutic strategy, with specific strains showing ability to modulate neurotransmitter levels, including GABA and serotonin, through direct microbial synthesis or host metabolic pathway regulation (Rathore et al., 2025). The vagus nerve serves as a critical anatomical conduit for gut-brain signaling, with studies showing that vagotomy abolishes many probiotic-mediated neurological effects in animal models (Hwang & Oh, 2025). Microbial metabolites such as SCFAs, including butyrate, propionate, and acetate, cross the blood-brain barrier and exert neuroprotective effects through histone deacetylase inhibition and anti-inflammatory actions (Wachamo and Gaultier, 2025). While preclinical evidence strongly supports probiotic modulation of the GBA, translation to human applications requires careful consideration of interspecies differences in gut microbiota composition and brain physiology. This growing body of literature establishes a solid foundation for investigating targeted probiotic interventions in neurological and psychiatric conditions.

Extensive research has characterized the mechanisms by which specific probiotic strains influence central nervous system function through the GBA. Lactobacillus and Bifidobacterium species have been particularly well-studied, demonstrating the ability to strengthen intestinal barrier function, reduce systemic lipopolysaccharide (LPS) levels, and decrease proinflammatory cytokine production (Bhatia et al., 2025) (Figure 11.2).

Clinical trials reveal that multispecies probiotic formulations often show superior efficacy compared to single-strain interventions, likely due to synergistic effects on microbial ecosystem stability and metabolic output (Gorzelanna et al., 2025). The HPA axis represents a key endocrine pathway modulated by probiotics, with certain strains shown to normalize cortisol levels and stress responses in both animal models and human subjects. Emerging evidence suggests probiotics may enhance neurogenesis and synaptic plasticity in hippocampal regions, potentially through brain-derived neurotrophic factor (BDNF) upregulation and microglial modulation (Bertollo et al., 2025). Gut microbiota composition has been linked to blood-brain barrier integrity, with probiotic administration shown to reduce

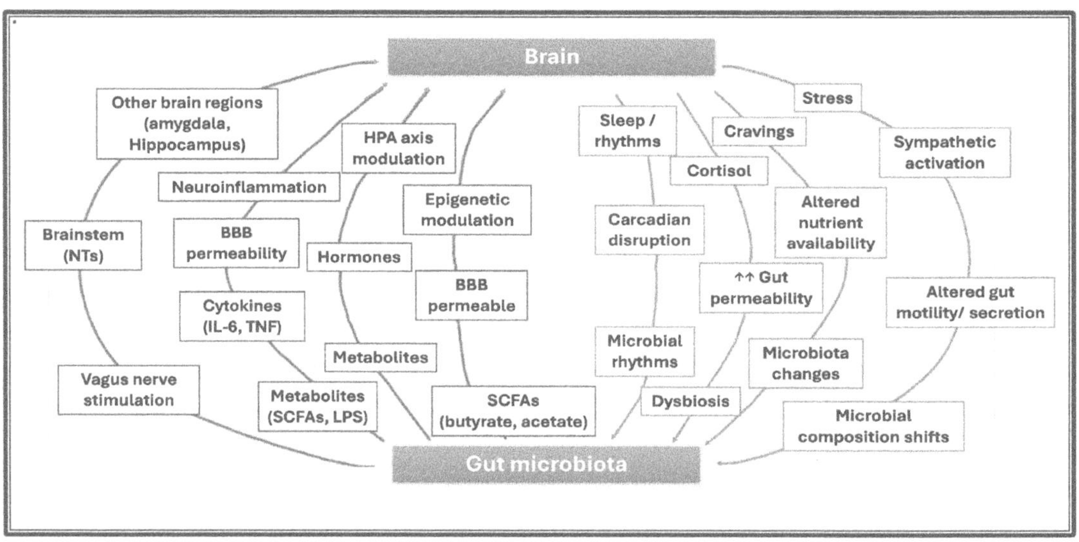

FIGURE 11.2 Two-way regulation: Gut microbiota and brain function.

permeability and potentially limit neurotoxic compound entry into the CNS. The timing of probiotic intervention appears crucial, with early-life administration demonstrating particularly robust effects on neurodevelopment and long-term brain function in animal studies (Ma et al., 2025). However, significant variability exists in reported outcomes across studies, reflecting differences in probiotic strains, dosages, administration protocols, and outcome measures. These findings highlight both the therapeutic potential and current limitations in our understanding of probiotic effects on the GBA.

The role of probiotics in neuropsychiatric disorders has gained increasing attention, with accumulating evidence supporting their use as adjunctive therapies. In depression, randomized controlled trials demonstrate that specific probiotic combinations can significantly reduce symptom severity, with effect sizes comparable to some conventional antidepressants (Elahinejad et al., 2025). Anxiety disorders show particular promise for probiotic intervention, with certain strains reducing both psychological and physiological markers of anxiety through GABAergic and serotonergic mechanisms (Butler et al., 2025). The gut microbiome's influence on neurodegenerative diseases has become a major research focus, with probiotics showing potential to slow disease progression in animal models of Parkinson's and Alzheimer's disease (Hasan et al., 2025). Autism spectrum disorder research reveals that probiotic administration may improve core behavioral symptoms, possibly through modulation of the microbiome-GBA and reduction of gastrointestinal distress commonly comorbid with ASD (Narula Khanna et al., 2025). Interesting findings in schizophrenia suggest probiotics may ameliorate some cognitive symptoms and potentially modulate antipsychotic medication efficacy through metabolic interactions (Slykerman et al., 2025). Chronic pain conditions, including migraine and fibromyalgia, have shown responsiveness to probiotic interventions, likely through anti-inflammatory mechanisms and descending pain pathway modulation (Ho et al., 2025). Despite these promising findings, significant challenges remain in establishing optimal strain combinations, dosages, and treatment durations for specific neurological conditions. The heterogeneity of study designs and patient populations underscores the need for larger, more standardized clinical trials to validate these preliminary observations (Table 11.2).

TABLE 11.2
Mechanisms of Action of Probiotics

CNS Effects	Mechanisms of Action of Probiotics	Reference
Anxiety reduction	• ↑ GABA receptor expression • Vagus nerve stimulation • ↓ Cortisol (HPA axis modulation) • Serotonin (5-HT) synthesis support	Nikolova et al. (2025)
Depression relief	• ↑ BDNF production • Tryptophan → serotonin conversion • ↓ Proinflammatory cytokines (IL-6, TNF-α) Dopamine precursor support	Zandifar et al. (2025)
Cognitive functions	• Neurogenesis via BDNF • Butyrate production → anti-inflammatory • Microglial maturation • ↓ Oxidative stress	Chen et al. (2021)
Adaptation to stress	• HPA axis regulation • ↓Cortisol release • Vagus nerve-mediated calming • Gut barrier strengthening	Wang et al. (2025)
Anti-inflammatory	• ↓ LPS translocation • ↑ Anti-inflammatory cytokines (IL-10) • ↓ TNF-α, IL-6 • SCFA production (butyrate, acetate)	Aparicio-Pascual et al. (2025)

Technical advancements in microbiome analysis have revolutionized our ability to study probiotic effects on the GBA at unprecedented resolution. High-throughput sequencing technologies now enable strain-level characterization of gut microbiota composition and functional potential, revealing intricate host-microbe interactions (Đokić et al., 2025). Metabolomic approaches have identified numerous microbially derived compounds that influence brain function, including tryptophan metabolites, secondary bile acids, and various neuroactive peptides (Interino et al., 2025). Advanced neuroimaging techniques demonstrate measurable probiotic-induced changes in brain activity patterns and functional connectivity, particularly in regions involved in emotional processing and stress response (Lee et al., 2025). Gnotobiotic animal models, where mice are colonized with defined microbial communities, have been instrumental in establishing causal relationships between specific bacteria and neurological outcomes. Sophisticated in vitro systems, including gut-on-a-chip and brain organoid technologies, are enabling mechanistic studies of host-microbe interactions at cellular and molecular levels (Metwaly et al., 2025). These technological advances have revealed the remarkable specificity of probiotic effects, where even subtle differences in bacterial strain or formulation can produce distinct neurological outcomes. Integration of multi-omics data through systems biology approaches is beginning to unravel the complex networks connecting gut microbial ecology with brain physiology and behavior. As these technologies continue to evolve, they promise to transform our understanding of probiotic mechanisms and accelerate the development of targeted microbial therapies for neurological disorders.

Critical examination of the current literature reveals several important challenges in probiotic research for GBA modulation. The field lacks consensus on optimal outcome measures for assessing probiotic effects on neurological function, with studies employing diverse behavioral, cognitive, and neurophysiological endpoints. Long-term safety data for chronic probiotic administration remains limited, particularly in vulnerable populations. The influence of host factors, including genetics, diet, lifestyle, and medication use, on probiotic efficacy requires systematic investigation to enable personalized treatment approaches. Most clinical trials to date have been relatively short-term, leaving open questions about the durability of probiotic effects and the necessity for ongoing administration. The pharmaceutical characteristics of probiotics, including stability, gastrointestinal survival, and colonization dynamics, present unique challenges for product development and quality control. Regulatory frameworks for probiotic claims related to neurological function remain underdeveloped compared to traditional pharmaceuticals, creating challenges for evidence-based clinical implementation. Economic considerations, including cost-effectiveness compared to conventional therapies and insurance coverage policies, may influence real-world adoption of probiotic interventions. Addressing these challenges will require coordinated efforts across academia, industry, and regulatory bodies to establish rigorous standards for probiotic research and clinical translation. Despite these hurdles, the accumulating evidence supports continued investigation of probiotics as a novel therapeutic modality for GBA disorders, with potential to complement or even replace current treatment paradigms in specific clinical contexts.

11.2.2 Research Gap

Despite the growing body of evidence supporting probiotic modulation of the GBA, several critical gaps remain. There is a lack of consensus on the optimal probiotic strains and dosages for specific neurological conditions, with many studies failing to account for interindividual variability in microbiome composition. Longitudinal data on the sustainability of probiotic effects are scarce, and the mechanisms underlying strain-specific outcomes are poorly understood. Additionally, most clinical trials have focused on single-strain probiotics, leaving the potential benefits of multi-strain formulations largely unexplored. The integration of multi-omics approaches, including metagenomics and metabolomics, into probiotic research remains limited, hindering a comprehensive understanding of host-microbe interactions. Addressing these gaps is essential for advancing the field and developing effective, personalized probiotic therapies.

11.3 METHODOLOGY

This chapter employs a systematic approach to review and analyze existing literature on probiotic interventions for GBA modulation. A comprehensive search of electronic databases, including PubMed, Scopus, and Web of Science, was conducted using predefined keywords related to probiotics, GBA, and neurological outcomes. Studies were selected based on inclusion criteria focusing on randomized controlled trials, meta-analyses, and mechanistic preclinical research. Data extraction prioritized information on probiotic strains, dosages, treatment durations, and measured neurological outcomes. Analytical methods included qualitative synthesis of mechanistic pathways and quantitative evaluation of clinical trial results where applicable. The methodology emphasizes transparency and reproducibility, with particular attention to study quality and potential biases in the existing literature.

11.4 ANALYSIS

The analysis reveals that probiotic interventions exhibit moderate but significant effects on mood and cognitive function, with effect sizes varying by strain and clinical population. Specific strains such as Lactobacillus rhamnosus and Bifidobacterium longum demonstrate consistent benefits in reducing anxiety and depressive symptoms, while others show limited efficacy. Mechanistic studies highlight the role of microbial metabolites in modulating neuroinflammation and neuronal activity, providing a plausible basis for observed clinical effects. However, substantial heterogeneity in study protocols and outcome measures complicates the interpretation of results across trials. Subgroup analyses suggest that baseline microbiome composition may influence treatment response, indicating the potential for personalized probiotic therapies. The analysis also identifies publication bias in the field, with positive results more likely to be reported than null findings.

11.5 FINDINGS

The findings indicate that probiotics hold promise as a noninvasive therapeutic strategy for GBA modulation, with particular relevance for mood disorders and neurodegenerative conditions. Strain-specific effects are evident, underscoring the importance of careful selection in clinical applications. The gut microbiome's baseline state emerges as a critical determinant of probiotic efficacy, suggesting that microbiome profiling could enhance treatment personalization. While safety and tolerability are generally high, long-term effects and potential interactions with conventional treatments require further investigation. The synthesis of available evidence supports the concept of psychobiotics but cautions against overgeneralization given the current limitations in research quality and consistency.

11.6 CONCLUSION

This chapter consolidates current knowledge on probiotic modulation of the GBA, highlighting both the potential and challenges of this therapeutic approach. While probiotics offer a promising avenue for addressing neurological disorders, significant gaps in understanding mechanisms and optimizing clinical protocols remain. The variability in individual responses necessitates a move toward personalized medicine strategies incorporating microbiome analysis. Future research must prioritize rigorous study designs, standardized outcome measures, and integration of multi-omics technologies to advance the field. Probiotics represent a valuable tool in the growing arsenal of microbiota-targeted therapies, but their full potential can only be realized through continued scientific exploration and translational efforts.

11.7 IMPLICATION

The implications of this research extend to clinical practice, where probiotics could serve as adjunct therapies for neurological and psychiatric conditions. Healthcare providers may consider incorporating probiotic interventions into treatment plans for patients with mood disorders or neurodegenerative diseases, particularly those unresponsive to conventional therapies. At a policy level, these findings advocate for updated regulatory frameworks to ensure accurate labeling and evidence-based claims for probiotic products. The research also underscores the need for interdisciplinary collaboration between microbiologists, neuroscientists, and clinicians to optimize probiotic development and application. From a public health perspective, increased awareness of diet-microbiome-brain interactions could lead to improved preventive strategies for mental health disorders.

11.8 FUTURE SCOPE

Future research should focus on elucidating the precise mechanisms by which specific probiotic strains influence neurological outcomes, employing advanced technologies such as gnotobiotic models and neural imaging. Large-scale, longitudinal clinical trials with standardized protocols are needed to establish efficacy across diverse populations. The development of microbiome-based predictive models could enable personalized probiotic recommendations, while exploration of synbiotic approaches may enhance therapeutic effects. Investigation of probiotic interactions with other interventions, including pharmacotherapies and dietary modifications, represents another critical direction. Emerging areas such as phage therapy and postbiotics warrant exploration as complementary approaches to probiotic interventions. Ultimately, the integration of GBA research into mainstream neurology and psychiatry holds transformative potential for understanding and treating brain disorders.

REFERENCES

Abdul Manan M. Progress in Probiotic Science: Prospects of Functional Probiotic-Based Foods and Beverages. Int J Food Sci. 2025;2025:5567567.

Abou Izzeddine N, Ahmad K, Bacha C, Jabbour M, Najjar M, Salhab S, Ghadieh HE, Kanaan A, Azar S, Khattar ZA, Harb F. The microbial guardians: Unveiling the Role of Gut Microbiota in Shaping Neurodegenerative Disease. IBRO Neurosci Rep. 2025;19:17–37.

Alipour S, Halimi H, Asri N, Rostami-Nejad M, Pishkar L, Houri H. Anti-Inflammatory effects of *Lactobacillus helveticus* and *Arthrospira platensis* on Colonic Cells Inflamed by Crohn's Disease-Associated *Escherichia coli*. BMC Mol Cell Biol. 2025;26(1):20.

Aparicio-Pascual D, Clemente-Suárez VJ, Tornero-Aguilera JF, Rubio-Zarapuz A. The Effect of Probiotic Supplementation on Cytokine Modulation in Athletes after a Bout of Exercise: A Systematic Review and Meta-Analysis. Sports Med Open. 2025;11(1):58.

Beikmohammadi M, Halimi S, Fard NA, Wen W. Therapeutic Potential of Probiotics: A Review of Their Role in Modulating Inflammation. Probiotics Antimicrob Proteins. 2025. doi: 10.1007/s12602-025-10609-z.

Bertollo AG, Santos CF, Bagatini MD, Ignácio ZM. Hypothalamus-Pituitary-Adrenal and Gut-Brain Axes in Biological Interaction Pathway of the Depression. Front Neurosci. 2025;19:1541075.

Bhatia A, Sharma D, Mehta J, Kumarasamy V, Begum MY, Siddiqua A, Sekar M, Subramaniyan V, Wong LS, Mat Rani NNI. Probiotics and Synbiotics: Applications, Benefits, and Mechanisms for the Improvement of Human and Ecological Health. J Multidiscip Healthc. 2025;18:1493–1510.

Butler MI, Kittel-Schneider S, Wagner-Skacel J, Mörkl S, Clarke G. The Gut Microbiome in Anxiety Disorders. Curr Psychiatry Rep. 2025;27(5):347–361.

Chandrasekaran P, Weiskirchen S, Weiskirchen R. Effects of Probiotics on Gut Microbiota: An Overview. Int J Mol Sci. 2024;25(11):6022.

Chen Y, Xu J, Chen Y. Regulation of Neurotransmitters by the Gut Microbiota and Effects on Cognition in Neurological Disorders. Nutrients. 2021;13(6):2099.

Cho M-Y, Eom J-H, Choi E-M, Yang S-J, Lee D, Kim YY, Kim H-S, Hwang I. Recent Advances in Therapeutic Probiotics: Insights from Human Trials. Clin Microbiol Rev. 2025;38(2):e0024024.

Dacaya P, Sarapis K, Moschonis G. The Role and Mechanisms of Probiotic Supplementation on Depressive Symptoms: A Narrative Review. Curr Nutr Rep. 2025;14(1):53.

Đokić J, Dinić M, Soković Bajić S, Bisenić A, Mitrović H, Jakovljević S, Radojević D, Brdarić E, Lukić J, Živković M, Tolinački M, Terzić-Vidojević A, Golić N. High-Throughput Workflow for Cultivation and Characterization of Gut Microbiota Strains with Anti-Inflammatory Properties and Metabolite Signature Associated with Gut-Brain Communication. Sci Rep. 2025;15(1):8741.

Duarte Luiz J, Manassi C, Magnani M, Cruz AGD, Pimentel TC, Verruck S. *Lactiplantibacillus plantarum* as a Promising Adjuvant for Neurological Disorders Therapy through the Brain-Gut Axis and Related Action Pathways. Crit Rev Food Sci Nutr. 2025;65(4):715–727.

Elahinejad V, Khorasanian AS, Tehrani-Doost M, Khosravi-Darani K, Mirsepasi Z, Effatpanah M, Askari-Rabori R, Tajadod S, Jazayeri S. Effects of Probiotics as Adjunctive Therapy to Fluoxetine on Depression Severity and Serum Brain-Derived Neurotrophic Factor, Cortisol, and Adrenocorticotropic Hormone in Patients with Major Depressive Disorder: A Randomized, Double-Blind, Placebo-Controlled Trial. Food Sci Nutr. 2025;13(4):e4698.

Feng J, Cen Q, Cui Y, Hu X, Li M, Wang L, Wei J, Sun N, Wang J, Zhang A. *Lactobacillus rhamnosus*: An Emerging Probiotic with Therapeutic Potential for Depression. Pharmacol Res. 2025;211:107541.

Gorzelanna Z, Mamrot A, Będkowska D, Bubak J, Miszczak M. Exploring the Potential of Novel Animal-Origin Probiotics as Key Players in One Health: Opportunities and Challenges. Int J Mol Sci. 2025;26(11):5143.

Grønbæk IMB, Halkjær SI, Mollerup S, Hansen EH, Paulsen SJ, Engel S, Theede K, Wilkens R, Boysen T, Petersen AM. The Effects of Probiotic Treatment with *Bifidobacterium breve*, Bif195 for Small Intestinal Crohn's Disease and the Gut Microbiome: Results from a Randomised, Double-Blind, Placebo-Controlled Trial. Gut Pathog. 2025;17(1):19.

Hasan A, Scuderi SA, Capra AP, Giosa D, Bonomo A, Ardizzone A, Esposito E. An Updated and Comprehensive Review Exploring the Gut–Brain Axis in Neurodegenerative Disorders and Neurotraumas: Implications for Therapeutic Strategies. Brain Sciences. 2025; 15(6):654.

Ho T, Elma Ö, Kocanda L, Brain K, Lam T, Kanhere T, Dong HJ. The Brain-Gut Axis and Chronic Pain: Mechanisms and Therapeutic Opportunities. Front Neurosci. 2025;19:1545997.

Hwang YK, Oh JS. Interaction of the Vagus Nerve and Serotonin in the Gut-Brain Axis. Int J Mol Sci. 2025;26(3):1160.

Interino N, Vitagliano R, D'Amico F, Lodi R, Porru E, Turroni S, Fiori J. Microbiota-Gut-Brain Axis: Mass-Spectrometry-Based Metabolomics in the Study of Microbiome Mediators-Stress Relationship. Biomolecules. 2025;15(2):243.

Lee SH, Han C, Shin C. IUPHAR Review: Microbiota-Gut-Brain Axis and Its Role in Neuropsychiatric Disorders. Pharmacol Res. 2025;216:107749.

Ma M, Li B, Qu Z, Liu S, Li S. Efficacy of Probiotics in Patients with Cognitive Impairment: A Systematic Review and Meta-Analysis. PLoS One. 2025;20(5):e0321567.

Mehta I, Juneja K, Nimmakayala T, Bansal L, Pulekar S, Duggineni D, Ghori HK, Modi N, Younas S. Gut Microbiota and Mental Health: A Comprehensive Review of Gut-Brain Interactions in Mood Disorders. Cureus. 2025;17(3):e81447.

Metwaly A, Kriaa A, Hassani Z, Carraturo F, Druart C; IHMCSA Consortium; Arnauts K, Wilmes P, Walter J, Rosshart S, Desai MS, Dore J, Fasano A, Blottiere HM, Maguin E, Haller D. A Consensus Statement on Establishing Causality, Therapeutic Applications and the Use of Preclinical Models in Microbiome Research. Nat Rev Gastroenterol Hepatol. 2025;22(5):343–356.

Narula Khanna H, Roy S, Shaikh A, Chhabra R, Uddin A. Impact of Probiotic Supplements on Behavioural and Gastrointestinal Symptoms in Children with Autism Spectrum Disorder: A Randomised Controlled Trial. BMJ Paediatr Open. 2025;9(1):e003045.

Nikolova VL, Cleare AJ, Young AH, Stone JM. Exploring the Mechanisms of Action of Probiotics in Depression: Results from a Randomized Controlled Pilot Trial. J Affect Disord. 2025;376:241–250.

O'Riordan KJ, Moloney GM, Keane L, Clarke G, Cryan JF. The Gut Microbiota-Immune-Brain Axis: Therapeutic Implications. Cell Rep Med. 2025;6(3):101982.

Petrut SM, Bragaru AM, Munteanu AE, Moldovan AD, Moldovan CA, Rusu E. Gut over Mind: Exploring the Powerful Gut-Brain Axis. Nutrients. 2025;17(5):842.

Rathore K, Shukla N, Naik S, Sambhav K, Dange K, Bhuyan D, Imranul Haq QM. The Bidirectional Relationship between the Gut Microbiome and Mental Health: A Comprehensive Review. Cureus. 2025;17(3):e80810.

Slykerman RF, Davies N, Vlckova K, O'Riordan KJ, Bassett SA, Dekker J, Schellekens H, Hyland NP, Clarke G, Patterson E. Precision Psychobiotics for Gut-Brain Axis Health: Advancing the Discovery Pipelines to Deliver Mechanistic Pathways and Proven Health Efficacy. Microb Biotechnol. 2025;18(1):e70079.

Tofani GSS, Leigh SJ, Gheorghe CE, Bastiaanssen TFS, Wilmes L, Sen P, Clarke G, Cryan JF. Gut Microbiota Regulates Stress Responsivity via the Circadian System. Cell Metab. 2025;37(1):138–153.e5.

Wachamo S, Gaultier A. The Emerging Role of Microbiota Derived SCFAs in Neurodegenerative Disorders. Brain Behav Immun Health. 2025;46:101012.

Wang D, Xu R, Liu S, Sun X, Zhang T, Shi L, Wang Y. Enhancing the Application of Probiotics in Probiotic Food Products from the Perspective of Improving Stress Resistance by Regulating Cell Physiological Function: A Review. Food Res Int. 2025;199:115369.

Xiao Y, Huang L, Zhao J, Chen W, Lu W. The Gut Core Microbial Species *Bifidobacterium longum*: Colonization, Mechanisms, and Health Benefits. Microbiol Res. 2025;290:127966.

Yassin LK, Nakhal MM, Alderei A, Almehairbi A, Mydeen AB, Akour A, Hamad MIK. Exploring the Microbiota-Gut-Brain Axis: Impact on Brain Structure and Function. Front Neuroanat. 2025;19:1504065.

Zandifar A, Badrfam R, Mohammaditabar M, Kargar B, Goodarzi S, Hajialigol A, Ketabforoush S, Heidari A, Fathi H, Shafiee A, Pourjafar H. The Effect of Prebiotics and Probiotics on Levels of Depression, Anxiety, and Cognitive Function: A Meta-Analysis of Randomized Clinical Trials. Brain Behav. 2025;15(3):e70401.

12 Stress Tolerance Mechanisms in Probiotic Formulations

Hassan Ali, Fatima Mujahid, Ahsen Taqveem, Hasan Ejaz, and Mohsin Khurshid

12.1 INTRODUCTION

The growing scientific and clinical interest in probiotics has established them as a viable alternative therapy for managing a range of diseases. Originating from the Latin ("pro" = for, "bios" = life), the term "probiotic" is formally defined by the FAO/WHO as "live microorganisms that, when administered in adequate amounts, confer a health benefit on the host." This concept traces back to Nobel laureate Elie Metchnikoff, who in 1908 linked the lactic acid bacteria (LAB) in fermented dairy to the longevity of Balkan populations, which is a groundbreaking yet long-overlooked hypothesis (Kitazawa et al., 2015; Organization, 2004). As pharmacologically active formulations, probiotics mediate host benefits through two primary mechanisms: (1) ecological antagonism against pathogens via competitive adhesion to intestinal epithelia and bacteriocin-mediated inhibition and (2) enzymatic biotransformation of dietary substrates into physiologically active compounds, particularly short-chain fatty acids (SCFAs: acetate [C2], propionate [C3], butyrate [C4]), which modulate host metabolism (Hills et al., 2019; Tirta et al., 2022). Meta-analyses demonstrate their significant efficacy in treating various gastroenterological conditions, including infectious diarrhea, antibiotic-associated diarrhea, and mild-to-moderate ulcerative colitis (de Simone, 2019). The pleiotropic health benefits of probiotics, encompassing gastrointestinal as well as systemic effects on metabolic, cardiovascular, immune, and neurocognitive functions, are summarized in Figure 12.1.

12.2 SYSTEMIC HEALTH BENEFITS OF PROBIOTICS

Contemporary preparations predominantly feature Generally Recognized as Safe (GRAS) strains of lactic acid-producing bacteria, with a taxonomic emphasis on *Lactobacillus* spp. (phylum Firmicutes) and *Bifidobacterium* spp. (phylum Actinobacteria), available in both pharmaceutical (Rx) and dietary supplement forms. Persistent quality assurance concerns center on three critical parameters: (a) genomic validation of strain identity, (b) accurate quantification of colony-forming units (CFUs) throughout shelf life, and (c) manufacturing process standardization, with observed heterogeneity in adherence to 21 CFR part 111 (cGMP for dietary supplements) and pharmaceutical-grade protocols (Goldenberg et al., 2017; Sánchez et al., 2017; Sarita et al., 2024). According to market analytics, the global probiotics industry is projected to reach USD 90.56 billion by 2025, expanding at a compound annual growth rate (CAGR) of 7.52% to attain USD 126.78 billion by 2030 (Mordor Intelligence). This growing demand across the food, pharmaceutical, and nutraceutical industries underscores the paramount importance of maintaining microbial viability and functional activity from production to consumption.

DOI: 10.1201/9781003399797-13

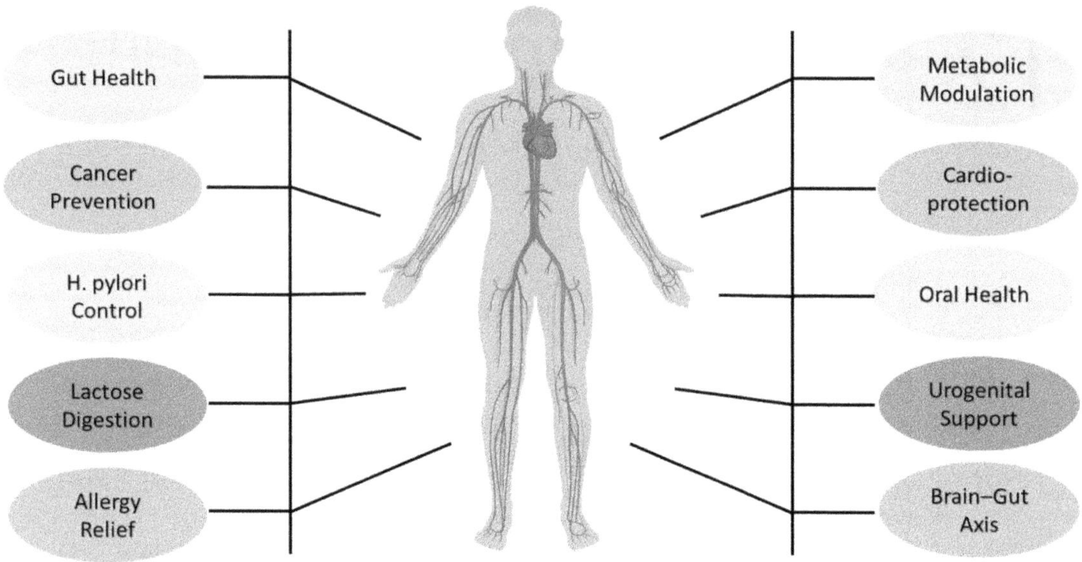

FIGURE 12.1 Systemic health benefits of probiotics.

A critical challenge in probiotic development is protecting these cultures from the multifaceted stressors encountered throughout their life cycle. During industrial processing, cells face variable fermentation conditions (pH/O_2 gradients), thermal stress (e.g., spray-drying at 150–200°C), and lyophilization-induced damage. Subsequent storage exposes them to oxidative stress and temperature fluctuations (Δ4°C/day), while gastrointestinal transit imposes extreme gastric acid (pH 1.5–3.0) and bile salt (0.3–0.5% w/v) challenges (Jannah et al., 2022). These cumulative stressors often reduce viable counts below the therapeutic threshold ($<10^6$ CFU/g), necessitating advanced stabilization strategies such as (1) selection of intrinsically adapted strains (e.g., *Lactobacillus acidophilus* with bile salt hydrolases), (2) microencapsulation (e.g., in alginate-chitosan matrices), and (3) optimized lyoprotectant formulations (e.g., trehalose-based cryoprotection). Ensuring that $\geq 10^7$ viable cells reach the intestinal niches remains essential for clinical efficacy, driving ongoing research into stress-resilient probiotic systems (Atanasov et al., 2023; Kathiriya et al., 2023; Sarita et al., 2024).

Therefore, overcoming these multifactorial stress barriers through advanced biotechnological and formulation strategies is essential to preserve probiotic viability, ensure functional delivery to the host, and achieve consistent therapeutic efficacy in clinical and nutritional applications.

12.2.1 Stress Tolerance Mechanisms in Probiotics

Probiotic microorganisms encounter a diverse array of physiological stresses throughout their industrial processing, storage, and gastrointestinal transit. These stressors disrupt cellular homeostasis by damaging proteins, membranes, nucleic acids, and metabolic pathways, thereby threatening viability and functional efficacy. Key challenges include thermal, osmotic, pH-related, oxidative, and bile salt stresses, each of which causes distinct cellular damage and triggers specific adaptive responses that are critical for probiotic survival and clinical performance (Corcoran et al., 2008; Mendonça et al., 2022). The intrinsic stress tolerance mechanisms of probiotics can be broadly categorized into eight major systems, as summarized in Figure 12.2.

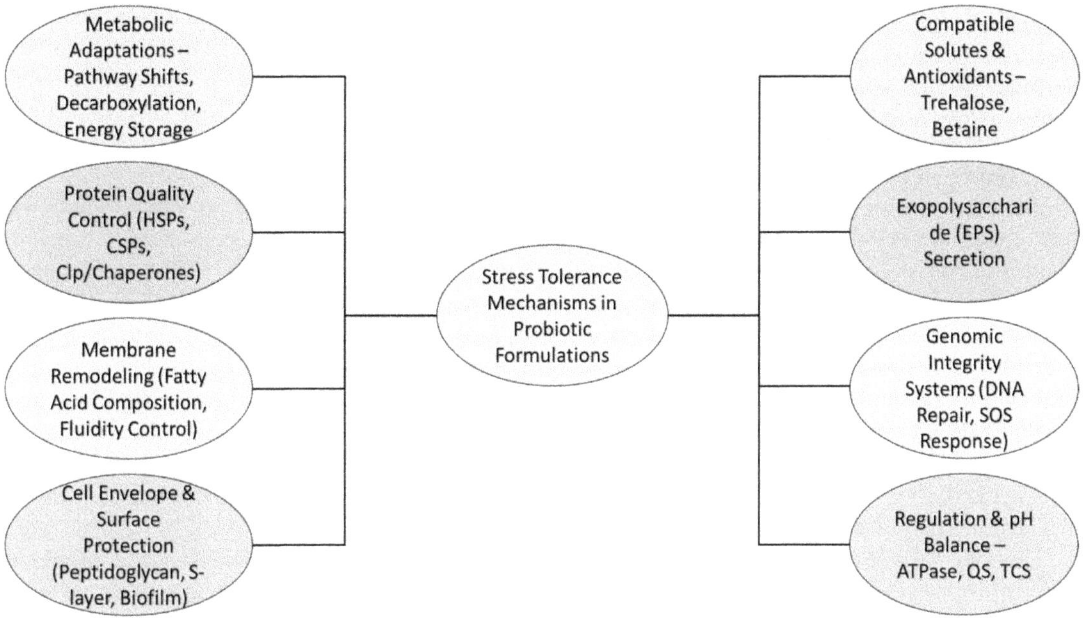

FIGURE 12.2 Intrinsic stress tolerance mechanisms of probiotics.

12.3 INTRINSIC STRESS TOLERANCE MECHANISMS OF PROBIOTICS

12.3.1 Thermal Stress

Thermal stress is one of the most critical technological challenges for probiotic microorganisms, as they are frequently exposed to elevated temperatures during industrial processing. Moderate heating (≈60–65°C) can destabilize non-covalent interactions, disrupt ribosomal function, and compromise cell envelope integrity, thereby impairing growth and protein synthesis. More extreme conditions, such as the transient exposure to temperatures up to 200°C during spray-drying, can induce irreversible denaturation of DNA, RNA, and proteins, ultimately causing a loss of metabolic activity and cell death. Elevated temperatures also alter membrane fluidity, increasing permeability and further compromising viability. Experimental studies on *Lactobacillus plantarum* have shown that heat shocks (e.g., 55°C for 15 min) significantly reduce acid tolerance and gastrointestinal survival. However, preadaptive conditioning at sublethal temperatures (e.g., 45°C for 30 min) can partially enhance resilience to subsequent stress exposure (Bove et al., 2013; Ferrando et al., 2016; Papadimitriou et al., 2016).

The adaptation to thermal stress relies on the induction of heat-responsive proteins and membrane remodeling processes that collectively safeguard cellular integrity. For instance, in *Lactobacillus kefiranofaciens* M1, exposure to elevated temperatures stimulates the synthesis of heat shock proteins (HSPs), phosphoenolpyruvate–protein phosphotransferase components, and molecular chaperones such as GroEL/GroES and DnaK/DnaJ. These are critical for protein folding, assembly, and translocation. The GroEL/GroES complex prevents irreversible protein aggregation during heat shock and is a central determinant of thermotolerance. Similar mechanisms operate in *Bifidobacterium longum*, where both large and small HSPs are strongly induced under stress. Small HSPs act not only as ATP-independent molecular chaperones but also as lipochaperones, interacting with lipid bilayers to stabilize membrane architecture during thermal perturbations. Complementing these protein-mediated defenses, LAB undergo lipidomic remodeling when subjected to heat stress. This involves an increased proportion of saturated and straight-chain fatty acids (C16:0, C18:0) to preserve optimal membrane fluidity and functionality. In some strains, a marked enrichment of cyclopropane fatty acids provides further stabilization under prolonged thermal exposure. Together, these adaptive responses enable probiotic cells to survive transient exposures to elevated temperatures (up to ~55–60°C) encountered during food processing, spray drying, and gastrointestinal transit, while maintaining metabolic functionality and viability (Capozzi et al., 2011; Chen et al., 2017; Hernández-Alcántara et al., 2018; Khaskheli et al., 2015; Russo et al., 2012).

Probiotic microorganisms deploy multiple strategies to counteract thermal stress, involving both protein-level defenses and structural membrane adjustments. For instance, in *Bifidobacterium bifidum*, brief pre-exposure to 42°C markedly enhances exopolysaccharide (EPS) secretion, a response associated with improved tolerance to downstream processing stresses like freeze-drying (Nguyen et al., 2014). Similarly, when *Lactobacillus casei* is exposed to sublethal heat (45–60°C), it exhibits differential expression of stress-responsive proteins. SDS-PAGE analysis has identified the overproduction of molecular chaperones, including DnaJ (40–55 kDa), GroEL (60 kDa), and DnaK (70 kDa), which act to stabilize protein folding and prevent aggregation under heat shock. In parallel, membrane lipid remodeling occurs, characterized by an increased proportion of saturated fatty acids (C16:0, C18:0) relative to unsaturated species, thereby supporting membrane stability at elevated temperatures (Bommasamudram et al., 2022).

The thermal stress encountered during food manufacturing poses a major barrier to probiotic survival, yet several LAB exhibit notable strain-dependent thermotolerance. *Lactobacillus fermentum* KGPMF28, for instance, can sustain active growth at 45°C for 24 h, underscoring its ability to proliferate under moderately elevated conditions. Likewise, *L. plantarum* strains Lp 813 and Lp 998 demonstrate enhanced resistance following thermal preconditioning at 45°C for 30 min, which increases their survival after a subsequent lethal exposure to 55°C for 15 min. This adaptive strategy not only improves thermotolerance by approximately two log units but also confers cross-protection against osmotic and oxidative stresses, highlighting the broad-spectrum benefits of sublethal heat adaptation. Certain *L. casei* strains further exemplify this resilience by retaining viability at 65°C, a temperature typically lethal to most microorganisms.

At the molecular level, these thermo-protective effects are associated with the induction of HSPs (e.g., GroEL, DnaK, small HSPs), remodeling of membrane fatty acid composition toward saturated and cyclopropane derivatives, and, in some cases, EPS secretion. These mechanisms collectively stabilize proteins, ribosomes, and lipid bilayers under heat stress (Ballini et al., 2023; Ferrando et al., 2015; Grujović et al., 2019). Key physiological outcomes, underlying molecular tolerance mechanisms, and thermal pre-adaptation strategies for probiotic survival are summarized in Table 12.1.

12.3.2 Cold Stress

Probiotics frequently encounter cold stress during storage in freeze-dried preparations or refrigerated food matrices, conditions that can markedly reduce their viability and functional performance. Low temperatures induce membrane rigidification, decrease enzymatic activity, and slow both RNA transcription and protein translation, ultimately leading to growth arrest. Freezing intensifies these effects through ice crystal formation, which physically disrupts membranes, and intracellular water crystallization, which elevates solute concentrations and generates osmotic stress. To withstand these challenges, probiotics employ diverse adaptive mechanisms. For instance, *Propionibacterium freudenreichii* remodels its membrane by increasing branched-chain fatty acids (BCFAs) derived from valine, leucine, and isoleucine, thereby sustaining fluidity at low temperatures. In *L. plantarum*, cold shock proteins (CSPs) such as CspP and CspL are upregulated; these proteins stabilize RNA and ensure continued transcription and translation during cold exposure. Some microorganisms also produce ice-binding proteins (IBPs), such as the β-solenoid-structured EfcIBP from *Euplotes focardii*, which inhibits ice recrystallization and enhances freeze-thaw survival by binding ice crystals through dual interaction surfaces. Proteomic analyses further reveal that cold-adapted *L. plantarum* overexpresses multiple stress-related proteins, including ATPases and chaperones, which collectively contribute to survival during repeated freeze-thaw cycles and long-term storage.

TABLE 12.1
Key Physiological Outcomes, Molecular Tolerance Mechanisms, and Thermal Pre-Adaptation Strategies for Probiotic Survival

Probiotic/Organism	Thermal Stress	Mechanism of Tolerance	Physiological Outcome	Ref.
Lactobacillus plantarum KLDS 1.0628	Pre-adaptation at 45°C for 1 h	Upregulation of heat shock proteins (HSPs), reduced SFA/USFA ratio in cell membrane	31.38-fold higher survival at 60°C, enhanced membrane fluidity, and cross-protection	Ma et al. (2021)
Tetragenococcus halophilus CGMCC 3792	Pre-adaptation at 45°C for 1.5 h	Upregulation of membrane transporters (ManM, OpuC, YidC), energy metabolism proteins (GpmA, AtpB, TpiA), and increased unsaturated fatty acids (C18:1)	7.24-fold higher survival under 10% ethanol stress, improved membrane integrity, and fluidity	Yang et al. (2021)
Enterococcus faecium HL7	Pre-adaptation at 52°C for 15 min	Increased saturated fatty acids (C16:0, C18:0), protein synthesis-dependent stress response	10^3–10^5× higher survival under H_2O_2 (0.01%), ethanol (20%), acid (pH 3), and alkali (pH 12) stress; 69.19% survival after freeze-drying (vs. 35.86% control)	Shin et al. (2019)
Lactobacillus kefiranofaciens M1	Heat adaptation (52°C for 30 min)	Upregulation of chaperones (DnaK, GroEL), ATP synthase (AtpA/B), and metabolic enzymes (triosephosphate isomerase, enolase)	52.5× higher heat survival (52°C); cross-protection against bile (4.8×), acid (3.5×), and cold stress (2.1×)	Chen et al. (2017)
Lactococcus lactis subsp. *Lactis* SD96 (evolved strain RD07)	Adaptive evolution at 41°C (400 generations)	Mutations in *purr* (G206S) reducing guanidine nucleotide pool; upregulated purine biosynthesis genes	Growth at 41°C in milk (unprecedented for mesophilic lactococci); 2–3× faster autolysis; enhanced cheese flavor development	Dorau et al. (2021)
Lactobacillus acidophilus NCFM (variant 1)	Gradual adaptation to 65°C (10–40 min cycles)	Downregulation of stress-sensitive proteins; potential HSP involvement	1000-fold greater survival at 65°C (1.7–1.9% vs. 0.0005% wild-type); 2-log improved survival at pH 2.0; enhanced gastric juice/salt tolerance; retained probiotic properties	Kulkarni et al. (2018)
Lactococcus lactis subsp. *cremoris*	Growth at 33°C (+3°C above optimal)	Upregulation of heat shock proteins (GroES/GroEL) and membrane adaptation	1.67× higher spray-drying survival (52.4% vs. 31.4% control); 0.7–1.5 log better heat resistance at 60°C	Hao et al. (2021)
Lactobacillus plantarum 998	Pre-adaptation at 45°C for 30 min	Heat shock protein upregulation; membrane composition modification	2-log higher thermal resistance (52°C); enhanced osmotolerance (up to 10% NaCl); maintained oxidative resistance	Ferrando et al. (2016)
Lactococcus lactis NZ9000 (UvrA+)	Heat challenge at 46°C for 6 h	Nucleotide excision repair (NER) system activation via UvrA overexpression	1.22× higher survival rate vs. control; Maintained DNA integrity under thermal stress	Moghaddam et al. (2017)
Lactobacillus acidophilus EG008 (ALE-derived)	65–75°C for 1 min	Non-synonymous SNP in *murD* (S435T) altering UDP-N-acetylmuramoyl-L-alanine-D-glutamate ligase structure	39.11% higher survival at 66°C; withstands 75°C (vs. WT limit of 65°C) while retaining probiotic traits	Jeon et al. (2021)

Notably, pre-adaptation to sub-lethal cold stress (e.g., 5°C for 6 h) markedly improves survival rates, with cold-conditioned *L. plantarum* achieving 78% viability post-freeze-thaw compared to 56% in non-adapted cells. Together, these adaptive responses, including membrane remodeling, CSP induction, IBP synthesis, and stress-protein upregulation, confer resilience to cold stress, supporting the stability and efficacy of probiotics in industrial formulations and clinical applications (Derzelle et al., 2002; Gagnaire et al., 2015; Mangiagalli et al., 2018; Song et al., 2014; Wang et al., 2025).

Proteomic investigations provide detailed insights into the cold-specific responses of probiotics. In *L. plantarum*, cold adaptation triggers the upregulation of at least 13 stress-related proteins, including ATPases and molecular chaperones, which collectively enhance resilience during repeated freeze-thaw cycles and extended storage. A similar deep-freeze adaptation is observed in *Lactobacillus delbrueckii* subsp. *Bulgaricus* LBB.B5, which, after several days at 4°C, shows strong induction of DNA repair proteins (AddB, UvrC, RecA) and the chaperone DnaJ compared to cells grown at 37°C. On a broader scale, *L. plantarum* K25 exhibits differential expression of over 500 proteins under cold stress (10°C vs. 37°C). This response includes increased production of DNA repair enzymes, transcriptional regulators, and ABC transporters, coupled with a downregulation of carbohydrate and fatty acid metabolism. This shift reflects a strategic reallocation of resources toward maintenance and energy conservation. The activation of two-component systems and quorum-sensing pathways further indicates that cold adaptation is a complex, coordinated process regulated at the systems level. Beyond cellular biochemistry, the physical parameter of freezing rate is a critical factor influencing viability. For example, rapid freezing in liquid nitrogen (–196°C) yields superior post-freeze-drying survival in LAB compared to slow freezing at –20°C, while yeasts display the opposite trend. These findings underscore the necessity of developing species-specific preservation strategies for optimal probiotic formulation design (Gaucher et al., 2019; Keto-Timonen et al., 2016; Liu et al., 2020; Polo et al., 2017; Prasad et al., 2003; Yin et al., 2017). Key physiological outcomes, molecular defense mechanisms, and cold pre-adaptation strategies contributing to probiotic survival are summarized in Table 12.2.

12.3.3 Osmotic Stress

Probiotic bacteria frequently encounter osmotic stress during cultivation, industrial processing, and transit through the gastrointestinal tract (GIT). Fluctuations in solute concentrations such as high salt in cheese, sugar in preserves, or brine in pickles create hyperosmotic conditions that trigger water efflux from cells. This results in cell shrinkage, reduced turgor pressure, and impaired physiological function. These disruptions compromise bacterial viability by altering the intracellular solute balance and destabilizing membrane and protein structures (Hickey et al., 2018; Waśko et al., 2013).

To mitigate osmotic stress, bacteria accumulate compatible solutes, which can be either imported from the environment or synthesized *de novo*. These solutes are typically neutral at physiological pH and can accumulate to high concentrations without interfering with cellular metabolism. Compatible solutes such as glycine betaine and trehalose are crucial for maintaining protein structure and membrane integrity under osmotic stress (Räsänen et al., 2004).

Specific bacterial responses illustrate this adaptive strategy. For example, *Lactobacillus rhamnosus* shows improved survival after freeze-drying when preconditioned with sublethal NaCl concentrations, a treatment that stimulates the expression of stress-responsive proteins, including GroEL, ClpB, and DnaK. In *Escherichia coli*, a model organism, the availability of glycine betaine under high salt pressure triggers a metabolic shift from aerobic respiration toward fermentative pathways. This adjustment helps preserve energy efficiency and cellular function even at reduced growth rates. Consequently, pre-adaptation to osmotic stress emerges as a practical strategy to enhance the viability of probiotics during industrial processes such as lyophilization and long-term storage (Metris et al., 2014; Meury, 1988; Waśko et al., 2013).

Probiotic bacteria exhibit remarkable adaptive responses to osmotic stress through multiple biochemical strategies. Research demonstrates that *P. freudenreichii* and *Lacticaseibacillus casei* significantly increase trehalose synthesis and secretion when exposed to osmotic stress (Cardoso et al., 2007). Similarly, lactobacilli modulate intracellular amino acid pools, particularly proline

TABLE 12.2
Cold Stress Tolerance Mechanisms and Physiological Outcomes in Selected Probiotic Strains

Probiotic/Organism	Cold Stress	Mechanism of Tolerance	Physiological Outcome	Ref.
Propionibacterium freudenreichii CIRM-BIA1T	Exposure to 4°C for 9 days (cheese-ripening simulation)	Upregulation of genes for glycogen synthesis (glgC, glgA, glgE), gluconeogenesis, and pyruvate metabolism; induction of cold-shock protein (cspB), stress-response regulators (dps, pspC), and proteomic chaperones (dnaK, groS, groL, clpB, hsp20)	Glycogen accumulation (peaking at ~120 h), 18-fold increase in trehalose synthesis, and continued production of flavor compounds (propionate, acetate, methylbutanoates) even at low temperature	Dalmasso et al. (2012); Dalmasso et al. (2012)
Lactobacillus acidophilus RD758	Cooling at moderate temperature (26–28°C) for 8 h before freezing to −20°C	Increased unsaturated/saturated fatty acid ratio, higher cycC19:0 fatty acid, synthesis of ClpP stress protein, pyruvate kinase, and putative glycoprotein endopeptidase	Reduced loss of acidification activity (0.64 min. day^{-1}); high residual acidification after 180 days of frozen storage (1011 min)	Wang et al. (2005)
Lactobacillus plantarum NC8 (cspL, cspP, cspC)	Temperature downshift (28°C → 8°C)	cspL: up to 17-fold induction; UB-like 5′UTR; RNA-binding motifs (RNP1/RNP2), cspP/cspC: slight or no induction; long 5′UTRs; RNA-binding motifs	cspL: enhanced cold adaptation, highest in late-exponential phase, cspP/cspC: maximal during lag phase; maintain translation at low temperature	Derzelle et al. (2000)
Lactobacillus plantarum WCFS1 (Hsp 18.5, Hsp 18.55, Hsp 19.3)	Exposure to 12°C	Overproduction of sHsps prevents protein misfolding and stabilizes membranes; regulated by CtsR/HrcA (Hsp 18.5) and σB-like (Hsp 18.55) promoters	Maintains viability and cell density during prolonged cold, enhancing cold adaptation	Fiocco et al. (2007)
Streptococcus thermophilus SFi39	Exposure to 20°C	ClpL protein is induced during cold stress, acting as a chaperone to prevent protein misfolding and maintain stress response	Cells maintain growth after temperature downshift, show reduced lag phase, and exhibit enhanced cryotolerance compared to the clpL mutant	Varcamonti et al. (2006)
Lactobacillus delbrueckii subsp. *bulgaricus*	0–25°C, sucrose 300–1800 mOsm L^{-1}	High and homogeneous membrane fluidity reduces rigid lipid domains, maintaining membrane integrity under cold and osmotic stress	Cells show enhanced freeze-thaw survival, reduced membrane disorganization, and preserved functionality compared to populations with rigid domains	Meneghel et al. (2017)
Lactobacillus paracasei, Pediococcus parvulus, Lactobacillus mali	−196°C	Cells were concentrated in lyoprotectants, which slowed metabolic activity and minimized ice crystal damage	High post-freeze-drying survival and maintained cell viability	Polo et al. (2017)
Lactobacillus plantarum L67	5°C for 6 h prior to freeze-thaw and storage	Upregulation of cold-shock protein genes (cspP, cspL) and 13 stress-related proteins, enhancing protein folding and cellular protection	Improved survival during freeze-thaw cycles and long-term storage, with 8% higher viability than control	Song et al. (2014)

and glutamate, as part of their osmotic adaptation process. These organisms employ specialized transport systems, such as the OpuABC (BusABC) transporter in *P. freudenreichii*, to accumulate compatible solutes like glycine betaine during hyperosmotic conditions (Gaucher et al., 2019). High solute concentrations in carrier matrices (e.g., sugars) can also induce the production of osmoprotective metabolites like mannitol, which enhances cell survival during industrial processing. Pre-exposure to sublethal osmotic stress activates stress-response proteins and stabilizes cellular structures, significantly improving viability during spray-drying and long-term storage.

EPS production constitutes another crucial protective mechanism. EPS forms a hydrated extracellular matrix around cells that maintains water balance and shields against osmotic fluctuations. Its synthesis is often upregulated under hyperosmotic conditions, contributing both to stress tolerance and desirable textural properties in probiotic formulations. Furthermore, osmotic conditioning using concentrated growth media has proven effective for enhancing biomass production and post-processing survival. This strategy reduces the need for costly supplements while improving long-term storage stability (Grosu-Tudor & Zamfir, 2014; Guérin et al., 2020; Song Huang et al., 2016; Nguyen et al., 2020; Peighambardoust et al., 2011).

Probiotic osmoadaptation integrates molecular-level adjustments with structural and physiological remodeling. Central to these adaptations are the accumulation of compatible solutes and the induction of stress-responsive proteins. For instance, *L. acidophilus* and *L. casei* activate the DnaK chaperone system and HtrS operon under high salt conditions to preserve protein conformation and membrane integrity. Similarly, *B. bifidum* CCFM16 exhibits a pronounced adaptive phenotype under extreme stress (1300 mOsm/kg), with evolved mutants achieving a threefold reduction in generation time via a proline-dominated osmolyte network. This adaptation involves metabolic reprogramming, including enhanced glutamate-to-proline conversion and reduced carbon flux that is an energetic trade-off that prioritizes stress protection over rapid proliferation (Palomino et al., 2016; Prasad et al., 2003; Zarzecka et al., 2022; Zhang et al., 2022).

Cellular resilience is further reinforced by surface layer (S-layer) remodeling, a key structural adaptation to osmotic stress. In *L. acidophilus* ATCC 4356, the proportion of SlpX within the S-layer increases from less than 10% to 40% under hyperosmotic conditions, a change that reinforces the cell envelope and enhances its barrier function. The industrial application of this knowledge is highly effective. Pre-adaptation to sublethal osmotic stress (e.g., 4% NaCl at pH 4.5) substantially improves survival during food processing. This is demonstrated by *L. rhamnosus* GG, which exhibits a markedly lower reduction in viability during yogurt storage (0.2 log) compared to non-adapted cells (0.5 log). Collectively, these findings confirm that controlled osmotic preconditioning orchestrates a multilevel defensive response—encompassing gene expression, metabolic reprogramming, and structural protein remodeling. This strategy offers a practical means to significantly enhance the stability, functionality, and industrial performance of probiotic formulations (Desmond et al., 2002; María Mercedes Palomino et al., 2016; Sagmeister et al., 2024; Settachaimongkon et al., 2015; Shi et al., 2024). Key physiological outcomes, molecular tolerance mechanisms, and osmotic pre-adaptation strategies are summarized in Table 12.3.

12.3.4 Acid Stress

LAB play a central role in fermentation by producing lactic acid, which acidifies their environment. The undissociated acid diffuses passively into the cytoplasm, where it dissociates into lactate ions and protons, thereby lowering the intracellular pH. This cytoplasmic acidification can damage DNA, enzymes, and structural proteins, impair metabolism, and ultimately reduce viability. The highly acidic conditions of the stomach and upper GIT present a further survival challenge, necessitating robust adaptive strategies for probiotic microbes (Pérez Montoro et al., 2018; Wu et al., 2012).

A primary biochemical defense involves amino acid decarboxylation reactions. These pathways consume intracellular protons and generate alkaline metabolites, which help neutralize cytoplasmic acidity while also producing ATP. For instance, the glutamate decarboxylase system in *Lactobacillus reuteri* and engineered histidine decarboxylation pathways in *Lactococcus lactis*

TABLE 12.3
Osmotic Stress Tolerance Mechanisms and Physiological Consequences in Selected Probiotic Strains

Probiotic/Organism	Osmotic Stress	Mechanism of Tolerance	Physiological Outcome	Ref.
Lactobacillus plantarum ATCC 14917	High salt (0.8M KCl)	Glycine-betaine acquisition via BusA-like transporters, intracellular K^+ accumulation, and enhanced proline biosynthesis	3-fold higher growth rate, maintains turgor, and cell integrity	Glaasker et al. (1996)
Propionibacterium freudenreichii (CIRM-BIA 129)	Hyperconcentrated sweet whey (30% dry matter)	Upregulation of OpuCA transporter (glycine-betaine uptake), ClpB chaperone, trehalose/polyphosphate synthesis	60.9% bile salt survival (vs. 2.6% control), 44.1% spray-drying viability	S. Huang et al. (2016)
Lactococcus lactis subsp. *Cremoris* MG1363	Hyperosmotic stress (0.5M NaCl)	BusR-mediated repression of glycine-betaine transporter BusA (OpuA)	6-fold induction of betaine uptake under osmotic stress	Romeo et al. (2003)
Lactobacillus plantarum P743	1M NaCl + 2mM betaine	Betaine uptake via BusA transporter (255 µmol/g DW)	26% survival after drying (vs. 11.7% control)	Kets et al. (1996)
Propionibacterium freudenreichii NIZO B365	2% NaCl + pH 4.7 + aerobic conditions	Trehalose biosynthesis via OtsAB pathway (400 mg/g protein)	Enhanced osmoprotection and carbon storage	Cardoso et al. (2004)
Lactobacillus fermentum IMDO 130101	Combined salt + temperature stress	Arginine deiminase (ADI) pathway activation	Ornithine production, ATP generation + pH homeostasis	Vrancken et al. (2009)
Lactobacillus casei ATCC 393	Hyperosmotic stress (NaCl-containing medium)	Membrane remodeling via increased cyc 19:0 fatty acid, AcylH3DG synthesis, and H4DG↑/H2DG↓ glycolipid ratio	Maintained membrane fluidity (DPH assay), increased hydrophobicity & bile sensitivity	Machado et al. (2004)
Lactobacillus casei BL23	High-salt (0.8 M NaCl)	Shorter LTA (31 vs. 42 Gro-P units), reduced D-alanylation (27% vs. 64%), downregulated *dltABCD/ltaS*	Increased cation binding, enhanced biofilm formation	Palomino et al. (2013)
Lactobacillus casei ATCC 393	1M NaCl	Reduced peptidoglycan cross-linking and altered PBP profiles	Cell enlargement and increased lytic sensitivity	Piuri et al. (2005)
Lactobacillus acidophilus IBB 801	NaCl (2%)	S-layer protein overexpression (~45 kDa)	Enhanced osmotic tolerance; cell envelope protection	Grosu-Tudor et al. (2016)

significantly enhance survival under extreme acid stress, enabling LAB to endure low-pH environments in fermented foods and during gastrointestinal transit (Trip et al., 2012; Wilson et al., 2014).

Complementing this metabolic response, LAB remodels its cell membrane to enhance resilience. Under acid stress, cells increase membrane fluidity by altering the ratio of unsaturated to saturated fatty acids. This restructuring improves membrane integrity, limits proton influx, and reduces acid-induced damage. These structural changes are crucial for supporting ongoing nutrient transport and energy transduction under acidic conditions. At the molecular level, acid stress triggers the upregulation of a suite of stress-related proteins. These include molecular chaperones, DNA repair enzymes, metabolic enzymes, elongation factors, ribosomal proteins, and ATP synthase subunits. This proteomic response functions to protect and repair cellular components, maintain efficient protein synthesis and folding, and sustain energy production. In summary, LAB employs an integrated defensive system that combines metabolic, structural, and proteomic strategies. By actively regulating intracellular pH, reinforcing membrane architecture, and activating specific stress-responsive proteins, they maintain viability and functionality in acidic environments. This multifaceted adaptability underscores their robustness and effectiveness as probiotics (Guan & Liu, 2020; He et al., 2025; Pérez Montoro et al., 2018; Sezgin & Tekin, 2023; Wu et al., 2012).

LAB employs a multi-faceted arsenal of strategies to survive acidic environments, complementing the core metabolic, membrane, and proteomic adaptations. One key biochemical strategy is malolactic fermentation (MLF), which involves the decarboxylation of malic acid into the milder lactic acid and carbon dioxide. This reaction consumes protons, thereby helping to neutralize acidity. The protective effect of MLF is demonstrated in *L. casei* ATCC 334, where the addition of malate significantly enhances survival at extremely low pH (Broadbent et al., 2010; Sumby et al., 2014; Torino et al., 2015).

Beyond metabolism, physical barrier formation is a crucial defense. Many LAB secrete EPSs under acidic stress to create a protective shield that limits proton influx. For instance, *Lactobacillus helveticus* ATCC 15807 produces increased amounts of EPS at pH 4.5 compared to neutral conditions, effectively shielding the cytoplasm from excessive acidification (Torino et al., 2005). LAB also actively regulate their intracellular pH through enzymatic activity. The F_0F_1 ATPase proton pump is central to this defense, expelling protons from the cell to maintain a viable cytoplasmic pH. Acid-adapted strains like *L. rhamnosus* GG upregulate this system when grown in acidic media, which is critical for ensuring survival under low-pH conditions (Koponen et al., 2012).

In addition to these primary mechanisms, LAB engages in supplementary acid stress responses. Quorum sensing via the LuxS system coordinates population-level adaptation, ClpE proteases degrade misfolded proteins resulting from acid-induced damage, and peptidoglycan biosynthesis modifications reinforce cell wall integrity. Together, these strategies synergistically enhance the resilience of LAB in acidic environments, reinforcing their effectiveness as probiotics and starter cultures (Pérez Montoro et al., 2018; Wang et al., 2018).

Probiotics combat acid stress through the production of stress proteins and molecular chaperones such as GroEL, GroES, DnaK, and Clp, as these are critical for preserving proper protein folding and preventing denaturation under low-pH conditions (Ferreira et al., 2013). This proteomic response is widespread; for instance, *L. plantarum* 423 significantly increases the abundance of HSPs when exposed to pH 2.5 for two hours, while *L. rhamnosus* GG exhibits broad changes in both protein expression and phosphorylation patterns under similar acidic stress (Heunis et al., 2014). Beyond protein protection, cells undergo structural adaptations. In *L. casei* ATCC 334, acid stress induces a shift in membrane fatty acid composition, notably increasing saturated and cyclopropane fatty acids. This remodeling reduces membrane proton permeability and enhances structural integrity. Concurrently, the Clp protease system is strongly upregulated in acid-challenged strains to ensure the efficient removal of misfolded or damaged proteins. These coordinated responses, which encompass chaperone induction, membrane stabilization, and protease activation, constitute a robust, integrated survival strategy that significantly enhances probiotic persistence in acidic environments (Broadbent et al., 2010). Key physiological outcomes, molecular tolerance mechanisms, and acidic pre-adaptation strategies are summarized in Table 12.4.

TABLE 12.4
Acidic Stress Tolerance Mechanisms and Physiological Consequences in Selected Probiotic Strains

Probiotic/Organism	Acidic Stress	Mechanism of Tolerance	Physiological Outcome	Ref.
Bifidobacterium longum NCIMB 8809 (wild-type) and acid-resistant mutant 8809dpH	Low pH (pH 2.0–4.8, gastric stress)	Overexpression of F_1F_0-ATPase (AtpA/AtpD) and metabolic rerouting (lactate, BCAA pathways); BSH downregulation	Maintains intracellular pH homeostasis; enhances survival in acidic GI tract	Sánchez et al. (2007)
Lactobacillus rhamnosus GG	Low pH (pH 2.0, gastric juice)	Glucose metabolism fuels F_1F_0-ATPase via glycolysis (GAPDH-dependent ATP production)	enhanced survival; proton extrusion maintains intracellular pH homeostasis	Corcoran et al. (2005)
Propionibacterium acidipropionici (mutant WSH1105)	High propionic acid (20 g/L)	Upregulation of ATP synthase (AtpA), GAPDH, and methylmalonyl-CoA epimerase; membrane fatty acid modification	Enhanced proton extrusion, redox balance, and PA conversion to succinyl-CoA	Guan et al. (2014)
Lactobacillus rhamnosus GG	Acid stress (pH 4.8 vs. 5.8)	Upregulation of F_1F_0-ATPase; pH-dependent protein phosphorylation (glycolytic enzymes)	Enhanced proton extrusion; metabolic flux modulation via phosphorylation	Koponen et al. (2012)
Lactobacillus reuteri	Acid stress (pH 2.5–3.5)	Glutamine deamidation (Gls3) and glutamate decarboxylation (GadB); arginine deiminase (ADI) pathway	Glutamine improves survival at pH 2.5; arginine conversion enhances resistance at pH 3.5; glutamate decarboxylation increases $\Delta\Psi$; coordinated amino acid conversions enhance ecological fitness in intestine and sourdough	Teixeira et al. (2014)
Lactobacillus bulgaricus L2	Freeze-drying stress	Increased cyclopropane fatty acid (cycC19:0) and cycC19:0/SFA ratio in membrane	Higher survival rates (up to 63.2%) at 30°C & pH 5.0; poor survival at 37°C due to low cycC19:0	Li et al. (2009)
Lactobacillus delbrueckii subsp. *Bulgaricus* CFL1	pH 5.25 (30 min., end of fermentation)	Cells modify membrane fatty acid composition and upregulate stress proteins while downregulating translation machinery.	9-fold higher freezing survival, better post-thaw acidification, and slower activity loss at −20°C	Streit et al. (2008)

(Continued)

TABLE 12.4 (Continued)
Acidic Stress Tolerance Mechanisms and Physiological Consequences in Selected Probiotic Strains

Probiotic/Organism	Acidic Stress	Mechanism of Tolerance	Physiological Outcome	Ref.
Bifidobacterium longum BBMN68	Pre-stress: pH 4.5 (2 h) Lethal stress: pH 3.5 (2 h)	Cells enhance ATP generation through glycolysis and amino acid metabolism, increase peptidoglycan synthesis (MurC, GalE1), elevate H^+-ATPase activity, and upregulate stress proteins (DnaK, GroEL) along with the translation factor TurfB	Maintained higher intracellular pH (6.2 vs. 5.3); 60× higher survival; sustained ATP/NH3 levels; strengthened cell wall	Jin et al. (2015)
Lactobacillus delbrueckii subsp. *bulgaricus*	Acid stress (pH 4.75) during exponential growth	Induction of heat shock proteins GroES, GroEL, and DnaK, which are involved in protein folding and protection under stress conditions	Enhanced cellular protection against acid damage, maintaining protein stability and function in acidic environments	Lim et al. (2000)
Lactobacillus delbrueckii subsp. *Bulgaricus* CFL1	Growth under non-controlled pH (natural acidification)	Increased production of heat shock proteins (Hsp70, GroEL, GroES) and glycolysis-related proteins	Enhanced resistance to spray-drying and heat stress (57°C), but no significant improvement during storage at 20°C	Silva et al. (2005)
Propionibacterium freudenreichii subsp. *shermanii*	Acid (pH 5.0)	Induction of OtsA–OtsB trehalose synthesis pathway (Trehalose-6-P synthase/phosphatase) with ADP-glucose; repression of TreS trehalose degradation pathway	Trehalose accumulation (~7-fold) stabilizes membranes/proteins and protects cells against acid damage	Filipa S. Cardoso et al. (2007)

12.3.5 Oxidative Stress

During industrial processing and gastrointestinal passage, probiotics are frequently exposed to aerobic conditions that impose oxidative stress. This is particularly challenging for oxygen-sensitive LAB and bifidobacteria. Oxygen toxicity primarily arises from its reduction at redox-active metal centers, such as those in heme-dependent cytochromes, which can generate reactive oxygen species (ROS), including superoxide, hydrogen peroxide, and hydroxyl radicals. These highly reactive molecules damage proteins, nucleic acids, and membrane lipids, thereby impairing metabolism, genomic stability, and membrane integrity, ultimately reducing cellular viability (Rezaïki et al., 2004; Yamamoto et al., 2005).

Many LAB are especially vulnerable because they lack canonical antioxidant enzymes like catalase and superoxide dismutase, limiting their ability to detoxify ROS. This deficiency often leads to the intracellular accumulation of hydrogen peroxide, which freely diffuses across membranes, exacerbating the oxidative burden and causing growth arrest or cell death. However, some species can mitigate this stress through heme-dependent mechanisms. For example, *L. plantarum* can express catalase when exogenous heme is available, and *L. lactis*, when supplied with heme, can assemble a functional respiratory chain. In this chain, cytochrome *bd* oxidase efficiently reduces oxygen to water, thereby preventing superoxide formation and significantly improving long-term survival (Kumar & Bandyopadhyay, 2005; Vido et al., 2004).

Although ROS can act as signaling molecules at low concentrations, probiotics lacking robust, intrinsic antioxidant systems are highly susceptible to their toxic effects. This vulnerability underscores oxidative stress as a critical determinant of probiotic survival during production and application (Cesselin et al., 2011). Key physiological outcomes, molecular tolerance mechanisms, and oxidative pre-adaptation strategies are summarized in Table 12.5.

12.3.6 Bile Stress

During gastrointestinal transit, probiotics must overcome multiple barriers, including salivary enzymes, gastric acidity, and bile salts. While exposure to saliva has a minimal impact on viability, the stomach presents a severe challenge due to its extreme acidity, which disrupts fundamental cellular processes. To survive this phase, probiotics employ strategies to regulate their intracellular pH, such as restricting proton entry, exporting protons via H^+ ATPase pumps, and consuming protons through metabolic pathways like the arginine deiminase, glutamate decarboxylase, and urease systems. These efforts are supported by additional protective measures, including membrane lipid remodeling, chaperone-mediated protein stabilization, and the activation of DNA repair mechanisms (Guan & Liu, 2020; Liu et al., 2024; Perez et al., 2018; Vila et al., 2019).

In the human intestine, probiotics are exposed to bile concentrations ranging from 0.05% to 2%, representing one of the most severe barriers to their survival. Acting as potent biological detergents, bile salts can damage DNA, induce protein misfolding, alter RNA structure, acidify the cytoplasm, and disrupt membrane integrity. To counteract these effects, probiotics activate a multifaceted defensive system. A central protective mechanism involves bile salt hydrolases (BSHs), which deconjugate bile acids to reduce their toxicity, while efflux transporters prevent their intracellular accumulation. Furthermore, modifications in membrane fatty acid composition and cell wall architecture are enacted to improve barrier integrity and resistance. At the molecular level, the stress response includes the upregulation of chaperones and proteases to repair damaged proteins, DNA repair systems to mitigate genotoxicity, and the F_0F_1-ATPase to maintain pH homeostasis. These coordinated responses are highly energetically demanding and are therefore supported by metabolic reprogramming that enhances ATP generation. Importantly, adaptation to bile stress not only ensures survival during intestinal transit but also confers cross-protection against acid and oxidative challenges. This highlights bile resistance as a critical determinant of probiotic

TABLE 12.5
Oxidative Stress Tolerance Mechanisms and Physiological Consequences in Selected Probiotic Strains

Probiotic/Organism	Acidic Stress	Mechanism of Tolerance	Physiological Outcome	Ref.
Lactobacillus casei Shirota	H_2O_2-induced oxidative and inflammatory stress	Modulation of Nrf-2/Keap1 signaling; Inhibition of the NF-κB inflammatory pathway	Prevented cellular ROS accumulation and membrane barrier disruption in human intestinal Caco-2/TC7 cells	Kong et al. (2020)
Lactobacillus plantarum (poxB poxF double mutant)	Aerobic stationary phase stress (endogenous H_2O_2 production)	Disruption of pyruvate oxidase (POX) activity, eliminating the major pathway for H_2O_2 and acetate production from lactate	Drastically improved cell viability and survival during prolonged stationary phase; prevented cell lysis	Goffin et al. (2006)
Lactobacillus plantarum CAUH2	5 mM H_2O_2 stress	Enhanced glycolysis for ATP; Upregulated Npr2, TrxR, Gpo; Maintained low free Fe2+; Activated DNA repair pathways; Upregulated Msr and Trx for protein repair	Improved survival and adaptive response to sublethal H_2O_2 exposure; Efficient H_2O_2 scavenging and reduced hydroxyl radical generation	Zhai et al. (2020)
Streptococcus thermophilus	Oxygen stress	Acquisition of small genomic islands via horizontal gene transfer, encoding functions for oxygen tolerance	Enhanced survival and fitness in the industrial dairy environment (e.g., milk)	Hols et al. (2005)
Lactobacillus acidophilus NCFM	Hydrogen Peroxide (H_2O_2)	Increased abundance of cysteine synthase (cysteine de novo synthesis); Elevated DNA repair enzymes	Accumulation of cysteine pool for redox homeostasis and protein stability; Reduction of oxidative damage to DNA and proteins	Calderini et al. (2017)
Bifidobacterium longum	Low Oxygen (5% O_2)	Sir2 enzyme deacetylates the transcription factor σH (SigH), upregulating the expression of antioxidant enzymes	Improved bacterial growth and survival under oxidative stress conditions	Guo et al. (2017)
Lactobacillus johnsonii/ gasseri (e.g., strains AL3, AL5)	Aerobic and Respiratory (O_2, H_2O_2, menadione, pyrogallol)	Activation of respiratory metabolism (using heme and menaquinone); Production of catalase (strain-specific); Possession of genes for thioredoxin reductase, SOD, and Dps-like peroxide resistance protein.	Increased biomass production; Enhanced tolerance to H_2O_2 and ROS generators (menadione, pyrogallol); Improved long-term survival in aerobic environments	Maresca et al. (2018)
Lactobacillus casei BL23	Hydrogen Peroxide (H_2O_2); Aeration	Heterologous expression of a manganese-dependent catalase (MnKat) from *L. plantarum*	Up to 320-fold increased survival after H_2O_2 exposure; 160-fold improved long-term survival in aerated stationary-phase cultures; Protection of co-cultured *L. bulgaricus* from H_2O_2	Rochat et al. (2006)
Lactococcus lactis	General Oxidative Stress (H_2O_2, O_2^-)	Multiple, overlapping genetic systems for resistance and control; include NADH oxidase/peroxidase, thioredoxin, and glutathione systems	Protection from molecular and metabolic damage; Maintenance of bacterial viability, reproducibility, and fermentative qualities under industrial conditions	Miyoshi et al. (2003)
Leuconostoc mesenteroides	General Oxygen Stress (Dissolved O_2)	Production of exopolysaccharide (EPS), which physically extrudes dissolved oxygen from the aqueous environment	Reduced intracellular ROS accumulation; Significant improvement in aerobic growth and survival; Protection of co-cultured oxygen-sensitive probiotics (*Lactobacillus*, *Bifidobacterium*)	Yan et al. (2018)

persistence, functionality, and overall efficacy in the GIT (Bezkorovainy, 2001; Desmond et al., 2004; He et al., 2019; Prete et al., 2020; Ruiz et al., 2013).

BSH serves as a primary defense enzyme in probiotics, catalyzing the deconjugation of bile salts into less toxic free bile acids and amino acids. This process not only reduces detergent toxicity but may also provide nutritional benefits to the host. Beyond enzymatic deconjugation, many strains employ active efflux systems to expel bile salts. The frequent genetic colocalization of these efflux pumps with BSH genes suggests a coordinated, evolutionarily conserved detoxification strategy. Bile salts also trigger a comprehensive cellular stress response. This includes the upregulation of ATP-dependent proteases and chaperones such as clpP, clpE, clpL, and clpX in *L. delbrueckii* UFV H2b20, which work to mitigate protein damage and maintain proteostasis. On a physiological level, prolonged exposure can drive remarkable adaptation. For instance, *L. lactis* mutants lacking LmrCD efflux pumps develop cholate resistance through altered surface hydrophilicity and enhanced biofilm formation enriched in extracellular polymeric substances. Collectively, these findings demonstrate that bile resistance is governed by a multifaceted network of defenses, including enzymatic modification, active transport, proteostasis maintenance, and lifestyle adaptation. These interconnected mechanisms are crucial for probiotic survival and persistence in the GIT (Dong et al., 2025; Ferreira et al., 2013; Zaidi et al., 2011). Key physiological outcomes, molecular tolerance mechanisms, and bile pre-adaptation strategies are summarized in Table 12.6.

12.3.7 Strategies to Enhance Probiotic Survival and Stress Resistance

A primary challenge in probiotic applications is ensuring a high number of viable cells survive both manufacturing processes and the harsh conditions of the GIT. To address this, several strategies have been developed to fortify probiotics against these stresses.

The Preconditioning (Stress Adaptation) strategy involves the controlled exposure of probiotic cells to mild, sublethal levels of stress such as heat, low pH, or high osmolarity to induce an adaptive response. This process primes the bacteria to withstand more severe subsequent challenges. Research demonstrates that heat-adapted cells of *Lactobacillus paracasei* showed a 700-fold improvement in survival during spray-drying and an 18-fold increase under heat stress. This enhanced resilience is linked to the increased production of protective chaperone proteins, such as the GroEL–GroES system. Similarly, pre-exposure to osmotic agents like NaCl or sucrose improves the growth and robustness of species like *L. acidophilus* and *L. casei* in challenging environments, a phenomenon known as cross-protection (Bisson et al., 2023; Mills et al., 2011).

Microencapsulation technology physically shields probiotic cells by entrapping them within protective biopolymer matrices, such as alginate, chitosan, or lipid-based hydrogels. Studies confirm that encapsulating *L. plantarum* in alginate-chitosan capsules significantly improves its viability during exposure to simulated gastric and intestinal fluids, as well as thermal processing. While core-shell alginate microgels are effective for protecting *B. longum*, it is noteworthy that a chitosan coating does not always provide an additional survival advantage. Recent advancements involve composite materials; for instance, incorporating magnesium oxide nanoparticles into alginate-gelatin matrices has proven effective at neutralizing acid and improving probiotic survival in low-pH conditions (Razavi et al., 2021; Trabelsi et al., 2013; Yeung et al., 2016).

The use of protective compounds is critical during drying processes like freeze-drying and spray-drying. Cryo- and lyo-protectants, including sugars like trehalose and sucrose, as well as proteins found in skim milk, stabilize cellular structures (e.g., membranes and proteins) against damage caused by freezing and dehydration. The internal accumulation of trehalose, often triggered by osmotic stress, is a key natural mechanism that confers resistance to desiccation and oxidative damage. The external addition of these protective agents to drying media is highly effective, with some formulations achieving survival rates exceeding 90% for sensitive species like *Streptococcus thermophilus* post-lyophilization (Gaucher et al., 2019; Leslie et al., 1995; Luft & Mazutti, 2025; Nezamdoost-Sani et al., 2023).

TABLE 12.6

Bile Stress Tolerance Mechanisms and Physiological Consequences in Selected Probiotic Strains

Probiotic/Organism	Bile Stress	Mechanism of Tolerance	Physiological Outcome	Ref.
Bifidobacterium animalis	Bile Salts (Ox bile)	Increased expression and activity of the F_1F_0-ATPase; Higher intracellular ATP reserve	Enhanced tolerance to low pH (pH 2.0); Better maintenance of intracellular pH homeostasis; Improved survival under bile stress	Sánchez et al. (2006)
Lactobacillus reuteri CRL 1098	Bile Salts	Changes in membrane lipid profile: decrease in phospholipids; lower saturated: unsaturated fatty acid ratio; appearance of unusual hydroxy and oxo fatty acids (C18:0,10-OH, C18:0,10-oxo); relative increase in glycolipids	Enhanced *in vitro* cholesterol removal. Changes did not confer improved tolerance to acid or cold stress	Taranto et al. (2003)
Lactobacillus acidophilus ATCC 4356	Bile Salts (0.01–0.1%)	Increased S-layer protein (Slp) production; Altered expression of the *slpA* gene; Changes in cell surface hydrophobicity	Shift in colony morphotype from Rough (R) to Smooth (S); Decreased maximum growth rate; Reduced H_2O_2 production; Altered adhesion properties	Khaleghi et al. (2010)
Lactobacillus salivarius Ren	Ox-bile (0.75 g/L)	Up-regulation of ABC transporters; Enhanced cell envelope barrier; Antioxidant production (γ-glutamylcysteine); Metabolic shift to maltose/glycerol	Improved bile extrusion; Reduced oxidative damage; Enhanced survival and gut adaptation	Wang et al. (2020)
Bifidobacterium longum	Bile Salts	Moderate induction of stress proteins (HtrA, DnaK, GroEL); Overlap with heat-shock response	Enhanced protein folding and degradation; Improved cellular stress management	Savijoki et al. (2005)
Lactobacillus plantarum KLDS1.0391	Bile Salts (0.3–2.0%)	Functional *luxS* gene and AI-2 quorum-sensing system	Enhanced acid and bile salt tolerance; Improved survival in simulated digestive juice; Increased adhesion to Caco-2 cells; Enhanced competitive exclusion of pathogens	Jia et al. (2018)
Lactobacillus plantarum; Lactobacillus fermentum	Bile Salts	Biofilm formation on abiotic surfaces; Enhanced spatial organization under bile stress	Increased resistance to high bile concentrations and gastrointestinal conditions; Enhanced inhibition of pathogen growth; Improved immunomodulatory effects (reduced TNF-α production)	Aoudia et al. (2016)
Lactobacillus acidophilus NCFM	Bile Salts	Multidrug resistance (MDR) transporters (LBA0552, LBA1429, LBA1446, LBA1679) for active extrusion	Enhanced bile survival; Reduced intracellular accumulation of toxins	Pfeiler & Klaenhammer (2009)
Lactobacillus reuteri ATCC 55730	Oxgall (Bile)	Induction of Clp chaperones (e.g., ClpL); Action of multidrug resistance transporters (e.g., lr1584); Putative esterase (lr1516) for cell envelope repair	Enhanced survival during bile shock; Resumption of growth in high bile concentrations; Cross-protection against acid stress	Whitehead et al. (2008)
Lactobacillus casei BL23	Bile Salts	Cell wall & fatty acid modification; Stress protein up-regulation; Altered carbohydrate metabolism; H^+-ATPase activation	Enhanced bile survival; Maintained cellular integrity	Alcántara & Zúñiga (2012)

This approach offers a targeted method to enhance innate probiotic stress tolerance by modifying microorganisms to overexpress specific protective genes, thereby directly amplifying their defense mechanisms. For example, engineering *L. paracasei* to overexpress the GroES–GroEL chaperone system resulted in a tenfold increase in survival after spray-drying and a twofold increase in resistance to freeze-drying. Future research is likely to explore the engineering of pathways involved in the transport and synthesis of protective solutes and osmolytes to further build intrinsic stress resistance (Mathipa & Thantsha, 2017; Mills et al., 2011; Singh et al., 2024).

The concerted application of these strategies provides a robust toolkit for significantly improving probiotic resilience. These approaches are essential for ensuring the successful delivery of therapeutically effective probiotic doses in functional foods and supplements, thereby overcoming the major barriers of industrial production and gastrointestinal digestion.

12.4 CONCLUSION

The functional efficacy of probiotics is fundamentally dependent on their survival through industrial processing, storage, and the harsh conditions of gastrointestinal transit. Throughout these stages, cells are challenged by a spectrum of stresses, including thermal, cold, osmotic, acid, oxidative, and bile that threaten their viability. In response, probiotics deploy a sophisticated array of intrinsic adaptive mechanisms. These include remodeling membrane fatty acid composition to maintain fluidity, accumulating compatible solutes like glycine betaine and proline, inducing universal stress proteins and molecular chaperones (e.g., GroEL, DnaK) to stabilize cellular proteins, and producing enzymes such as BSHs to detoxify harmful compounds. These innate responses underscore the remarkable biological resilience of probiotic microorganisms.

Complementing these natural defenses, a suite of technological strategies has been developed to enhance probiotic robustness. Preconditioning with sublethal stress primes cellular defense pathways, often conferring valuable cross-protection. Microencapsulation within biopolymer matrices offers a physical barrier against gastric acidity and bile salts. The strategic use of cryo- and lyoprotectants, such as trehalose, mitigates damage during freeze-drying and storage, ensuring long-term stability. Furthermore, advanced genetic engineering techniques provide precision tools to amplify key stress response pathways, paving the way for designing next-generation probiotics with superior resilience.

In conclusion, ensuring the delivery of a therapeutically effective dose of probiotics necessitates an integrated approach that synergizes the selection of intrinsically robust strains with external protective interventions. The rational and combined application of preconditioning, encapsulation, protective formulations, and genetic optimization is therefore essential. This multidisciplinary strategy provides a clear path toward achieving consistent clinical efficacy and reliable industrial performance for probiotic products.

REFERENCES

Alcántara, C., & Zúñiga, M. (2012). Proteomic and transcriptomic analysis of the response to bile stress of *Lactobacillus casei* BL23. *Microbiology (Reading)*, *158*(Pt 5), 1206–1218. https://doi.org/10.1099/mic.0.055657-0

Aoudia, N., Rieu, A., Briandet, R., Deschamps, J., Chluba, J., Jego, G., Garrido, C., & Guzzo, J. (2016). Biofilms of *Lactobacillus plantarum* and *Lactobacillus fermentum*: Effect on stress responses, antagonistic effects on pathogen growth and immunomodulatory properties. *Food Microbiol*, *53*(Pt A), 51–59. https://doi.org/10.1016/j.fm.2015.04.009

Atanasov, N., Trifonova, E., Evstatieva, Y., & Nikolova, D. (2023). Effect of two lyoprotectants on the survival rate and storage stability of freeze-dried probiotic lactic acid bacterial strains. *Journal of Chemical Technology and Metallurgy*, *58*(6), 1003–1010. https://doi.org/10.59957/jctm.v58i6.1

Ballini, A., Charitos, I. A., Cantore, S., Topi, S., Bottalico, L., & Santacroce, L. (2023). About functional foods: The probiotics and prebiotics state of art. *Antibiotics*, *12*(4), 635. https://doi.org/10.3390/antibiotics12040635

Bezkorovainy, A. (2001). Probiotics: Determinants of survival and growth in the gut. *Am J Clin Nutr, 73*(2 Suppl), 399s–405s. https://doi.org/10.1093/ajcn/73.2.399s

Bisson, G., Maifreni, M., Innocente, N., & Marino, M. (2023). Application of pre-adaptation strategies to improve the growth of probiotic lactobacilli under food-relevant stressful conditions [10.1039/D2FO03215E]. *Food & Function, 14*(4), 2128–2137. https://doi.org/10.1039/D2FO03215E

Bommasamudram, J., Muthu, A., & Devappa, S. (2022). Effect of sub-lethal heat stress on viability of lacticaseibacillus casei n in spray-dried powders. *LWT, 155*, 112904. https://doi.org/z10.1016/j.lwt.2021.112904

Bove, P., Russo, P., Capozzi, V., Gallone, A., Spano, G., & Fiocco, D. (2013). *Lactobacillus plantarum* passage through an oro-gastro-intestinal tract simulator: Carrier matrix effect and transcriptional analysis of genes associated to stress and probiosis. *Microbiological Research, 168*(6), 351–359. https://doi.org/10.1016/j.micres.2013.01.004

Broadbent, J. R., Larsen, R. L., Deibel, V., & Steele, J. L. (2010). Physiological and transcriptional response of *Lactobacillus casei* ATCC 334 to acid stress. *Journal of Bacteriology, 192*(9), 2445–2458. https://doi.org/10.1128/jb.01618-09

Calderini, E., Celebioglu, H. U., Villarroel, J., Jacobsen, S., Svensson, B., & Pessione, E. (2017). Comparative proteomics of oxidative stress response of *Lactobacillus acidophilus* NCFM reveals effects on DNA repair and cysteine de novo synthesis. *Proteomics, 17*(5). https://doi.org/10.1002/pmic.201600178

Capozzi, V., Weidmann, S., Fiocco, D., Rieu, A., Hols, P., Guzzo, J., & Spano, G. (2011). Inactivation of a small heat shock protein affects cell morphology and membrane fluidity in *Lactobacillus plantarum* WCFS1. *Res Microbiol, 162*(4), 419–425. https://doi.org/10.1016/j.resmic.2011.02.010

Cardoso, F. S., Castro, R. F., Borges, N., & Santos, H. (2007). Biochemical and genetic characterization of the pathways for trehalose metabolism in *Propionibacterium freudenreichii*, and their role in stress response. *Microbiology (Reading), 153*(Pt 1), 270–280. https://doi.org/10.1099/mic.0.29262-0

Cardoso, F. S., Gaspar, P., Hugenholtz, J., Ramos, A., & Santos, H. (2004). Enhancement of trehalose production in dairy propionibacteria through manipulation of environmental conditions. *International Journal of Food Microbiology, 91*(2), 195–204. https://doi.org/10.1016/s0168-1605(03)00387-8

Cesselin, B., Derré-Bobillot, A., Fernandez, A., Lamberet, G., Lechardeur, D., Yamamoto, Y., Pedersen, M. B., Garrigues, C., Gruss, A., & Gaudu, P. (2011). Responses of lactic acid bacteria to oxidative stress. In E. Tsakalidou, & K. Papadimitriou (Eds.), *Stress responses of lactic acid bacteria* (pp. 111–127). Springer US. https://doi.org/10.1007/978-0-387-92771-8_6

Chen, M.-J., Tang, H.-Y., & Chiang, M.-L. (2017). Effects of heat, cold, acid and bile salt adaptations on the stress tolerance and protein expression of kefir-isolated probiotic *Lactobacillus kefiranofaciens* M1. *Food Microbiology, 66*, 20–27. https://doi.org/10.1016/j.fm.2017.03.020

Corcoran, B. M., Stanton, C., Fitzgerald, G. F., & Ross, R. P. (2005). Survival of probiotic lactobacilli in acidic environments is enhanced in the presence of metabolizable sugars. *Applied and Environmental Microbiology, 71*(6), 3060–3067. https://doi.org/10.1128/aem.71.6.3060-3067.2005

Corcoran, B. M., Stanton, C., Fitzgerald, G. F., & Ross, R. P. (2008). Life under stress: The probiotic stress response and how it may be manipulated. *Current Pharmaceutical Design, 14*(14), 1382–1399. https://doi.org/10.2174/138161208784480225

Dalmasso, M., Aubert, J., Briard-Bion, V., Chuat, V., Deutsch, S. M., Even, S., Falentin, H., Jan, G., Jardin, J., Maillard, M. B., Parayre, S., Piot, M., Tanskanen, J., & Thierry, A. (2012). A temporal-omic study of *Propionibacterium freudenreichii* CIRM-BIA1 adaptation strategies in conditions mimicking cheese ripening in the cold. *PLoS One, 7*(1), e29083. https://doi.org/10.1371/journal.pone.0029083

Dalmasso, M., Aubert, J., Even, S., Falentin, H., Maillard, M. B., Parayre, S., Loux, V., Tanskanen, J., & Thierry, A. (2012). Accumulation of intracellular glycogen and trehalose by *Propionibacterium freudenreichii* under conditions mimicking cheese ripening in the cold. *Applied and Environmental Microbiology, 78*(17), 6357–6364. https://doi.org/10.1128/aem.00561-12

de Simone, C. (2019). The unregulated probiotic market. *Clinical Gastroenterology and Hepatology, 17*(5), 809–817. https://doi.org/10.1016/j.cgh.2018.01.018

Derzelle, S., Hallet, B., Ferain, T., Delcour, J., & Hols, P. (2002). Cold shock induction of the cspL Gene in *Lactobacillus plantarum* involves transcriptional regulation. *Journal of Bacteriology, 184*(19), 5518–5523. https://doi.org/10.1128/jb.184.19.5518-5523.2002

Derzelle, S., Hallet, B., Francis, K. P., Ferain, T., Delcour, J., & Hols, P. (2000). Changes in cspL, cspP, and cspC mRNA abundance as a function of cold shock and growth phase in *Lactobacillus plantarum*. *Journal of Bacteriology, 182*(18), 5105–5113. https://doi.org/10.1128/jb.182.18.5105-5113.2000

Desmond, C., Fitzgerald, G. F., Stanton, C., & Ross, R. P. (2004). Improved stress tolerance of GroESL-overproducing *Lactococcus lactis* and probiotic *Lactobacillus paracasei* NFBC 338. *Applied and Environmental Microbiology, 70*(10), 5929–5936. https://doi.org/10.1128/aem.70.10.5929-5936.2004

Desmond, C., Stanton, C., Fitzgerald, G. F., Collins, K., & Paul Ross, R. (2002). Environmental adaptation of probiotic lactobacilli towards improvement of performance during spray drying. *International Dairy Journal*, *12*(2), 183–190. https://doi.org/10.1016/S0958-6946(02)00040-7

Dong, Z., Yang, S., Tang, C., Li, D., Kan, Y., & Yao, L. (2025). New insights into microbial bile salt hydrolases: From physiological roles to potential applications [Review]. *Frontiers in Microbiology*, *16*(2025). https://doi.org/10.3389/fmicb.2025.1513541

Dorau, R., Chen, J., Liu, J., Ruhdal Jensen, P., & Solem, C. (2021). Adaptive laboratory evolution as a means to generate *Lactococcus lactis* strains with improved thermotolerance and ability to autolyze. *Applied and Environmental Microbiology*, *87*(21), e0103521. https://doi.org/10.1128/aem.01035-21

Ferrando, V., Quiberoni, A., Reinheimer, J., & Suárez, V. (2016). Functional properties of *Lactobacillus plantarum* strains: A study in vitro of heat stress influence. *Food Microbiology*, *54*, 154–161. https://doi.org/10.1016/j.fm.2015.10.003

Ferrando, V., Quiberoni, A., Reinhemer, J., & Suárez, V. (2015). Resistance of functional *Lactobacillus plantarum* strains against food stress conditions. *Food Microbiology*, *48*, 63–71. https://doi.org/10.1016/j.fm.2014.12.005

Ferreira, A. B., De Oliveira, M. N. V., Freitas, F. S., Alfenas-Zerbini, P., Da Silva, D. F., De Queiroz, M. V., Borges, A. C., & De Moraes, C. A. (2013). Increased expression of clp genes in *Lactobacillus delbrueckii* UFV H2b20 exposed to acid stress and bile salts. *Beneficial Microbes*, *4*(4), 367–374. https://doi.org/10.3920/bm2013.0022

Fiocco, D., Capozzi, V., Goffin, P., Hols, P., & Spano, G. (2007). Improved adaptation to heat, cold, and solvent tolerance in *Lactobacillus plantarum*. *Applied Microbiology and Biotechnology*, *77*(4), 909–915. https://doi.org/10.1007/s00253-007-1228-x

Gagnaire, V., Jardin, J., Rabah, H., Briard-Bion, V., & Jan, G. (2015). Emmental cheese environment enhances *Propionibacterium freudenreichii* stress tolerance. *PLoS One*, *10*(8), e0135780. https://doi.org/10.1371/journal.pone.0135780

Gaucher, F., Bonnassie, S., Rabah, H., Marchand, P., Blanc, P., Jeantet, R., & Jan, G. (2019). Review: Adaptation of beneficial propionibacteria, lactobacilli, and bifidobacteria improves tolerance toward technological and digestive stresses. *Frontiers in Microbiology*, *10*, 841. https://doi.org/10.3389/fmicb.2019.00841

Glaasker, E., Konings, W. N., & Poolman, B. (1996). Osmotic regulation of intracellular solute pools in *Lactobacillus plantarum*. *Journal of Bacteriology*, *178*(3), 575–582. https://doi.org/10.1128/jb.178.3.575-582.1996

Goffin, P., Muscariello, L., Lorquet, F., Stukkens, A., Prozzi, D., Sacco, M., Kleerebezem, M., & Hols, P. (2006). Involvement of pyruvate oxidase activity and acetate production in the survival of *Lactobacillus plantarum* during the stationary phase of aerobic growth. *Applied and Environmental Microbiology*, *72*(12), 7933–7940. https://doi.org/10.1128/aem.00659-06

Goldenberg, J. Z., Yap, C., Lytvyn, L., Lo, C. K., Beardsley, J., Mertz, D., & Johnston, B. C. (2017). Probiotics for the prevention of clostridium difficile-associated diarrhea in adults and children. *Cochrane Database of Systematic Reviews*, *12*(12), Cd006095. https://doi.org/10.1002/14651858.CD006095.pub4

Grosu-Tudor, S.-S., & Zamfir, M. (2014). Exopolysaccharide production by selected lactic acid bacteria isolated from fermented vegetables.

Grosu-Tudor, S. S., Brown, L., Hebert, E. M., Brezeanu, A., Brinzan, A., Fadda, S., Mozzi, F., & Zamfir, M. (2016). S-layer production by *Lactobacillus acidophilus* IBB 801 under environmental stress conditions. *Applied Microbiology and Biotechnology*, *100*(10), 4573–4583. https://doi.org/10.1007/s00253-016-7355-5

Grujović, M., Mladenović, K. G., Nikodijević, D. D., & Čomić, L. R. (2019). Autochthonous lactic acid bacteria—Presentation of potential probiotics application. *Biotechnology Letters*, *41*(11), 1319–1331. https://doi.org/10.1007/s10529-019-02729-8

Guan, N., & Liu, L. (2020). Microbial response to acid stress: Mechanisms and applications. *Applied Microbiology and Biotechnology*, *104*(1), 51–65. https://doi.org/10.1007/s00253-019-10226-1

Guan, N., Shin, H. D., Chen, R. R., Li, J., Liu, L., Du, G., & Chen, J. (2014). Understanding of how *Propionibacterium acidipropionici* respond to propionic acid stress at the level of proteomics. *Scientific Reports*, *4*, 6951. https://doi.org/10.1038/srep06951

Guérin, M., Silva, C. R.-D., Garcia, C., & Remize, F. (2020). Lactic acid bacterial production of exopolysaccharides from fruit and vegetables and associated benefits. *Fermentation*, *6*(4), 115. https://doi.org/10.3390/fermentation6040115

Guo, Q., Li, S., Xie, Y., Zhang, Q., Liu, M., Xu, Z., Sun, H., & Yang, Y. (2017). The NAD(+)-dependent deacetylase, *Bifidobacterium longum* Sir2 in response to oxidative stress by deacetylating SigH (σ(H)) and FOXO3a in *Bifidobacterium longum* and HEK293T cell respectively. *Free Radical Biology and Medicine*, *108*, 929–939. https://doi.org/10.1016/j.freeradbiomed.2017.05.012

Hao, F., Fu, N., Ndiaye, H., Woo, M. W., Jeantet, R., & Chen, X. D. (2021). Thermotolerance, survival, and stability of lactic acid bacteria after spray drying as affected by the increase of growth temperature. *Food and Bioprocess Technology*, *14*(1), 120–132. https://doi.org/10.1007/s11947-020-02571-1

He, J., Wang, W., Wu, Z., Pan, D., Guo, Y., Cai, Z., & Lian, L. (2019). Effect of *Lactobacillus reuteri* on intestinal microbiota and immune parameters: Involvement of sex differences. *Journal of Functional Foods*, *53*, 36–43. https://doi.org/10.1016/j.jff.2018.12.010

He, X., Cui, Y., Jia, Q., Zhuang, Y., Gu, Y., Fan, X., & Ding, Y. (2025). Response mechanisms of lactic acid bacteria under environmental stress and their application in the food industry. *Food Bioscience*, *64*, 105938. https://doi.org/10.1016/j.fbio.2025.105938

Hernández-Alcántara, A. M., Wacher, C., Llamas, M. G., López, P., & Pérez-Chabela, M. L. (2018). Probiotic properties and stress response of thermotolerant lactic acid bacteria isolated from cooked meat products. *LWT*, *91*, 249–257. https://doi.org/10.1016/j.lwt.2017.12.063

Heunis, T., Deane, S., Smit, S., & Dicks, L. M. (2014). Proteomic profiling of the acid stress response in *Lactobacillus plantarum* 423. *Journal of Proteome Research*, *13*(9), 4028–4039. https://doi.org/10.1021/pr500353x

Hickey, C. D., Fallico, V., Wilkinson, M. G., & Sheehan, J. J. (2018). Redefining the effect of salt on thermophilic starter cell viability, culturability and metabolic activity in cheese. *Food Microbiology*, *69*, 219–231. https://doi.org/10.1016/j.fm.2017.08.015

Hills, R. D. Jr., Pontefract, B. A., Mishcon, H. R., Black, C. A., Sutton, S. C., & Theberge, C. R.. (2019). Gut microbiome: Profound implications for diet and disease. *Nutrients*, *11*(7). https://doi.org/10.3390/nu11071613

Hols, P., Hancy, F., Fontaine, L., Grossiord, B., Prozzi, D., Leblond-Bourget, N., Decaris, B., Bolotin, A., Delorme, C., Dusko Ehrlich, S., Guédon, E., Monnet, V., Renault, P., & Kleerebezem, M. (2005). New insights in the molecular biology and physiology of *Streptococcus thermophilus* revealed by comparative genomics. *FEMS Microbiology Reviews*, *29*(3), 435–463. https://doi.org/10.1016/j.femsre.2005.04.008

Huang, S., Cauty, C., Dolivet, A., Le Loir, Y., Chen, X. D., Schuck, P., Jan, G., & Jeantet, R. (2016). Double use of highly concentrated sweet whey to improve the biomass production and viability of spray-dried probiotic bacteria. *Journal of Functional Foods*, *23*, 453–463. https://doi.org/10.1016/j.jff.2016.02.050

Huang, S., Rabah, H., Jardin, J., Briard-Bion, V., Parayre, S., Maillard, M. B., Le Loir, Y., Chen, X. D., Schuck, P., Jeantet, R., & Jan, G. (2016). Hyperconcentrated sweet whey, a new culture medium that enhances *Propionibacterium freudenreichii* stress tolerance. *Applied and Environmental Microbiology*, *82*(15), 4641–4651. https://doi.org/10.1128/aem.00748-16

Jannah, S. R., Rahayu, E. S., Yanti, R., Suroto, D. A., & Wikandari, R. (2022). Study of viability, storage stability, and shelf life of probiotic instant coffee *Lactiplantibacillus plantarum* subsp. *Plantarum* dad-13 in vacuum and nonvacuum packaging at different storage temperatures. *Int J Food Sci*, *2022*, 1663772. https://doi.org/10.1155/2022/1663772

Jeon, S., Kim, H., Choi, Y., Cho, S., Seo, M., & Kim, H. (2021). Complete genome sequence of the newly developed *Lactobacillus acidophilus* strain with improved thermal adaptability. *Frontiers in Microbiology*, *12*, 697351. https://doi.org/10.3389/fmicb.2021.697351

Jia, F. F., Zheng, H. Q., Sun, S. R., Pang, X. H., Liang, Y., Shang, J. C., Zhu, Z. T., & Meng, X. C. (2018). Role of luxS in stress tolerance and adhesion ability in *Lactobacillus plantarum* KLDS1.0391. *BioMed Research International*, *2018*, 4506829. https://doi.org/10.1155/2018/4506829

Jin, J., Qin, Q., Guo, H., Liu, S., Ge, S., Zhang, H., Cui, J., & Ren, F. (2015). Effect of pre-stressing on the acid-stress response in *Bifidobacterium* revealed using proteomic and physiological approaches. *PLoS One*, *10*(2), e0117702. https://doi.org/10.1371/journal.pone.0117702

Kathiriya, M. R., Vekariya, Y. V., & Hati, S. (2023). Understanding the probiotic bacterial responses against various stresses in food matrix and gastrointestinal tract: A review. *Probiotics Antimicrob Proteins*, *15*(4), 1032–1048. https://doi.org/10.1007/s12602-023-10104-3

Keto-Timonen, R., Hietala, N., Palonen, E., Hakakorpi, A., Lindström, M., & Korkeala, H. (2016). Cold shock proteins: A minireview with special emphasis on csp-family of enteropathogenic yersinia. *Frontiers in Microbiology*, *7*, 1151. https://doi.org/10.3389/fmicb.2016.01151

Kets, E., Teunissen, P., & de Bont, J. (1996). Effect of compatible solutes on survival of lactic acid bacteria subjected to drying. *Applied and Environmental Microbiology*, *62*(1), 259–261. https://doi.org/10.1128/aem.62.1.259-261.1996

Khaskheli, G. B., Zuo, F., Yu, R., & Chen, S. (2015). Overexpression of small heat shock protein enhances heat- and salt-stress tolerance of *Bifidobacterium longum* NCC2705. *Current Microbiology*, *71*(1), 8–15. https://doi.org/10.1007/s00284-015-0811-0

Kitazawa, H., Alvarez, S., Suvorov, A., Melnikov, V., Villena, J., & Sánchez, B. (2015). Recent advances and future perspective in microbiota and probiotics. *BioMed Research International*, *2015*, 275631. https://doi.org/10.1155/2015/275631

Kong, Y., Olejar, K. J., On, S. L. W., & Chelikani, V. (2020). The potential of *Lactobacillus* spp. For modulating oxidative stress in the gastrointestinal tract. *Antioxidants (Basel)*, *9*(7). https://doi.org/10.3390/antiox9070610

Koponen, J., Laakso, K., Koskenniemi, K., Kankainen, M., Savijoki, K., Nyman, T. A., de Vos, W. M., Tynkkynen, S., Kalkkinen, N., & Varmanen, P. (2012). Effect of acid stress on protein expression and phosphorylation in *Lactobacillus rhamnosus* GG. *J Proteomics*, *75*(4), 1357–1374. https://doi.org/10.1016/j.jprot.2011.11.009

Kulkarni, S., Haq, S. F., Samant, S., & Sukumaran, S. (2018). Adaptation of *Lactobacillus acidophilus* to thermal stress yields a thermotolerant variant which also exhibits improved survival at pH 2. *Probiotics Antimicrob Proteins*, *10*(4), 717–727. https://doi.org/10.1007/s12602-017-9321-7

Kumar, S., & Bandyopadhyay, U. (2005). Free heme toxicity and its detoxification systems in human. *Toxicology Letters*, *157*(3), 175–188. https://doi.org/10.1016/j.toxlet.2005.03.004

Leslie, S. B., Israeli, E., Lighthart, B., Crowe, J. H., & Crowe, L. M. (1995). Trehalose and sucrose protect both membranes and proteins in intact bacteria during drying. *Applied and Environmental Microbiology*, *61*(10), 3592–3597. https://doi.org/10.1128/aem.61.10.3592-3597.1995

Li, C., Zhao, J.-L., Wang, Y.-T., Han, X., & Liu, N. (2009). Synthesis of cyclopropane fatty acid and its effect on freeze-drying survival of *Lactobacillus bulgaricus* L2 at different growth conditions. *World Journal of Microbiology and Biotechnology*, *25*(9), 1659–1665. https://doi.org/10.1007/s11274-009-0060-0

Lim, E. M., Ehrlich, S. D., & Maguin, E. (2000). Identification of stress-inducible proteins in *Lactobacillus delbrueckii* subsp. bulgaricus. *Electrophoresis*, *21*(12), 2557–2561. https://doi.org/10.1002/1522-2683(20000701)21:12<2557::Aid-elps2557>3.0.Co;2-b

Liu, S., Ma, Y., Zheng, Y., Zhao, W., Zhao, X., Luo, T., Zhang, J., & Yang, Z. (2020). Cold-stress response of probiotic *Lactobacillus plantarum* K25 by iTRAQ proteomic analysis. *Journal of Microbiology and Biotechnology*, *30*(2), 187–195. https://doi.org/10.4014/jmb.1909.09021

Liu, Z., Zhao, X., & Bangash, H. I. (2024). Expression of stress responsive genes enables *Limosilactobacillus reuteri* to cross-protection against acid, bile salt, and freeze-drying [Original research]. *Frontiers in Microbiology*, *15*. https://doi.org/10.3389/fmicb.2024.1437803

Luft, L., & Mazutti, M. A. (2025). Freeze and spray drying technologies to produce solid microbial formulations for sustainable agriculture. *Processes*, *13*(7), 2188. https://doi.org/10.3390/pr13072188

Ma, J., Xu, C., Liu, F., Hou, J., Shao, H., & Yu, W. (2021). Stress adaptation and cross-protection of *Lactobacillus plantarum* KLDS 1.0628. *CyTA – Journal of Food*, *19*(1), 72–80. https://doi.org/10.1080/19476337.2020.1859619

Machado, M. C., López, C. S., Heras, H., & Rivas, E. A. (2004). Osmotic response in *Lactobacillus casei* ATCC 393: biochemical and biophysical characteristics of membrane. *Arch Biochem Biophys*, *422*(1), 61–70. https://doi.org/10.1016/j.abb.2003.11.001

Mangiagalli, M., Sarusi, G., Kaleda, A., Bar Dolev, M., Nardone, V., Vena, V. F., Braslavsky, I., Lotti, M., & Nardini, M. (2018). Structure of a bacterial ice binding protein with two faces of interaction with ice. *FEBS Journal*, *285*(9), 1653–1666. https://doi.org/10.1111/febs.14434

Maresca, D., Zotta, T., & Mauriello, G. (2018). Adaptation to aerobic environment of *Lactobacillus johnsonii/gasseri* strains [Original research]. *Frontiers in Microbiology*, *9*. https://doi.org/10.3389/fmicb.2018.00157

Mathipa, M. G., & Thantsha, M. S. (2017). Probiotic engineering: Towards development of robust probiotic strains with enhanced functional properties and for targeted control of enteric pathogens. *Gut Pathogens*, *9*, 28. https://doi.org/10.1186/s13099-017-0178-9

Mendonça, A. A., Pinto-Neto, W. P., da Paixão, G. A., Santos, D. D. S., De Morais, M. A. Jr., & De Souza, R. B. (2022). Journey of the probiotic bacteria: Survival of the fittest. *Microorganisms*, *11*(1). https://doi.org/10.3390/microorganisms11010095

Meneghel, J., Passot, S., Cenard, S., Réfrégiers, M., Jamme, F., & Fonseca, F. (2017). Subcellular membrane fluidity of *Lactobacillus delbrueckii* subsp. bulgaricus under cold and osmotic stress. *Applied Microbiology and Biotechnology*, *101*(18), 6907–6917. https://doi.org/10.1007/s00253-017-8444-9

Metris, A., George, S. M., Mulholland, F., Carter, A. T., & Baranyi, J. (2014). Metabolic shift of *Escherichia coli* under salt stress in the presence of glycine betaine. *Applied and Environmental Microbiology*, *80*(15), 4745–4756. https://doi.org/10.1128/aem.00599-14

Meury, J. (1988). Glycine betaine reverses the effects of osmotic stress on DNA replication and cellular division in *Escherichia coli*. *Archievs of Microbiology*, *149*(3), 232–239. https://doi.org/10.1007/bf00422010

Mills, S., Stanton, C., Fitzgerald, G. F., & Ross, R. P. (2011). Enhancing the stress responses of probiotics for a lifestyle from gut to product and back again. *Microbial Cell Factories*, *10*(Suppl 1), S19. https://doi.org/10.1186/1475-2859-10-s1-s19

Miyoshi, A., Rochat, T., Gratadoux, J. J., Le Loir, Y., Oliveira, S. C., Langella, P., & Azevedo, V. (2003). Oxidative stress in *Lactococcus lactis*. *Genetics and Molecular Research*, *2*(4), 348–359.

Moghaddam, T. K., Zhang, J., & Du, G. (2017). UvrA expression of *Lactococcus lactis* NZ9000 improve multiple stresses tolerance and fermentation of lactic acid against salt stress. *Journal of Food Science and Technology*, *54*(3), 639–649. https://doi.org/10.1007/s13197-017-2493-z

Mordor Intelligence. *Probiotics Market Size & Share Analysis – Growth Trends and Forecast (2025–2030)*. Retrieved August 16, 2025 from https://www.mordorintelligence.com/industry-reports/probiotics-market

Nezamdoost-Sani, N., Khaledabad, M. A., Amiri, S., & Mousavi Khaneghah, A. (2023). Alginate and derivatives hydrogels in encapsulation of probiotic bacteria: An updated review. *Food Bioscience*, *52*, 102433. https://doi.org/10.1016/j.fbio.2023.102433

Nguyen, H. T., Razafindralambo, H., Blecker, C., N'Yapo, C., Thonart, P., & Delvigne, F. (2014). Stochastic exposure to sub-lethal high temperature enhances exopolysaccharides (EPS) excretion and improves *Bifidobacterium bifidum* cell survival to freeze–drying. *Biochemical Engineering Journal*, *88*, 85–94. https://doi.org/10.1016/j.bej.2014.04.005

Nguyen, P. T., Nguyen, T. T., Bui, D. C., Hong, P. T., Hoang, Q. K., & Nguyen, H. T. (2020). Exopolysaccharide production by lactic acid bacteria: The manipulation of environmental stresses for industrial applications. *AIMS Microbiology*, *6*(4), 451–469. https://doi.org/10.3934/microbiol.2020027

Organization, W. H. (2004). *Food and Agriculture Organization of the United Nations* (Vol. 2).

Palomino, M. M., Allievi, M. C., Gründling, A., Sanchez-Rivas, C., & Ruzal, S. M. (2013). Osmotic stress adaptation in *Lactobacillus casei* BL23 leads to structural changes in the cell wall polymer lipoteichoic acid. *Microbiology (Reading)*, *159*(Pt 11), 2416–2426. https://doi.org/10.1099/mic.0.070607-0

Palomino, M. M., Waehner, P. M., Fina Martin, J., Ojeda, P., Malone, L., Sánchez Rivas, C., Prado Acosta, M., Allievi, M. C., & Ruzal, S. M. (2016). Influence of osmotic stress on the profile and gene expression of surface layer proteins in *Lactobacillus acidophilus* ATCC 4356. *Applied Microbiology and Biotechnology*, *100*(19), 8475–8484. https://doi.org/10.1007/s00253-016-7698-y

Papadimitriou, K., Alegría, Á, Bron, P. A., de Angelis, M., Gobbetti, M., Kleerebezem, M., Lemos, J. A., Linares, D. M., Ross, P., Stanton, C., Turroni, F., van Sinderen, D., Varmanen, P., Ventura, M., Zúñiga, M., Tsakalidou, E., & Kok, J. (2016). Stress physiology of lactic acid bacteria. *Microbiology and Molecular Biology Reviews*, *80*(3), 837–890. https://doi.org/10.1128/mmbr.00076-15

Peighambardoust, S. H., Golshan Tafti, A., & Hesari, J. (2011). Application of spray drying for preservation of lactic acid starter cultures: A review. *Trends in Food Science & Technology*, *22*(5), 215–224. https://doi.org/10.1016/j.tifs.2011.01.009

Perez, H. A., Bustos, A. Y., Taranto, M. P., Frías, M. D. A., & Ledesma, A. E. (2018). Effects of lysozyme on the activity of ionic of fluoroquinolone species. *Molecules*, *23*(4), 741. https://www.mdpi.com/1420-3049/23/4/741

Pérez Montoro, B., Benomar, N., Caballero Gómez, N., Ennahar, S., Horvatovich, P., Knapp, C. W., Gálvez, A., & Abriouel, H. (2018). Proteomic analysis of *Lactobacillus pentosus* for the identification of potential markers involved in acid resistance and their influence on other probiotic features. *Food Microbiology*, *72*, 31–38. https://doi.org/10.1016/j.fm.2017.11.006

Pfeiler, E. A., & Klaenhammer, T. R. (2009). Role of transporter proteins in bile tolerance of *Lactobacillus acidophilus*. *Applied and Environmental Microbiology*, *75*(18), 6013–6016. https://doi.org/10.1128/aem.00495-09

Piuri, M., Sanchez-Rivas, C., & Ruzal, S. M. (2005). Cell wall modifications during osmotic stress in *Lactobacillus casei*. *Journal of Applied Microbiology*, *98*(1), 84–95. https://doi.org/10.1111/j.1365-2672.2004.02428.x

Polo, L., Mañes-Lázaro, R., Olmeda, I., Cruz-Pio, L. E., Medina, Á, Ferrer, S., & Pardo, I. (2017). Influence of freezing temperatures prior to freeze-drying on viability of yeasts and lactic acid bacteria isolated from wine. *Journal of Applied Microbiology*, *122*(6), 1603–1614. https://doi.org/10.1111/jam.13465

Prasad, J., McJarrow, P., & Gopal, P. (2003). Heat and osmotic stress responses of probiotic *Lactobacillus rhamnosus* HN001 (DR20) in relation to viability after drying. *Applied and Environmental Microbiology*, *69*(2), 917–925. https://doi.org/10.1128/aem.69.2.917-925.2003

Prete, R., Long, S. L., Gallardo, A. L., Gahan, C. G., Corsetti, A., & Joyce, S. A. (2020). Beneficial bile acid metabolism from *Lactobacillus plantarum* of food origin. *Scientific Reports*, *10*(1), 1165. https://doi.org/10.1038/s41598-020-58069-5

Räsänen, L. A., Saijets, S., Jokinen, K., & Lindström, K. (2004). Evaluation of the roles of two compatible solutes, glycine betaine and trehalose, for the *Acacia senegal*–Sinorhizobium symbiosis exposed to drought stress. *Plant and Soil*, *260*(1), 237–251. https://doi.org/10.1023/B:PLSO.0000030181.03575.e1

Razavi, S., Janfaza, S., Tasnim, N., Gibson, D. L., & Hoorfar, M. (2021). Nanomaterial-based encapsulation for controlled gastrointestinal delivery of viable probiotic bacteria. *Nanoscale Advances*, *3*(10), 2699–2709. https://doi.org/10.1039/d0na00952k

Rezaïki, L., Cesselin, B., Yamamoto, Y., Vido, K., van West, E., Gaudu, P., & Gruss, A. (2004). Respiration metabolism reduces oxidative and acid stress to improve long-term survival of *Lactococcus lactis*. *Molecular Microbiology*, *53*(5), 1331–1342. https://doi.org/10.1111/j.1365-2958.2004.04217.x

Rochat, T., Gratadoux, J. J., Gruss, A., Corthier, G., Maguin, E., Langella, P., & van de Guchte, M. (2006). Production of a heterologous nonheme catalase by *Lactobacillus casei*: An efficient tool for removal of H_2O_2 and protection of *Lactobacillus bulgaricus* from oxidative stress in milk. *Applied and Environmental Microbiology*, *72*(8), 5143–5149. https://doi.org/10.1128/aem.00482-06

Romeo, Y., Obis, D., Bouvier, J., Guillot, A., Fourçans, A., Bouvier, I., Gutierrez, C., & Mistou, M. Y. (2003). Osmoregulation in *Lactococcus lactis*: BusR, a transcriptional repressor of the glycine betaine uptake system BusA. *Molecular Microbiology*, *47*(4), 1135–1147. https://doi.org/10.1046/j.1365-2958.2003.03362.x

Ruiz, L., Margolles, A., & Sánchez, B. (2013). Bile resistance mechanisms in *Lactobacillus* and *Bifidobacterium*. *Frontiers in Microbiology*, *4*, 396. https://doi.org/10.3389/fmicb.2013.00396

Russo, P., De la Luz Mohedano, M., Capozzi, V., De Palencia, P. F., López, P., Spano, G., & Fiocco, D. (2012). Comparative proteomic analysis of *Lactobacillus plantarum* WCFS1 and ΔctsR mutant strains under physiological and heat stress conditions. *International Journal of Molecular Sciences*, *13*(9), 10680–10696. https://www.mdpi.com/1422-0067/13/9/10680

Sagmeister, T., Gubensäk, N., Buhlheller, C., Grininger, C., Eder, M., Đordić, A., Millán, C., Medina, A., Murcia, P. A. S., Berni, F., Hynönen, U., Vejzović, D., Damisch, E., Kulminskaya, N., Petrowitsch, L., Oberer, M., Palva, A., Malanović, N., Codée, J., & Pavkov-Keller, T. (2024). The molecular architecture of *Lactobacillus* s-layer: Assembly and attachment to teichoic acids. *Proceedings of the National Academy of Sciences of the United States of America*, *121*(24), e2401686121. https://doi.org/10.1073/pnas.2401686121

Sánchez, B., Champomier-Vergès, M. C., Collado Mdel, C., Anglade, P., Baraige, F., Sanz, Y., de los Reyes-Gavilán, C. G., Margolles, A., & Zagorec, M. (2007). Low-pH adaptation and the acid tolerance response of *Bifidobacterium longum* biotype longum. *Applied and Environmental Microbiology*, *73*(20), 6450–6459. https://doi.org/10.1128/aem.00886-07

Sánchez, B., de los Reyes-Gavilán, C. G., & Margolles, A. (2006). The F1F0-ATPase of *Bifidobacterium animalis* is involved in bile tolerance. *Environmental Microbiology*, *8*(10), 1825–1833. https://doi.org/10.1111/j.1462-2920.2006.01067.x

Sánchez, B., Delgado, S., Blanco-Míguez, A., Lourenço, A., Gueimonde, M., & Margolles, A. (2017). Probiotics, gut microbiota, and their influence on host health and disease. *Molecular Nutrition & Food Research*, *61*(1). https://doi.org/10.1002/mnfr.201600240

Sarita, B., Samadhan, D., Hassan, M. Z., & Kovaleva, E. G. (2024). A comprehensive review of probiotics and human health-current prospective and applications. *Frontiers in Microbiology*, *15*, 1487641. https://doi.org/10.3389/fmicb.2024.1487641

Savijoki, K., Suokko, A., Palva, A., Valmu, L., Kalkkinen, N., & Varmanen, P. (2005). Effect of heat-shock and bile salts on protein synthesis of *Bifidobacterium longum* revealed by [35S]methionine labelling and two-dimensional gel electrophoresis. *FEMS Microbiology Letters*, *248*(2), 207–215. https://doi.org/10.1016/j.femsle.2005.05.032

Settachaimongkon, S., van Valenberg, H. J. F., Winata, V., Wang, X., Nout, M. J. R., van Hooijdonk, T. C. M., Zwietering, M. H., & Smid, E. J. (2015). Effect of sublethal preculturing on the survival of probiotics and metabolite formation in set-yoghurt. *Food Microbiology*, *49*, 104–115. https://doi.org/10.1016/j.fm.2015.01.011

Sezgin, E., & Tekin, B. (2023). Molecular evolution and population genetics of glutamate decarboxylase acid resistance pathway in lactic acid bacteria [original research]. *Frontiers in Genetics*, *14*. https://doi.org/10.3389/fgene.2023.1027156

Shi, Z., Li, X., Fan, X., Zeng, X., Zhang, T., Wu, Z., Wu, X., & Pan, D. (2024). The SlpX protein plays a crucial role in the intestinal juice tolerance of *Lactobacillus acidophilus* CICC6074. *Food Bioscience*, *59*, 103865. https://doi.org/10.1016/j.fbio.2024.103865

Shin, Y., Kang, C. H., Kim, W., & So, J. S. (2019). Heat adaptation improved cell viability of probiotic *Enterococcus faecium* HL7 upon various environmental stresses. *Probiotics Antimicrob Proteins*, *11*(2), 618–626. https://doi.org/10.1007/s12602-018-9400-4

Silva, J., Carvalho, A. S., Ferreira, R., Vitorino, R., Amado, F., Domingues, P., Teixeira, P., & Gibbs, P. A. (2005). Effect of the pH of growth on the survival of *Lactobacillus delbrueckii* subsp. bulgaricus to stress conditions during spray-drying. *Journal of Applied Microbiology*, *98*(3), 775–782. https://doi.org/10.1111/j.1365-2672.2004.02516.x

Singh, B., Mal, G., Kalra, R. S., & Marotta, F. (2024). Genetic engineering and genome editing of probiotic microbes. In B. Singh, G. Mal, R. S. Kalra, & F. Marotta (Eds.), *Probiotics as live biotherapeutics for veterinary and human health, volume 2: Functional foods and post-genomics* (pp. 521–538). Springer Nature Switzerland. https://doi.org/10.1007/978-3-031-65459-6_24

Song, S., Bae, D. W., Lim, K., Griffiths, M. W., & Oh, S. (2014). Cold stress improves the ability of *Lactobacillus plantarum* L67 to survive freezing. *International Journal of Food Microbiology, 191*, 135–143. https://doi.org/10.1016/j.ijfoodmicro.2014.09.017

Streit, F., Delettre, J., Corrieu, G., & Béal, C. (2008). Acid adaptation of *Lactobacillus delbrueckii* subsp. bulgaricus induces physiological responses at membrane and cytosolic levels that improves cryotolerance. *Journal of Applied Microbiology, 105*(4), 1071–1080. https://doi.org/10.1111/j.1365-2672.2008.03848.x

Sumby, K. M., Grbin, P. R., & Jiranek, V. (2014). Implications of new research and technologies for malolactic fermentation in wine. *Applied Microbiology and Biotechnology, 98*(19), 8111–8132. https://doi.org/10.1007/s00253-014-5976-0

Taranto, M. P., Fernandez Murga, M. L., Lorca, G., & de Valdez, G. F. (2003). Bile salts and cholesterol induce changes in the lipid cell membrane of *Lactobacillus reuteri. Journal of Applied Microbiology, 95*(1), 86–91. https://doi.org/10.1046/j.1365-2672.2003.01962.x

Teixeira, J. S., Seeras, A., Sanchez-Maldonado, A. F., Zhang, C., Su, M. S., & Gänzle, M. G. (2014). Glutamine, glutamate, and arginine-based acid resistance in *Lactobacillus reuteri. Food Microbiology, 42*, 172–180. https://doi.org/10.1016/j.fm.2014.03.015

Tirta, G. D., Martin, L., Bani, M. D., Kho, K., Pramanda, I. T., Pui, L. P., How, Y. H., Lim, C. S. Y., & Devanthi, P. V. P. (2022). Spray drying encapsulation of pediococcus acidilactici at different inlet air temperatures and wall material ratios. *Foods, 12*, 1. https://doi.org/10.3390/foods12010165

Torino, M. I., Font de Valdez, G., & Mozzi, F. (2015). Biopolymers from lactic acid bacteria. Novel applications in foods and beverages. *Frontiers in Microbiology, 6*, 834. https://doi.org/10.3389/fmicb.2015.00834

Torino, M. I., Mozzi, F., & Font de Valdez, G. (2005). Exopolysaccharide biosynthesis by *Lactobacillus helveticus* ATCC 15807. *Applied Microbiology and Biotechnology, 68*(2), 259–265. https://doi.org/10.1007/s00253-004-1865-2

Trabelsi, I., Bejar, W., Ayadi, D., Chouayekh, H., Kammoun, R., Bejar, S., & Ben Salah, R. (2013). Encapsulation in alginate and alginate coated-chitosan improved the survival of newly probiotic in oxgall and gastric juice. *International Journal of Food Microbiology, 61*, 36–42. https://doi.org/10.1016/j.ijbiomac.2013.06.035

Trip, H., Mulder, N. L., & Lolkema, J. S. (2012). Improved acid stress survival of *Lactococcus lactis* expressing the histidine decarboxylation pathway of *Streptococcus thermophilus* CHCC1524. *Journal of Biological Chemistry, 287*(14), 11195–11204. https://doi.org/10.1074/jbc.M111.330704

Varcamonti, M., Arsenijevic, S., Martirani, L., Fusco, D., Naclerio, G., & De Felice, M. (2006). Expression of the heat shock gene clpL of *Streptococcus thermophilus* is induced by both heat and cold shock. *Microb Cell Fact, 5*, 6. https://doi.org/10.1186/1475-2859-5-6

Vido, K., le Bars, D., Mistou, M.-Y., Anglade, P., Gruss, A., & Gaudu, P. (2004). Proteome analyses of heme-dependent respiration in *Lactococcus lactis*: Involvement of the proteolytic system. *Journal of Bacteriology, 186*(6), 1648–1657. https://doi.org/10.1128/jb.186.6.1648-1657.2004

Vila, T., Rizk, A. M., Sultan, A. S., & Jabra-Rizk, M. A. (2019). The power of saliva: Antimicrobial and beyond. *PLoS Pathog, 15*(11), e1008058. https://doi.org/10.1371/journal.ppat.1008058

Vrancken, G., Rimaux, T., Wouters, D., Leroy, F., & De Vuyst, L. (2009). The arginine deiminase pathway of *Lactobacillus fermentum* IMDO 130101 responds to growth under stress conditions of both temperature and salt. *Food Microbiology, 26*(7), 720–727. https://doi.org/10.1016/j.fm.2009.07.006

Wang, C., Cui, Y., & Qu, X. (2018). Mechanisms and improvement of acid resistance in lactic acid bacteria. *Archives of Microbiology, 200*(2), 195–201. https://doi.org/10.1007/s00203-017-1446-2

Wang, G., Zhai, Z., Ren, F., Li, Z., Zhang, B., & Hao, Y. (2020). Combined transcriptomic and proteomic analysis of the response to bile stress in a centenarian-originated probiotic *Lactobacillus salivarius* ren. *Food Research International, 137*, 109331. https://doi.org/10.1016/j.foodres.2020.109331

Wang, J., Wu, P., Dhital, S., Yu, A., & Chen, X. D. (2025). Impact of freezing and freeze drying on *Lactobacillus rhamnosus* GG survival: Mechanisms of cell damage and the role of pre-freezing conditions and cryoprotectants. *Foods, 14*(10). https://doi.org/10.3390/foods14101817

Wang, Y., Delettre, J., Guillot, A., Corrieu, G., & Béal, C. (2005). Influence of cooling temperature and duration on cold adaptation of *Lactobacillus acidophilus* RD758. *Cryobiology, 50*(3), 294–307. https://doi.org/10.1016/j.cryobiol.2005.03.001

Waśko, A., Polak-Berecka, M., & Gustaw, W. (2013). Increased viability of probiotic *Lactobacillus rhamnosus* after osmotic stress. *Acta Alimentaria, 42*(4), 520–528. https://doi.org/10.1556/aalim.42.2013.4.7

Whitehead, K., Versalovic, J., Roos, S., & Britton, R. A. (2008). Genomic and genetic characterization of the bile stress response of probiotic *Lactobacillus reuteri* ATCC 55730. *Applied and Environmental Microbiology, 74*(6), 1812–1819. https://doi.org/10.1128/aem.02259-07

Wilson, C. M., Loach, D., Lawley, B., Bell, T., Sims, I. M., O'Toole, P. W., Zomer, A., & Tannock, G. W. (2014). *Lactobacillus reuteri* 100-23 modulates urea hydrolysis in the murine stomach. *Applied and Environmental Microbiology, 80*(19), 6104–6113. https://doi.org/10.1128/aem.01876-14

Wu, C., Zhang, J., Wang, M., Du, G., & Chen, J. (2012). *Lactobacillus casei* combats acid stress by maintaining cell membrane functionality. *J Ind Microbiol Biotechnol, 39*(7), 1031–1039. https://doi.org/10.1007/s10295-012-1104-2

Yamamoto, Y., Poyart, C., Trieu-Cuot, P., Lamberet, G., Gruss, A., & Gaudu, P. (2005). Respiration metabolism of group b streptococcus is activated by environmental haem and quinone and contributes to virulence. *Molecular Microbiology, 56*(2), 525–534. https://doi.org/10.1111/j.1365-2958.2005.04555.x

Yan, M., Wang, B. H., Xu, X., der Meister, T. Jr., Tabγač, H. T., Hwang, F. F., & Liu, Z. (2018). Extrusion of dissolved oxygen by exopolysaccharide from leuconostoc mesenteroides and its implications in relief of the oxygen stress. *Frontiers in Microbiology, 9*, 2467. https://doi.org/10.3389/fmicb.2018.02467

Yang, H., Yao, S., Zhang, M., & Wu, C. (2021). Heat adaptation induced cross protection against ethanol stress in *Tetragenococcus halophilus*: Physiological characteristics and proteomic analysis. *Frontiers in Microbiology, 12*, 686672. https://doi.org/10.3389/fmicb.2021.686672

Yeung, T. W., Üçok, E. F., Tiani, K. A., McClements, D. J., & Sela, D. A. (2016). Microencapsulation in alginate and chitosan microgels to enhance viability of *Bifidobacterium longum* for Oral delivery. *Frontiers in Microbiology, 7*, 494. https://doi.org/10.3389/fmicb.2016.00494

Yin, X., Salemi, M. R., Phinney, B. S., Gotcheva, V., Angelov, A., & Marco, M. L. (2017). Proteomes of *Lactobacillus delbrueckii* subsp. bulgaricus LBB.B5 incubated in milk at optimal and low temperatures. *mSystems, 2*(5). https://doi.org/10.1128/mSystems.00027-17

Zaidi, A. H., Bakkes, P. J., Krom, B. P., van der Mei, H. C., & Driessen, A. J. (2011). Cholate-stimulated biofilm formation by *Lactococcus lactis* cells. *Applied and Environmental Microbiology, 77*(8), 2602–2610. https://doi.org/10.1128/aem.01709-10

Zarzecka, U., Chajęcka-Wierzchowska, W., Zakrzewski, A., Zadernowska, A., & Fraqueza, M. J. (2022). High pressure processing, acidic and osmotic stress increased resistance to aminoglycosides and tetracyclines and the frequency of gene transfer among strains from commercial starter and protective cultures. *Food Microbiology, 107*, 104090. https://doi.org/10.1016/j.fm.2022.104090

Zhai, Z., Yang, Y., Wang, H., Wang, G., Ren, F., Li, Z., & Hao, Y. (2020). Global transcriptomic analysis of *Lactobacillus plantarum* CAUH2 in response to hydrogen peroxide stress. *Food Microbiology, 87*, 103389. https://doi.org/10.1016/j.fm.2019.103389

Zhang, Y., Mao, B., Tang, X., Liu, X., Zhao, J., Zhang, H., Cui, S., & Chen, W. (2022). Integrative genome and metabolome analysis reveal the potential mechanism of osmotic stress tolerance in *Bifidobacterium bifidum*. *LWT, 159*, 113199. https://doi.org/10.1016/j.lwt.2022.113199

13 Personalized Microbiome-Based Nutritional Strategies

Samy Selim, Yaseen Galali, Muhammad Naeem Riaz, Mohammad Al Hemairy, Lia Gallo-Urrego, and Sajid Iqbal

13.1 INTRODUCTION

Imagine a time when your diet is a finely tuned plan tailored especially for you, rather than merely a set of guidelines. Not only for your taste receptors or calorie count, but also for the trillions of microscopic occupants of your gut. This is the fascinating frontier of personalized microbiome-based dietary strategies, not science fiction; it's also rapidly altering our perceptions of health and disease. For a long period, recommendations on diet have mostly taken a "one-size-fits-all approach." Although general recommendations on good eating have surely helped to improve public health, they sometimes overlook a vital point.

Our bodies respond to food in different ways depending on how complexly our genes, lifestyle, and surroundings interact. And our gut microbiome (Abeltino et al., 2024; Bermingham et al., 2024) is the silent but strong conductor at the core of this complex dance. Consider your gut microbiome as a busy inner city full of trillions of microscopic residents—bacteria, fungi, viruses, and other small organisms (Baker et al., 2024; Heilbronner et al., 2021). This is not just a haphazard collection; rather, it is a dynamic, linked ecosystem that influences practically all aspects of our health. From our food digestion and nutrient absorption to our immune system's strength to even our mood and brain activity, these microscopic residents are always at work (Hou et al., 2022). Our gut microbiome's state is two-way, a really bidirectional conversation between what we consume (Abeltino et al., 2024; Bermingham et al., 2024). The foods we eat directly determine which bacteria flourish and which fight in our gut. For some helpful bacteria, for example, the fiber in fruits and vegetables functions as a feast; it helps them generate molecules that are quite beneficial for humans, including short-chain fatty acids (SCFAs) (Mentella et al., 2020). These bacteria return by helping us break down food, extract energy, and even synthesize important vitamins. Therefore, a diet that makes one person feel great could leave another feeling lazy, just because their internal microbial populations differ (Pillai et al., 2024; Song & Shin, 2022). This natural uniqueness is exactly the reason the conventional, all-encompassing approach to diet usually fails. Numerous studies have shown that a diet that helps one person lose weight or control blood sugar may have little to no effect on another. Previously unknown, this variability is now increasingly linked to the particular composition and function of every person's gut microbiome (Pillai et al., 2024; Song & Shin, 2022).

Personalized microbiome-based nutrition has great actual promise. The continuous loop between our dietary choices, the gut microbiome, and our well-being. What we eat directly shapes our gut's microbial community, which in turn affects our health. Our health status can then influence our food preferences, completing the cycle (Figure 13.1). It's about creating a dietary plan as unique as your fingerprint, surpassing general guidance. Understanding the particular microbial terrain within your gut, together with other personal elements like your genes and lifestyle, will help us to create eating plans that support a healthy microbial balance, maximize how your body absorbs food, and finally result in improved health and a reduced risk of chronic diseases (Abeltino et al., 2024; Bermingham et al., 2024). From sophisticated sequencing methods that map out your microbial world to strong computing tools and artificial intelligence (AI) that can make sense of

 DOI: 10.1201/9781003399797-14

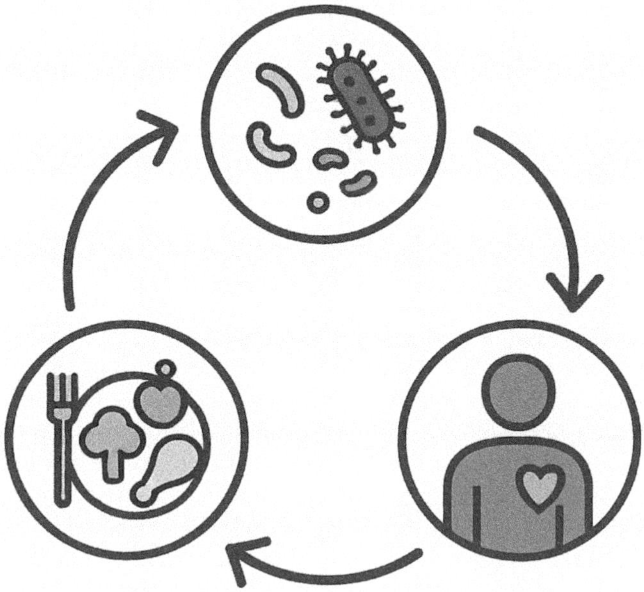

FIGURE 13.1 The interconnected cycle of diet, gut microbes, and health.

all that complicated data, amazing technological advancements drive this revolutionary approach (Acharjee & Choudhury, 2022; Ferrario & Gedrich, 2024). Personalized nutrition is ready to revolutionize preventive medicine and enable each of us to take a more exact, informed, and effective approach to our own well-being as we keep discovering the secrets of the microbiome.

13.2 FUNDAMENTALS OF THE GUT MICROBIOTA AND ITS INFLUENCE ON HEALTH

We must first explore the amazing realm of the gut microbiota itself before we can really value the power of a tailored diet. Far more than a collection of bacteria, this dynamic and sophisticated community of microorganisms is a vital organ system in its own right, continuously interacting with our bodies and affecting a great range of physiological processes. Gaining its great influence on our health requires an awareness of its composition, basic purposes, and what happens when it deviates from equilibrium. A very varied ecosystem, the human gut microbiome is a busy city of microbial life (Baker et al., 2024). Although the most plentiful and well-known occupants are bacteria, this community also comprises archaea (single-celled organisms unique from bacteria), viruses (including bacteriophages infecting bacteria), fungi (the "mycobiome"), and even protozoa. Every one of these groups helps the gut environment to be generally dynamic and healthy (Abeltino et al., 2024).

Two bacterial phyla usually define the scene in a healthy adult gut: Bacteroidota (formerly Bacteroidetes) and Bacillota (formerly Firmicutes). The great majority of the microbial population is made up by them taken together (Heilbronner et al., 2021). Individual differences in the balance between these two groups can be greatly influenced by elements including diet, age, and even genes. While Bacillota is a varied group with many members known for generating important compounds like butyrate, a vital short-chain fatty acid with great anti-inflammatory properties, bacteria are especially adept at breaking down complex carbohydrates.

Though fewer than bacteria, the other microbial occupants are quite important. Particularly bacteriophages, viruses are quite plentiful and can greatly influence the bacterial community by selectively attacking particular bacterial species. Fungi, including Saccharomyces species and Candida, also live in the gut and interact with the bacterial flora, affecting health in different ways (Baker et al., 2024). By eating hydrogen generated by bacteria, archaea—especially methanogens—help to preserve a balanced gut environment and support the fermenting process. Generally speaking, a great degree of variation in the gut microbiome is regarded as a sign of good health. Often more resilient, better able to adapt to changes, and able to perform a greater spectrum of metabolic activities, a more varied community supports general well-being.

13.3 FUNCTIONS OF A HEALTHY GUT MICROBIOTA

Far from passive, a healthy and balanced gut microbiota actively performs a multitude of absolutely vital tasks (Mentella et al., 2020) metabolic activities (e.g., vitamin synthesis, SCFA production). The ability of the gut flora to ferment indigestible dietary components, mostly complex carbohydrates like dietary fiber, defines one of its most important functions. Our human enzymes cannot break down fiber; thus, it passes to the large intestine, where our gut bacteria can begin their work. SCFAs, including acetate, propionate, and butyrate, are produced during fermentation. These SCFAs are absolutely vital; they give the cells lining our colon energy, help control gut movement, improve mineral absorption, and have systemic effects on our metabolism and inflammation level [1, 3]. Particularly as a superstar, butyrate is essential for preserving the health of our colon lining and has anti-inflammatory and even anti-cancer effects. Beyond SCFAs, gut bacteria also synthesize vitamins, including vitamin K and several B vitamins (like biotin, folate, and B12), which our bodies cannot create enough of on their own. They also help break down foreign molecules (like some medications), convert dietary polyphenols into more active forms, and handle bile acids.

13.4 IMMUNE SYSTEM MODULATION

The development and fine-tuning of our immune system are much aided by our gut bacteria. From the time we are born, the colonization of our gut by bacteria helps to "educate" our immune cells, so promoting the development of both our immediate (innate) and long-term (adaptive) immune defenses (Hou et al., 2022). A good microbiota helps to maintain the balance of our immune responses, reducing too much inflammation and guaranteeing our ability to fight off dangerous pathogens. Particularly important for preventing autoimmune reactions and preserving immune tolerance are specific microbial species and the molecules they generate, which can affect the activity of different immune cells, including regulatory T cells (Tregs). Further defending against invaders, the microbiota also generates antimicrobial compounds and strengthens the protective barrier of our gut lining.

Consider your intestinal barrier as a very selective gatekeeper separating the contents of your gut (with its great microbial count and food particles) from the rest of your body. Maintaining a strong and intact barrier depends on the gut bacteria. Microbes boost mucus production, which covers the intestinal lining and provides protection (Hou et al., 2022). They also promote the synthesis of "tight junction proteins," which resemble microscopic seals between the gut's lining of cells. These seals stop dangerous compounds, including bacterial toxins (like lipopolysaccharide, LPS), from "leaking" out of the gut and into the bloodstream—a condition sometimes known as "leaky gut." Some microbial byproducts, such as butyrate, also give these intestinal cells necessary energy to enable their growth and self-healing process. We term this disturbance of the delicate balance of the gut flora "gut dysbiosis." This imbalance can result in a decrease in beneficial bacteria, a loss of microbial diversity, or an overgrowth of maybe dangerous or proinflammatory species. Linked to a wide spectrum of human diseases, dysbiosis emphasizes the need for a balanced gut ecosystem in preserving health (Hou et al., 2022; Xia et al., 2023).

13.5 METABOLIC DISORDERS

Evidence linking gut dysbiosis to metabolic diseases, including obesity and type 2 diabetes, is mounting. All of which contribute to metabolic dysfunction: changes in the gut microbiota can affect how much energy we extract from our diet, how our bodies store fat, how we process glucose, and even help to cause systemic inflammation (Pillai et al., 2024). People with diabetes and obesity, for instance, sometimes show less SCFA-producing bacteria and more inflammatory-promoting microbes. The bacteria can also influence the synthesis of gut hormones, controlling insulin sensitivity and appetite.

13.6 INFLAMMATORY DISEASES

A hallmark of inflammatory bowel diseases (IBD), including Crohn's disease and ulcerative colitis, is dysbiosis. Usually, IBD results in a reduction in microbial diversity and a changed balance between inflammatory-causing and anti-inflammatory-reducing microorganisms, causing chronic inflammation and gut lining damage (Mentella et al., 2020). Beyond IBD, gut dysbiosis has also been linked to other inflammatory diseases, including rheumatoid arthritis and psoriasis, implying that imbalances in the gut can have broad effects on inflammation all around the body.

New studies suggest a major part of the gut bacteria play in cardiovascular health. Certain gut bacteria can convert dietary elements, including L-carnitine—found in red meat and eggs—and choline into a molecule known as trimethylamine (TMA). Increased levels of trimethylamine N-oxide (TMAO) have been linked to an elevated risk of atherosclerosis (hardening of the arteries) and other cardiovascular problems (Pillai et al., 2024); this TMA is then converted in the liver to TMAO. On the other hand, good flora can generate helpful molecules that really prevent heart disease.

Through a sophisticated network called the gut-brain axis, our gut and brain are in continuous conversation. Through many signaling pathways involving nerves, hormones, and the immune system, the gut microbiota is a major modulator of this axis, affecting everything from brain development to our mood and behavior (Hou et al., 2022). Though the precise cause-and-effect relationships are still being worked out, dysbiosis has been linked to a variety of neurological and psychiatric disorders, including anxiety, depression, autism spectrum disorder, and Parkinson's disease.

In cancer, the gut flora has a multifarious influence on its development, course, and even the efficacy of treatment. Particularly in the colon, dysbiosis-driven chronic inflammation can fuel tumor development. Some microbial waste can either promote or inhibit the spread of cancer. Moreover, the gut flora emphasizes its relevance in thorough cancer treatment since it can influence the efficacy and side effects of immunotherapy and chemotherapy. Fundamentally, gut bacteria are an indispensable part of human life since they carry important metabolic and immune roles. Disruption of this microbial ecosystem can have far-reaching effects and lead to a great variety of diseases. This knowledge emphasizes especially the possibilities of therapeutic approaches meant to restore and preserve a healthy gut microbial balance, especially customized diets.

Diet is the one external element most likely to shape the composition and activity of our gut flora (Mentella et al., 2020; Song & Shin, 2022). The foods we consume provide the necessary fuel and building blocks that define which kinds of microorganisms grow in our gut. Developing really successful individualized dietary plans absolutely depends on knowing exactly how various dietary components affect these microbial populations. Our main sources of energy, carbohydrates, proteins, and fats, not only feed us but also significantly affect our gut bacterial count (Table 13.1). But it's more about the particular forms and ratios of these macronutrients that make all the difference than about the count. Arguably, the most important dietary change for the gut flora is carbohydrates. Simple sugars are rapidly absorbed in our small intestine; complex carbohydrates—especially dietary fiber and resistant starches—make it all the way to the colon essentially undigested. Once there, they truly become a feast for a wide range of gut bacteria.

TABLE 13.1
Dietary Modulators and Their Impact on Gut Microbiota

Dietary Component	Examples/Sources	Impact on Gut Microbiota	Key Metabolites/Effects
		Carbohydrates	
Fiber/Prebiotics	Whole grains, fruits, vegetables, legumes, inulin, FOS, GOS	Increases diversity, promotes beneficial bacteria (*Bifidobacterium*, *Lactobacillus*)	Short-chain fatty acids (SCFAs: acetate, propionate, butyrate), improved gut barrier, anti-inflammatory
Refined Sugars/ Processed Carbs	Sugary drinks, white bread, processed snacks	Decreases diversity, favors less beneficial/pro-inflammatory bacteria	Reduced SCFA production, increased inflammation
		Proteins	
Animal-Based	Red meat, dairy	Can increase *Bacteroides*, potentially harmful metabolites	Branched-chain fatty acids, ammonia, hydrogen sulfide (pro-inflammatory)
Plant-Based	Legumes, nuts, seeds, plant-based proteins	Promotes beneficial bacteria, more favorable metabolic profile	Beneficial SCFAs, diverse microbial community
		Fats	
Saturated Fats	Processed foods, some animal fats	Proinflammatory microbiota, increased intestinal permeability	Increased LPS, altered bile acid metabolism
Unsaturated Fats	Olive oil, avocados, fatty fish, nuts	Supports diverse and healthy microbiota, anti-inflammatory	Anti-inflammatory compounds, improved gut barrier
		Bioactive Compounds	
Polyphenols	Fruits, vegetables, tea, coffee	Metabolized by gut bacteria, enhanced bioavailability	Antioxidant, anti-inflammatory properties

By definition, dietary fiber is carbohydrate polymers our human enzymes cannot break down in the small intestine. Whole grains, fruits, lentils, vegetables, seeds, and nuts [3] abound in it. SCFAs, including acetate, propionate, and butyrate, are produced when particular gut bacteria—including those from the Bacteroidota and Bacillota phyla—ferment these fibers. These SCFAs are absolutely vital for gut health; they give the cells lining our colon energy, preserve the integrity of our gut barrier, and act systemically to reduce inflammation [1, 3]. Consistently linked to improved health outcomes, a diet high in varied fibers promotes a more varied and resilient gut flora.

Prebiotics, a unique kind of dietary fiber designed especially to stimulate the growth and/or activity of beneficial gut microorganisms, especially *Bifidobacterium* and *Lactobacillus* species, follow then. Natural prebiotics abound in foods including chicory root, garlic, onions, and bananas. Including prebiotics in your diet can increase the counts of these health-promoting bacteria, thereby enhancing metabolic performance and immune control. Conversely, diets heavy in highly processed carbohydrates and refined sugars—typical of the modern Western diet—tend to lower microbial variety. They support the proliferation of less advantageous or even dangerous bacteria, which would lower SCFA generation and raise proinflammatory compounds. This dysbiosis is that imbalance greatly increases the risk of chronic diseases.

The gut flora is shaped in part by our diet's protein content and source as well. Although most proteins are broken down and absorbed in the small intestine, some undigested proteins can pass via the colon. Gut bacteria there ferment it, generating different metabolites. While some of these can be bad for our health, others are helpful. Diets heavy in animal proteins, for instance, have been shown to boost the abundance of Bacteroides species and generate possibly toxic metabolites, including branched-chain fatty acids, ammonia, and hydrogen sulfide. These substances have been connected to a higher risk of some diseases and can aggravate gut inflammation. On the other hand, plant-based protein diets are linked with a more favorable metabolic profile and usually encourage the development of helpful bacteria, including *lactobacilli* and *bifidobacteria*. Often found alongside plant-based proteins, the fiber content boosts these benefits even more.

Our gut bacteria are also much influenced by dietary fats. Commonly found in Western diets, saturated fats have been linked to a proinflammatory microbial profile, enhanced intestinal permeability (that "leaky gut" phenomenon), and higher levels of circulating LPSs, which can aggravate metabolic problems, including obesity. Additionally, changing bile acid metabolism by saturated fats influences the growth of some bacterial species. On the other hand, diets heavy in unsaturated fats—especially monounsaturated and polyunsaturated fatty acids found in olive oil, avocados, nuts, and fatty fish—tend to support a more varied and healthy gut flora. These fats help to produce anti-inflammatory molecules and promote the growth of helpful bacteria, thus supporting gut barrier function and lowering inflammation systemically.

Beyond the main macronutrients, micronutrients (vitamins and minerals), and many bioactive compounds, including polyphenols, also affect the gut flora. Not readily absorbed in the small intestine, polyphenols abound in fruits, vegetables, tea, and coffee. In the colon, gut bacteria break them down. Their bioavailability and bioactivity can be improved by this microbial transformation, thus improving their effects on host health, including anti-inflammatory and antioxidant ones.

Examining whole dietary patterns offers a more complete picture of their combined impact on the gut flora than concentrating on specific nutrients. For example, the Mediterranean diet—which emphasizes fruits, vegetables, whole grains, legumes, nuts, olive oil, and modest consumption of fish and chicken—is repeatedly connected to a varied and healthy gut flora. This diet encourages the development of anti-inflammatory microbial profiles and SCFA-producing bacteria, supporting their well-documented advantages in avoiding some cancers, metabolic syndrome, and cardiovascular disease. By contrast, the Western diet, which is marked by high levels of processed foods, refined sugars, saturated fats, and animal proteins, usually results in a less varied microbiota, an increase in proinflammatory bacteria, and decreased SCFA generation. Rising obesity, type 2 diabetes, and inflammatory diseases all have a clear correlation with this dietary pattern.

13.7 PROBIOTICS AND POSTBIOTICS

Although diet mostly shapes the gut flora already present, probiotics and postbiotics provide more direct means to change its composition and action. Live microorganisms called probiotics help the host by means of appropriate doses. Yogurt, kefir, sauerkraut, and kimchi are among the fermented foods that abound in them; dietary supplements also contain them. In order to help gut health and general well-being, probiotics can momentarily colonize the gut, change the immune system, generate good metabolites, and fight bad bacteria.

Conversely, postbiotics are soluble elements created by microorganisms either during fermentation or following microbial breakdown (such as metabolites, components of cell walls, and enzymes). Furthermore, providing health advantages for the host are these nonliving microbial elements. Particularly for people with compromised immune systems, they can have immunomodulating, anti-inflammatory, and antimicrobial effects, thus providing a safer and more consistent substitute for live probiotics. All things considered, our dietary decisions have quite a strong influence on the makeup and purpose of our gut flora. Understanding the particular effects of various macronutrients, micronutrients, and overall dietary patterns, as well as the function of probiotics and postbiotics, helps us to create very customized nutritional plans meant to maximize the gut microbiome for better health results.

13.8 METHODOLOGIES FOR NUTRITIONAL PERSONALIZATION AND MICROBIOME ANALYSIS

Complex scientific tools are necessary to really personalize dietary recommendations depending on the microbiome. Together with strong computational techniques, the fast developments in high-throughput sequencing technologies have totally changed our ability to describe the gut microbiome and translate its complex data into useful, pragmatic insights. The main approaches applied in microbiome analysis are discussed in this part, together with their application in creating customized dietary recommendations. It outlines the standard sequence of steps in developing personalized nutrition plans based on the gut microbiome. It begins with sample collection and laboratory evaluation, followed by bioinformatics processing and AI analysis, ultimately leading to individualized dietary guidance tailored to a person's unique microbial profile (Figure 13.2).

Modern microbiome research is mostly based on high-throughput sequencing, which lets scientists fully find and count the microbial populations in a given sample. Two main strategies are often used: 16S rRNA sequence analysis. A rather popular and reasonably affordable approach for microbial community profiling is 16S rRNA gene sequencing (Abeltino et al., 2024). Found in every bacterium and archaeon, the 16S ribosomal RNA gene is a universal marker. It comprises variable regions (which vary between species), interspersed with highly conserved areas that are shared by all species. Often down to the genus or even species level, researchers can identify and classify many microbial taxa by sequencing these varied areas. This approach shows the relative abundance of many groups, so offering a decent summary of "who is there" in the microbial population. It does, however, provide scant details regarding "what they can do"—their particular functional capacity.

Shotgun metagenomic sequencing, on the other hand, sequences all DNA in a sample, including DNA from fungi, archaea, bacteria, and viruses. This method offers a far more complete picture of the microbial population. By means of gene analysis, researchers can deduce the metabolic pathways, enzymes, and other functional capabilities of the whole microbial community, enabling not only exact taxonomic identification (who is there) but also thorough functional profiling (what they can do). This is especially important for tailored nutrition since it provides a better knowledge of the

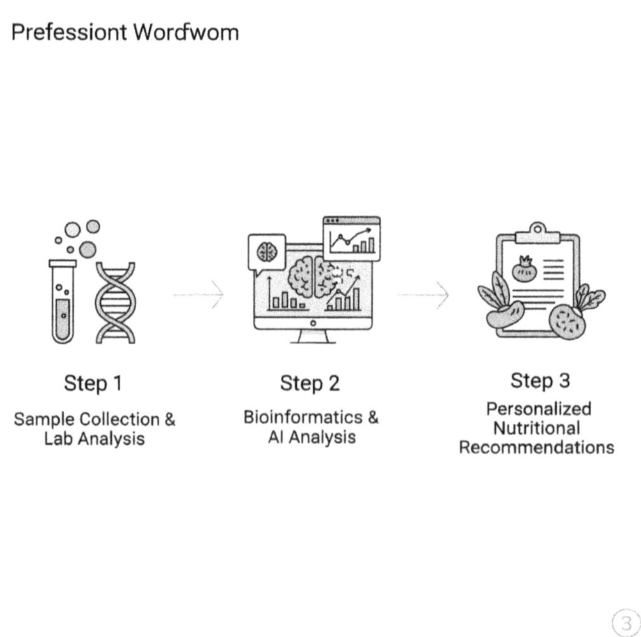

FIGURE 13.2 Workflow for personalized microbiome analysis and nutritional recommendation.

functional potential of the microbiome by revealing the possibility for particular bacteria to generate either beneficial or harmful metabolites in reaction to dietary interventions.

13.9 MICROBIAL FUNCTION

Although sequence technologies reveal the genetic potential of the microbiome, metabolomics and proteomics give vital information on its actual activity and direct influence on the host. Offering a real-time functional readout of the microbiome, these "omics" techniques track the end products of microbial metabolism and the proteins they express. The broad study of small molecules, or metabolites, within a biological sample—such as stool, blood, or urine—is known as metabolomics. Direct outcomes of host metabolic activities, as well as microbial ones, are these metabolites. Analyzing the metabolome helps scientists to pinpoint particular molecules created by bacteria, including SCFAs, bile acids, and several amino acid derivatives—all important mediators of the interactions between the gut flora and the host. Important biomarkers for both health and disease states, changes in the metabolomic profile can indicate changes in microbial function in response to diet. Metabolomics clarifies precisely how a person's microbiome handles particular nutrients and generates molecules that affect their health outcomes, guiding personalized nutrition.

The broad study of proteins, the main functional molecules in cells, is known as proteomics. Analyzing the proteome of a microbial community helps scientists to understand the active metabolic routes and functional mechanisms the microorganisms now employ. Although less often used in standardized personalized nutrition settings than sequencing and metabolomics, proteomics can offer important complementary information about the dynamic functional state of the microbiome. High-throughput sequencing and "omics" technologies produce a simply enormous volume and complexity of data that demand sophisticated bioinformatics and computational tools for efficient analysis and interpretation. Absolutely necessary tools for cleaning raw sequencing data, fixing mistakes, and filtering out any human DNA possibly present in the sample constitute quality control and preprocessing, and using large reference databases, taxonomic groups are assigned microbial genetic sequences.

13.10 AI AND MI IN PREDICTING DIETARY RESPONSES

By allowing us to forecast how a person will react to various diets based on their unique microbiome and other personal data (Acharjee & Choudhury, 2022; Ferrario & Gedrich, 2024), AI and machine learning (ML) are really transforming personalized nutrition. Because it cannot consider the great variation in how each person's body absorbs food and how their gut microbiome responds, conventional dietary advice often falls short. AI/ML models are meant to transcend.

Based on their gut microbiome, AI models can forecast a person's blood sugar reaction following particular food consumption. This enables very customized dietary advice meant to stabilize blood sugar levels (Abavisani et al., 2024). Finding Microbiological Biomarkers: ML techniques can identify particular microbial species or functional pathways acting as biomarkers for specific dietary reactions or disease risks (Ferrario & Gedrich, 2024). Improving dietary interventions, whether it's weight control, lowering inflammation, or enhancing metabolic health, AI can examine the complex interactions among diet, microbiome, and human physiology to propose the most successful dietary interventions for particular health goals. Customized Diet Suggestions: Beyond broad recommendations, AI can suggest particular foods or combinations of foods most likely to support an individual's particular microbiome profile, thus optimizing positive health effects.

All of this microbiome research and computational interpretation is ultimately meant to translate complicated scientific data into sensible, easily followed, customized dietary recommendations. Usually, this procedure consists of several important phases: microbiome profiling starts with gathering from the individual a biological sample, typically a stool sample. Following 16S rRNA gene sequencing or shotgun metagenomics, this sample helps to define the gut microbial composition and functional potential. Crucially, then, this microbial data is merged with other pertinent personal information. This covers genetic data, metabolic markers (such as blood glucose and lipid profiles), lifestyle choices (such as exercise patterns and stress levels), and the particular health objectives of the person.

13.11 CLINICAL APPLICATIONS OF PERSONALIZED MICROBIOME

Nutrition profound insights gained from understanding the intricate relationship between the gut microbiome, diet, and host health have opened up a vast landscape of clinical applications for personalized microbiome-based nutritional strategies. This approach holds immense promise for the prevention, management, and even treatment of a wide array of health conditions, offering a powerful shift from generalized dietary advice to highly tailored interventions. Obesity and metabolic syndrome represent significant global health challenges, and it's clear that individual responses to weight loss interventions vary dramatically.

By considering the distinct microbial profiles that affect how our bodies absorb energy from food, store fat, and control metabolism, personalized microbiome-based nutrition presents a revolutionary solution (Bermingham et al., 2024). Studies on gut microbial compositions have revealed that some of them are connected to a greater inclination for weight gain or resistance against attempts at weight loss (Pillai et al., 2024). Higher ratios of Bacillota (Firmicutes) to Bacteroidota (Bacteroidetes), for instance, have been linked repeatedly to higher energy extraction from the diet. Investigating a person's microbiome helps one to create customized plans to balance this equilibrium, promote the growth of good bacteria (such as *Akkermansia muciniphila*, connected with better metabolic health), and increase dietary fiber intake to increase feelings of fullness and lower calorie absorption.

Modified treatments might center on particular prebiotics or probiotics that support a better metabolic profile, thus enabling more efficient and sustainable weight control. Chronic inflammation of the gastrointestinal tract defines IBD, which comprise ulcerative colitis and Crohn's disease. Gut dysbiosis is a main characteristic of several disorders. Customized diets seek to restore microbial balance, lower inflammation, and help IBD sufferers' gut barrier to remain intact. Although low-FODMAP diets or specific carbohydrate diets have shown some promise, their efficacy varies greatly among individuals. While lowering proinflammatory species, microbiome analysis can direct the choice of dietary components that are well-tolerated and encourage the growth of anti-inflammatory bacteria (such as fecal bacterium parasitize, a main producer of butyrate). Modified approaches may also include focused probiotic treatments or even fecal microbiota transplantation (FMT), carefully selected based on the individual's particular microbial deficits and inflammatory markers, to help re-establish a healthy gut ecosystem and maybe cause remission or prevent relapses.

Still, the leading cause of death globally is cardiovascular disease (CVD). Particularly in relation to the synthesis of metabolites like TMAO, the gut microbiome's influence on CVD development is under increasing focus. Certain gut bacteria break down dietary choline and L-carnitine, which are found in foods including red meat and eggs, and elevated TMAO levels have been linked to an increased risk of atherosclerosis (Pillai et al., 2024). Based on their microbiome profile, personalized dietary plans can find people more likely to produce TMAO. This enables recommendations to change their diet to lower the consumption of TMAO precursors or to change their gut flora to lower the bacteria generating TMA. On the other hand, promoting the expansion of bacteria that generate useful metabolites, such as SCFAs, can help to improve lipid profiles, lower blood pressure, and increase endothelial function, so lowering the risk of CVD.

Crucially controlling our immune system, the gut microbiome affects local immune reactions within the gut as well as systemic reactions all around the body. Among the several autoimmune diseases, including rheumatoid arthritis, multiple sclerosis, and type 1 diabetes, dysbiosis has been linked to their development and course. Aiming to rebalance the gut flora, personalized nutritional interventions can help to induce immune tolerance and lower autoimmune inflammation. To raise

SCFA generation, this may mean consuming more fermentable fibers, thus improving the function of regulatory T cells (Tregs), vital cells for preserving immune tolerance. Alternatively, particular probiotic strains well known to alter immune systems could be added. Knowing a person's immune status and microbial profile enables very focused dietary and microbial treatments to reduce symptoms and maybe change the course of autoimmune diseases.

Known as the gut-brain axis, the bidirectional communication network between the gut and the brain emphasizes how significantly the gut microbiome shapes mental health and neurological performance. Disorders, including anxiety, depression, and autism spectrum disorder, have been linked to dysbiosis. Personalized microbiome-based dietary plans can investigate the possibilities of "psychobiotics" (probiotics especially connected to mental health benefits) and dietary interventions encouraging the generation of neuroactive compounds by gut bacteria (e.g., precursors of serotonin, GABA, and dopamine). Customized diets high in prebiotics and particular amino acids can help to support a good gut-brain axis, improving mood, cognitive ability, and stress tolerance. Research in this field is fast-growing and creating fresh opportunities for complementary therapies in mental healthcare. In the realm of cancer, the gut microbiome plays a multifaceted role, influencing everything from the development and progression of tumors to the effectiveness and side effects of cancer therapies, particularly immunotherapy.

Personalized nutritional strategies in oncology support aim to optimize the gut microbiome to enhance treatment responses, reduce adverse effects, and improve overall patient outcomes. For instance, certain microbial compositions have been linked to better responses to immune checkpoint inhibitors, a type of cancer immunotherapy. Dietary interventions, including specific kinds of fiber or focused probiotics, can help to create a gut environment that supports anti-tumor immunity and lowers treatment-related inflammation or gastrointestinal discomfort. This developing area has great potential to combine microbial control with nutritional modification into all-encompassing cancer treatments.

Fundamentally, in clinical environments, personalized microbiome-based nutrition is quickly moving from a theoretical idea to a useful and potent tool. By leveraging advanced analytical techniques and a deeper understanding of the intricate interactions between host and microbes, clinicians can now offer more precise and effective dietary interventions, uniquely tailored to each individual's biological landscape. This promises to optimize health and disease management across a broad spectrum of conditions, marking a significant leap forward in precision medicine.

13.12 CHALLENGES AND LIMITATIONS

Although tailored microbiome-based dietary plans hold great promise, several major obstacles and limits need to be carefully addressed before this method can be widely used and completely realized. These challenges call for coordinated effort among many stakeholders in scientific, technical, ethical, and legal spheres.

The present lack of standardization in microbiome research techniques presents one of the main difficulties. Variations in how samples are obtained, DNA is extracted, sequencing platforms are used (e.g., 16S rRNA versus shotgun metagenomics), and bioinformatics pipelines are applied can result in inconsistencies. Comparing data across several studies and labs becomes quite challenging this way. Such variation makes it difficult to create big, strong, interoperable datasets—which are absolutely required for creating dependable predictive models and broad insights. Advancement of the field depends on common protocols, strict quality control procedures, and shared data repositories being established.

Personalized nutrition mostly depends on the gathering and analysis of very sensitive personal data, including an individual's unique microbiome profile, genetic information, comprehensive dietary habits, and health records. This begs for serious ethical questions about informed permission, data privacy, and security. Ensuring responsible handling of patient data, anonymizing where relevant, and shielding from illegal access comes first. Moreover, ethical issues surround the possibility of discrimination depending on a person's genetic predispositions or microbiome profile. Furthermore, it is difficult to properly present complicated scientific data to the public to prevent misinterpretation, unnecessary anxiety, or support of unproven treatments.

Although many studies have found convincing links between particular microbial profiles and health effects or dietary responses, the generalizability and repeatability of these results remain major challenges. The gut microbiome is quite dynamic and affected by many different elements; it is difficult to separate the exact influence of particular dietary changes. Many of the present studies are carried out on rather small groups of individuals; their results might not be directly relevant to many populations with different genetic backgrounds, lifestyles, and dietary choices. Larger, well-designed, longitudinal studies spanning many populations are desperately needed to validate first results and create strong, generalizable principles for customized nutrition in order to overcome this.

Comprehensive microbiome testing, especially advanced techniques like shotgun metagenomics and metabolomics, can be rather costly right now and might not be covered by health insurance. This greatly reduces its availability to a larger population, aggravating current health inequalities. Personalized microbiome-based nutrition must become a widely accepted method if the expenses of testing are to drop significantly, and these services must become more generally available and included in regular healthcare procedures. More medical professionals who are qualified to analyze intricate microbiome data and translate it into sensible, evidence-based dietary recommendations also need training.

Modern microbiome research contrasts greatly with its useful application in clinical environments. Many doctors lack the tools and knowledge required to properly include complicated microbiome data into regular patient care. The evidence base for particular tailored dietary recommendations is still developing; thus, more thorough clinical studies are required to unequivocally show quantifiable health benefits. To enable the seamless transfer of research results into practical clinical strategies, developing user-friendly interfaces, decision-support tools, and unambiguous clinical guidelines for healthcare providers will be absolutely vital. The fast speed of innovation in personalized microbiome-based nutrition has exceeded the evolution of clear rules and regulations. Important issues regarding the control of direct-to-consumer microbiome testing kits, the tailored dietary recommendations given by different companies, and the health claims made about their advantages surface. Protecting consumers from false information and guaranteeing the responsible development of the field depend on well-defined regulations for product safety, efficacy, and marketing claims. This also covers defining the suitable roles of several healthcare experts in providing individualized dietary recommendations. Dealing with these complex issues calls for cooperative efforts among researchers, doctors, businesses, legislators, and regulatory agencies. Unlocking the full potential of tailored microbiome-based dietary plans and properly including them in both preventive and therapeutic healthcare systems will depend on overcoming these constraints.

13.13 FUTURE DIRECTIONS AND EMERGING TECHNOLOGIES

Personalized microbiome-based nutritional strategies are a dynamic and fast-changing field driven ahead by technological developments and a growing knowledge of the complex interactions between our bodies and their microbial occupants. Many fascinating future directions and new technologies have great potential to help to further improve the range of this transforming method. Personalized nutrition is about to see even more central and sophisticated involvement from AI and ML. Apart from their present uses in glycemic response prediction, future AI models will combine an even more extensive range of various datasets. This will comprise real-time physiological data streamed from wearable sensors, thorough genetic information (genomics), detailed metabolic profiles (metabolomics), and longitudinal microbiome data—tracking changes over time. This integration will allow the creation of very complex predictive models able to forecast individual responses to a wider spectrum of dietary interventions with hitherto unheard-of accuracy. Discovery of new microbial biomarkers, identification of hitherto unidentified diet-microbe-host interactions, and optimization of intricate dietary recommendations for highly specific health objectives will also depend much on AI. Moreover, explainable artificial intelligence (XAI) will be quite important. XAI seeks to make AI decisions open and understandable, fostering trust in these AI-driven recommendations and enabling both doctors and people to precisely understand the justification behind the tailored advice.

Originally applied mostly to human cells, the revolutionary CRISpen-Cas gene-editing technology is now under investigation for its ability to precisely alter the gut microbiome (Amen et al., 2025; Brödel et al., 2024). Targeting and eradicating particular unwelcome bacterial species or genes—for example, those involved in the synthesis of toxic compounds like TMAO—this innovative approach could on the other hand, be used to bring useful genes into already existing microorganisms, thus improving their functional capacity. Moving beyond general dietary changes to direct, surgical-like microbial engineering, CRISpen-based microbiome editing presents the exciting possibility of highly targeted and exact modulation of the gut ecosystem while still in very early phases of research.

Synthetic biology is either redesigning current natural biological systems for particular, practical uses or designing and building new biological parts, devices, and systems (Nazir et al., 2024). Within the framework of the microbiome, this could result in the creation of "engineered probiotics," or microorganisms especially meant to carry out intended roles in the gut. Probiotics might be designed, for example, to generate specific therapeutic compounds, effectively break down toxic metabolites, or even provide targeted treatments right to the gut. This method provides a great degree of control and specificity, hence maybe resulting in stronger and more predictable treatments than conventional, naturally occurring probiotics. The journey into the intricate world of the human gut microbiome has profoundly reshaped our understanding of health and disease, ushering in a new era of personalized nutrition (Abeltino et al., 2024; Nazir et al., 2025).

Personalized nutrition combined with wearable sensors and sophisticated digital health systems will allow real-time monitoring of our physiological reactions to diet and generate continuous feedback loops (Nazir et al., 2025). Wearable devices provide priceless real-time data by tracking a range of criteria, including glucose levels, heart rate variability, sleep patterns, and activity levels. Subsequently, this data can be easily combined with dietary information and microbiome. Then, using AI, digital health systems can create dynamic, flexible dietary recommendations that change depending on a person's real-time physiological state and response. This makes tailored nutrition far more sensitive and efficient, thus producing a really unique and dynamic health management system that develops with you. Large-scale, longitudinal studies tracking people over long periods of time will take front stage in future research. These studies will compile thorough information on diet, lifestyle, health effects, and—above all—microbiota dynamics across time. This will offer priceless practical data on the long-term safety and effectiveness of tailored microbiome-based dietary plans. Beyond basic correlational results, these kinds of studies are crucial for establishing clear causal links and for spotting strong biomarkers and predictive elements fit for clinical practice use. Developing new foods and ingredients especially suited for microbiome modulation is likely to benefit from precision fermentation, a technique using microorganisms to generate particular ingredients. Highly specific prebiotics, postbiotics, and other bioactive compounds created by this technology can then be added to foods or supplements to exactly target intended microbial roles. This provides completely new opportunities for developing functional foods meant from the ground up to support a healthy gut microbiome and maximizes personal health. These future directions point clearly toward progressively exact, dynamic, and integrated approaches to personalized microbiome-based nutrition. A really preventive and personalized healthcare system will be made possible as these technologies develop and our scientific knowledge deepens our capacity to customize dietary interventions to each individual's unique biological environment.

13.14 CONCLUSION

Clearly underlining their transforming potential in optimizing human health, this chapter has investigated the fundamental ideas, the innovative approaches, the expanding clinical applications, and the fascinating future trajectories of personalized microbiome-based nutritional strategies. We now know that the gut microbiota is a dynamic and powerful ecosystem that significantly affects host metabolism, immunity, and even neurological activity; it is not only a group of commensal organisms. Diet and microbiome have a bidirectional relationship whereby our food choices significantly influence our microbial environment, and, hence, our response to the foods we eat determines our unique microbial composition (Bermingham et al., 2024; Song & Shin, 2022). This natural inter-individual variability emphasizes the limits of conventional, broad dietary recommendations and the great need for really individualized solutions.

Together with metabolomics, proteomics, and sophisticated bioinformatics tools, advanced high-throughput sequencing technologies have given us hitherto unheard-of understanding of the composition and functional capacity of the gut microbiome. The smooth integration of AI and ML is improving our capacity to understand these complicated datasets, facilitating the prediction of individual dietary responses and the creation of very customized nutritional recommendations. From a theoretical idea to a useful and powerful reality, these approaches are fast bringing personalized nutrition.

Personalized microbiome-based diets have great promise across a broad spectrum of health issues and are fast finding clinical use. From optimizing weight management and mitigating metabolic syndrome to lowering inflammation in IBD, preventing cardiovascular disease, modulating immune responses in autoimmune conditions, supporting mental health via the gut-brain axis, and even improving oncology treatments, the potential for focused dietary interventions guided by microbiome insights is truly broad. This method addresses underlying biological processes rather than only treating symptoms, providing a more exact and efficient means of disease prevention and management.

The road ahead does not, however, present without difficulties. Methodically addressed are issues including the need for data standardization, important ethical concerns about patient privacy, demand for strong reproducibility and generalizability of results, and the present high cost and availability of testing techniques. Moreover, public confidence depends on clearly establishing regulatory frameworks and translating difficult research into useful clinical practice. Looking ahead, the field will be further transformed by newly developed technologies, including even more advanced AI and predictive modeling, revolutionary CRISpen-based microbiome editing, creative synthetic biology approaches, and flawless interaction with real-time wearable sensors. These developments will open the path for a really preventive and individualized healthcare system by enabling even more exact, dynamic, and integrated methods for microbiome modulation.

Finally, tailored microbiome-based dietary plans offer a fresh and very exciting horizon for the control of health and disease. Accepting the inherent complexity and uniqueness of the human-microbe relationship will open hitherto unheard-of chances to maximize health outcomes, avoid chronic diseases, and enable people to participate more actively and wisely in their own well-being. Unquestionably, health is personal; the gut microbiome is at its very core.

REFERENCES

Abavisani, M., Khoshrou, A., Foroushan, S. K., Ebadpour, N., & Sahebkar, A. (2024). Deciphering the gut microbiome: The revolution of artificial intelligence in microbiota analysis and intervention. Current Research in Biotechnology, 7, 100211. https://doi.org/10.1016/j.crbiot.2024.100211

Abeltino, A., Hatem, D., Serantoni, C., Riente, A., De Giulio, M. M., De Spirito, M., De Maio, F., & Maulucci, G. (2024). Unraveling the gut microbiota: Implications for precision nutrition and personalized medicine. Nutrients, 16(22), 3806. https://doi.org/10.3390/nu16223806

Acharjee, A., & Choudhury, S. P. (2022). Artificial intelligence-based personalized nutrition and prediction of irritable bowel syndrome patients. Therapeutic Advances in Gastroenterology, 15, 17562848221145612. https://doi.org/10.1177/17562848221145612

Amen, R. A., Hassan, Y. M., Essmat, R. A., Ahmed, R. H., Azab, M. M., Shehata, N. R., Elgazzar, M. M., & El-Sayed, W. M. (2025). Harnessing the microbiome: CRISPR-based gene editing and antimicrobial peptides in combating antibiotic resistance and cancer. Probiotics and Antimicrobial Proteins. https://doi.org/10.1007/s12602-025-10573-8

Baker, J. L., Mark Welch, J. L., Kauffman, K. M., McLean, J. S., & He, X. (2024). The oral microbiome: Diversity, biogeography and human health. Nature Reviews Microbiology, 22(2), 89–104. https://doi.org/10.1038/s41579-023-00963-6

Bermingham, K. M., Linenberg, I., Polidori, L., Asnicar, F., Arrè, A., Wolf, J., Badri, F., Bernard, H., Capdevila, J., Bulsiewicz, W. J., Gardner, C. D., Ordovas, J. M., Davies, R., Hadjigeorgiou, G., Hall, W. L., Delahanty, L. M., Valdes, A. M., Segata, N., Spector, T. D., & Berry, S. E. (2024). Effects of a personalized nutrition program on cardiometabolic health: A randomized controlled trial. Nature Medicine, 30, 1888–1897. https://doi.org/10.1038/s41591-024-02951-6

Brödel, A. K., Charpenay, L. H., Galtier, M., Fuche, F. J., Terrasse, R., Poquet, C., Havránek, J., Pignotti, S., Krawczyk, A., Arraou, M., Prevot, G., Spadoni, D., Yarnall, M. T. N., Hessel, E. M., Fernandez-Rodriguez, J., Duportet, X., & Bikard, D. (2024). In situ targeted base editing of bacteria in the mouse gut. Nature, 632(8026), 877–884. https://doi.org/10.1038/s41586-024-07681-w

Ferrario, P. G., & Gedrich, K. (2024). Machine learning and personalized nutrition: A promising liaison? European Journal of Clinical Nutrition, 78, 74–76. https://doi.org/10.1038/s41430-023-01350-3

Heilbronner, S., Krismer, B., Brötz-Oesterhelt, H., & Peschel, A. (2021). The microbiome-shaping roles of bacteriocins. Nature Reviews Microbiology, 19(11), 726–739. https://doi.org/10.1038/s41579-021-00569-w

Hou, K., Wu, Z.-X., Chen, X.-Y., Wang, J.-Q., Zhang, D., Xiao, C., Zhu, D., Koya, J. B., Wei, L., Li, J., & Chen, Z.-S. (2022). Microbiota in health and diseases. Signal Transduction and Targeted Therapy, 7(1), 135. https://doi.org/10.1038/s41392-022-00974-4

Mentella, M. C., Scaldaferri, F., Pizzoferrato, M., Gasbarrini, A., & Miggiano, G. A. D. (2020). Nutrition, IBD and gut microbiota: A review. Nutrients, 12(4), 944. https://doi.org/10.3390/nu12040944

Nazir, A., Hussain, F. H. N., Raza, A., & Edison, L. K. (2024). Advancing microbiota therapeutics: The role of synthetic biology in engineering microbial communities for precision medicine. Frontiers in Bioengineering and Biotechnology, 12, 1511149. https://doi.org/10.3389/fbioe.2024.1511149

Nazir, A., Hussain, F. H. N., Raza, A., & Edison, L. K. (2025). Microbiome-driven therapeutics: From gut health to precision medicine. Gastrointestinal Disorders, 7(1), 7. https://doi.org/10.3390/gidisord7010007

Pillai, S. S., Gagnon, C. A., Foster, C., & Ashraf, A. P. (2024). Exploring the gut microbiota: Key insights into its role in obesity, metabolic syndrome, and type 2 diabetes. The Journal of Clinical Endocrinology & Metabolism, 109(11), 2709–2719. https://doi.org/10.1210/clinem/dgae499

Song, E.-J., & Shin, J.-H. (2022). Personalized diets based on the gut microbiome as a target for health maintenance: From current evidence to future possibilities. Journal of Microbiology and Biotechnology, 32(12), 1497–1505. https://doi.org/10.4014/jmb.2209.09050

Xia, C., Su, J., Liu, C., Mai, Z., Yin, S., Yang, C., & Fu, L. (2023). Human microbiomes in cancer development and therapy. MedComm, 4(2), e221. https://doi.org/10.1002/mco2.221

Index

Note: Page numbers in *italics* refer to figures and **bold** refer to tables

For Product Safety Concerns and Information please contact our EU
representative GPSR@taylorandfrancis.com
Taylor & Francis Verlag GmbH, Kaufingerstraße 24, 80331 München, Germany

www.ingramcontent.com/pod-product-compliance
Ingram Content Group UK Ltd.
Pitfield, Milton Keynes, MK11 3LW, UK
UKHW052221190926
475702UK00006B/1138

* 9 7 8 1 0 3 2 5 0 8 2 2 1 *